Chemistry and Technology of Natural and Synthetic Dyes and Pigments

Edited by Ashis Kumar Samanta, Nasser S. Awwad and Hamed Majdooa Algarni

Published in London, United Kingdom

IntechOpen

Supporting open minds since 2005

Chemistry and Technology of Natural and Synthetic Dyes and Pigments
http://dx.doi.org/10.5772/intechopen.83199
Edited by Ashis Kumar Samanta, Nasser S. Awwad and Hamed Majdooa Algarni

Contributors
Shekh Md. Mamun Kabir, Joonseok Koh, Sharad Rohidas Rohidas Patil, Amol S. Chaudhary, Tushar A. Anil Shinde, Suprit Borse, Kamlesh Sonawane, Madhuri V. Kakde, Deepti Pargai, Shahnaz Jahan, MANISHA GAHLOT GAHLOT, Pubalina Samanta, Ahipa T.N., Deepak Devadiga, Virendra Kumar Gupta, Prof Nasser S Awwad, Adel A. El-Zahhar, Jamila A.M. Alasmary, Ashis Kumar Samanta, Anowar Hossain, Anupama Mishra, Sapna Gautam, M. Kavyashree, Deepali Singhee

First published in London, United Kingdom, 2020 by IntechOpen
IntechOpen is the global imprint of INTECHOPEN LIMITED, registered in England and Wales, registration number: 11086078, 5 Princes Gate Court, London, SW7 2QJ, United Kingdom
Printed in Croatia

British Library Cataloguing-in-Publication Data
A catalogue record for this book is available from the British Library

Additional hard and PDF copies can be obtained from orders@intechopen.com

Chemistry and Technology of Natural and Synthetic Dyes and Pigments
Edited by Ashis Kumar Samanta, Nasser S. Awwad and Hamed Majdooa Algarni
p. cm.
Print ISBN 978-1-78985-997-3
Online ISBN 978-1-78985-998-0
eBook (PDF) ISBN 978-1-83968-758-7

For EU product safety concerns:
IN TECH d.o.o., Prolaz Marije Krucifikse Kozulić 3, 51000 Rijeka, Croatia, info@intechopen.com or visit our website at intechopen.com.

Meet the editors

Prof (Dr) Ashis Kumar Samanta is presently a Professor and Former HOD of the Department of Jute and Fibre Technology, Institute of Jute Technology (IJT), University of Calcutta. He is also the Chairman of the BIS TXD-07 Committee for Speciality Chemicals and Dyestuff as well as the Chairman and Board of Studies/Experts for the Bachelor of Fashion and Apparel Design (BFAD) Course, University of Calcutta. He was the Professor In-Charge (Academic) at the Institute of Jute Technology (IJT) for 2 years (Aug 2009 to July 2011). He was a member of the Governing Council of IJT for 3 terms, a member of the Research Advisory Committee (RAC) of ICAR-NIRJAFT, Kolkata (2010-2012), a member of the Board of Studies for the B.Des. Course, Viswabharati (2011-13), a member of the PhD Committee in Textile Technology, CU (2009-2017) and Former Chairman/Convener and Co-Chairman, Textile Engg. Divn. WBSC of IE(I) (1998-2018). He is a member of AASCIT and AATCC, USA. He also served as Former Asst. Director in MANTRA, Surat, 1997-98. He is listed in the global directory of the ISO-TC-38 Committee as expert member from BIS and has attended all plenary meetings of ISO TC 38 since 2015 to present 5 NWIP on the testing of natural dyes. He is the recipient of the "Bharat Nirman Award-2014" for special contributions in the Jute and Textile Technology Education, Research and Training. He received the Eminent Engineer Award in Sept-2019, the "Triguna Charan Sen Award" twice in 1995-96 and & in 2013-14, the "Subject Paper Award" in 2006; and the "Special Merit Award" in 2007 all from the Institution of Engineers (India). He is the reviewer of many international journals such as TRJ, JNF, JIT, JTI, IJFTR and JFP, Coloration Tech, Jclp, J Bio-Res & Tech and Ind. JNF. He is an editorial board member of five reputed national and international textile journals. His main research areas are chemical processing and finishing of textiles including natural dyeing, nano-technology and PCM smart materials, where he has contributed more than 300 publications including 130 research papers (70 international and 60 national) besides 50 technical articles and 101 conference papers, 8 international book chapters and 11 training manuals and 3 books. Presently, he has edited this book on "Chemistry and Technology of Natural and Synthetic Dyes and Pigments". So far, he has guided twelve PhD students in chemical processing of Ttxtiles and is presently guiding another 3 research scholars. He was PI/Project leader of 16 R & D & 2 HRD projects in his career.

Dr. Nasser S. Awwad has a PhD in inorganic and radiochemistry (2000) from Ain Shams University and a post-doctorate degree at Sandia National Labs, New Mexico, USA, 2004. Nasser Awwad was an Associate Professor of radiochemistry in 2006 and Professor of inorganic and radiochemistry in 2011 at the Egyptian Atomic Energy Authority. He has been a Professor at King Khalid University, Abha, KSA from 2011 to now. He has published two chapters in the following books "Natural Gas - Extraction to End Use" and "Advances in Petrochemicals". He has been the editor for six books about: uranium, new trends in nuclear sciences, dyes in industry and lanthanides, and nuclear power plants. In addition, he has published 84 papers in ISI journals. He supervised 4 PhD and 16 MSc students in the field of radioactive and wastewater treatment. He participated in 25 international conferences in South Korea, USA, Lebanon, KSA, Egypt and India. He participated in 6 large projects with KACST at KSA and Sandia National Labs at USA on the conditioning of radioactive sealed sources and wastewater treatment. He has been the leader of many research groups about the utilization of nanomaterials for treatment of inorganic and organic pollutants and has also been a member of some research groups. He is a member of the Arab Society of Forensic Sciences and Forensic Medicine and is a member of the Egyptian Society for Nuclear Sciences and its applications. He is on the editorial board of the Journal of Energy and Environmental Research and Technology. He was a member of the organizing committee of the 9th World Congress and Expo on Nanotechnology & Material Science. He was the chairman of a session on Reuse and Recycling of Materials and their products held at India, Kerala, Kottayam 2009. He is a rapporteur of the Permanent Committee for Nuclear and Radiological Protection at King Khalid University and a member of the Committee for the Development of International Cooperation Management at KKU.

Dr. Hamed Majdooa Algarni is the Dean of Scientific Research at King Khalid University (KKU). He received his MSc in physics in 2007 from KKU, Saudi Arabia, and earned his PhD in 2014 from the University of York, the United Kingdom. Dr. Algarni received the Kathleen Mary Stott Prize for excellence in scientific research for the best PhD in 2013 from the University of York, UK. He received an academic distinction award (three times) from the Royal Embassy of Saudi Arabia, Cultural Bureau, London. He specializes in condensed matter physics, nano-science and nanotechnology, which includes growth, properties, and applications of nanostructured and biomaterials and optoelectronic devices. He has authored and co-authored 90 research articles in highly reputed journals.

Contents

Preface

Dyeing is an ancient art for textiles and clothing. In ancient times, when no synthetic dyes were available, primitive dyeing using natural dyes was used for dyeing textiles and clothing. With the discovery of the synthetic dye 'Mauve' by W. H. Perkin in 1856 and thereafter on its subsequent commercialization, production and application, synthetic dyes conquered the market for their obvious ready use and lower cost and hence the use of natural colorants declined sharply. However, in recent times, the detrimental effects of some azo-based synthetic dyes (releasing carcinogenic and toxic amines and their hazardous effect on the environment and harmful effect on human skin) have caused a ban of a few such synthetic dyes. Hence, the following three approaches for textile dyeing have started increasing: (i) dyeing and finishing textiles with natural dyes and finishes, (ii) using eco-safe synthetic dyes and chemicals, and (iii) effective effluent treatment for synthetic dyeing as useful viable option.

There is limited information available in the form of a consolidated book on natural dyes, pigments, and finishes and production of herbal textiles as well as eco-friendly dyeing chemicals and eco-safe synthetic dyeing and removal of toxic materials and colours from such dyeing effluent etc. Moreover, for more and more commercial exploitation of natural dyes and finishes exploring their functional properties, standardization of natural dyeing techniques need to be adopted for its revival as an alternative to eco-safe synthetic dyeing. Hence, this book was planned and finally published entitled "Chemistry and Technology of Natural and Synthetic Dyes and Pigments" under IntechOpen access publishing media.

This book includes the following chapters: an introductory editorial chapter on bio-mordants, bio-dyes and bio-finishes; review of natural dyes and pigments and its application, pantone-like shade generation with natural colorants, colour-based natural dyes and pigments, printing with natural dyes and pigments, functional property and functional finishes with natural dyes and pigments, eco-safe synthetic dyes and chemicals, and miscellaneous review on dyed textiles and clothing including natural dye-based herbal textiles.

The academic editor acknowledges the constant support of all the administrative and policy making executives, Author Service Managers and the editorial and production team of IntechOpen publisher without whom this book project could not have seen the light of the day and could not be completed in time.

Both the IntechOpen publisher and the Academic Editor jointly acknowledge all the contributors as authors, whose chapters have been finally accepted after peer review /revision/ modification as required.

The academic editor and IntechOpen Publisher hope that this book is found to be useful for dyers of the textile industry as well as to the future researchers in this field.

Ashis Kumar Samanta
University of Calcutta,
India

Nasser S. Awwad and Hamed Majdooa Algarni
King Khalid University,
Saudi Arabia

Section 1

Introduction

Chapter 1

Bio-Dyes, Bio-Mordants and Bio-Finishes: Scientific Analysis for Their Application on Textiles

Ashis Kumar Samanta

Abstract

This chapter deals with testing/characterization and other scientific analyses and process standardization for the application of bio-dyes, bio-mordants, and bio-finishing agents on cotton textiles. Few case studies on selective natural/bio-dyes in terms of Fourier-transform infrared spectroscopy (FTIR), high-performance liquid chromatography (HPLC), ultraviolet–visible (UV–VIS) spectroscopy, and differential scanning calorimetry (DSC) analysis have been discussed. Based on these characterizations, few test protocols have been established for the identification of each natural dye from dyed textiles, as standards of Bureau of Indian Standards (BIS). One such case study on madder is shown. Exemplary case studies on the standardization of extraction, suitable mordanting, and dyeing process variables for the application of specific natural dyes on cotton have also been provided. Few case studies on improving color fastness to washing and sunlight by suitable post-treatment with natural agents have been discussed. Case studies on different post-treatments/simultaneous dyeing and finishing treatments with bio-dyes and bio-finishing agents for improving antibacterial properties and UV protection factor (UPF) have also been explained.

Keywords: bio-mordants/mordants, bio-dyes/natural dyes, bio-finishes/natural finishes, cotton, colour fastness, dyeing process variables, natural antibacterial finish, natural UV-protective finish, madder, tesu

1. Introduction

Natural dyes are the coloring matters derived from natural sources. The production of synthetic dyes involves many chemical reactions with petrochemical-based dye intermediates, which are high energy consuming and deliver hazardous toxic chemicals to the environment [1–5]. This pollution problem of production of synthetic dyes has led many researchers worldwide to reinvestigate the methods of producing eco-friendly natural dyed textile products for future eco-textiles. Natural fiber-based textiles are biochemically processed with enzymes and dyed with natural dyes using bio-mordant and later finished with natural finishing agents in a eco-friendly mode; these whole processing chain and such textile products developed are considered as sustainable [6–8]. A sustainable textile product thus should begin and end its life cycle as smoothly as possible without harming not only human beings but also the flora and fauna of the environment.

The sustainability of textile and clothing consists of a fairly long supply chain, which starts from fiber formation and ends at the apparel production, use and disposal at the end, covering the entire life cycle analysis of textile products, where nearly half of this entire supply chain occupies the said production process life cycle and the rest is left to the consumers in terms of its usage and disposal.

A sustainable textile product can be defined as one that is created, produced, transported, used and disposed of with due consideration to environmental impacts, social aspects and economic implications, thereby satisfying all three pillars of sustainability, and is expected to create the minimum possible or very least environmental and social impacts throughout its entire life cycle. In a specific product life cycle of any textiles, it uses different materials and consumes different forms of energy and many other processing inputs to produce or generate the desirable textile products as desirable outputs in different stages of their manufacture, packaging, despatch, use and final disposal. To make this life cycle sustainable for any textile goods, all these inputs and outputs are to be environment-friendly and eco-safe particularly to the living bodies on earth including humans for the entire life cycle, i.e. it should be noncarcinogenic and has to utilize renewable/recycled input streams with the least energy consumption.

When synthetic dyes were not known to humankind, dyeing of textiles was dependant on natural colors only. Later, when synthetic dyes were made available at ease and were started their commercial selling at low cost under different classes as ready-made dye powders to use with low cost and processing advantages, large-scale textile manufacturers and even small-scale dyers had been shifted to use more and more of synthetic dyes and pigments. But, with the enhancement of knowledge on environmental concern, people started to search for eco-friendly dyeing approaches as alternatives, as synthetic dyes are made from a non-renewable source of petrochemical base with chances of generating toxic chemicals as the by-product, which are not environment-friendly. Hence, the importance of eco-friendly dyeing of textile goods has started re-examination with the growing interest of consumers towards the use of eco-friendly natural dyed fiber-based textile products, which thus have been revived now needing more and more scientific information and analysis on those.

Moreover, the century-old natural dyeing processes had never died or discarded fully, and it is being practised still in different corners of the world and India is not an exception. Considering the niche market of eco-friendly dyed and finished natural fiber-based or organic textiles, recently, most of the textile dyers-cum-exporters are showing fresh interest for using natural dyes and natural finishing agents for textile dyeing, printing and finishing, if it can bring much more value addition (than using synthetic dyes), as a sustainable textile processing for eco-textile products. Natural dyes in addition of being eco-friendly produce very uncommon, soothing and soft shades in comparison with synthetic dyes [9]. For a successful commercial use of natural dyes for any particular natural fibre-based textiles, there has to be a standardization of process variables [9, 10] for extraction, mordanting and dyeing of that particular fiber-natural dye system with its testing and characterization [11], identification of important ingredients and other scientific analysis and process standardization for application of such bio-/natural dyes, bio-/natural mordants and bio-/natural finishing agents on natural fiber-based textiles. To obtain newer compound shades using a mixture of natural dyes needs a test of compatibility [12, 13]. The scientific determination of the dyeing rate, dyeing kinetics/thermodynamics and other physico-chemical parameters [14] of dyeing of a specific natural dye for specific textiles should be derived to establish appropriate standardized scientific dyeing techniques/procedures with the optimization of dyeing process variables and to decide the required after-treatment process necessary for obtaining an acceptable color fastness behavior with reproducible and uniform color yield. The status of natural dyeing up to 2001 [4], a comprehensive review on scientific studies on natural dyes up to 2009 [5, 6] including the characterization and scientific analysis for natural dyes, eco-friendly natural dyeing with bio-mordants [15], use of sonicator/ultrasonic assistance [16, 17] in natural dyeing and enzyme-based preparation and natural dyeing of textiles [18] and functional properties of selective natural dyes like UV protection [19, 20] and antibacterial properties [21, 22] are available in the literature.

However, the major problem of natural dyes encountered is their sufficient availability due to difficulty in collection, cost, poor color fastness and standardized methods of their application. Till date, there are limited studies available in the literature on scientific evaluation and testing, characterization, identification of important ingredients and process standardization related to optimization of dyeing process variables, mechanism of dye-mordant-fiber fixation, role of different mordants and mordanting assistants/additives and chemical pre-treatments/post-treatments/modifications for natural dyeing with evaluation of dyeing rate, dyeing kinetics and to develop compound shades (to overcome shade limitations) by use of binary or ternary mixture. Hence, there is a need for precise scientific and technological knowledge and development of systematic scientific methods of dyeing textiles with natural dyes and natural finishing agents.

Natural colorants have the following advantages as compared to synthetic colorants, but natural dyes have some disadvantages too.

The advantages [6–9] of natural colorants and natural finishing agents:

1. Eco-friendliness: Natural dyes are less toxic, less polluting, less health hazardous, noncarcinogenic and non-toxic. Most of the natural dyes are considered to be eco-friendly as these are obtained from renewable resources as compared to synthetic dyes (derived from non-renewable petroleum resources and synthesized in an intermediate route involving many chemical hazards).

2. Soothing to the human eye: They have harmonizing colors, are gentle, soft and subtle and create a restful effect producing a soothing shade.

3. Biodegradability: Unlike non-renewable basic raw materials for synthetic dyes, the natural dyes of plant sources are usually agro-renewable/vegetable-based products and at the same time are biodegradable.

4. Availability of a wide range of colors: Natural dyestuff also can produce a wide range of colors by a mix-and-match system. A very small variation in the dyeing technique or the use of different mordants with the same dye or different concentrations of mordants on the same dye can create a variety of new shades.

5. Functional benefits of natural dyes/finishing agents towards wearers: Many natural dyes and natural finishing agents can be used as UV-protective and antibacterial materials. Natural textiles dyed with suitable natural colorants and finished with specific natural finishing agents can thus provide protection from UV rays, microbes or even mosquito bites. Natural dyes, e.g. myrobalan, turmeric, madder/manjistha (MJ), Arjuna, safflower, etc., possess a medical curative property as Ayurvedic medicine.

Despite so many advantages, natural dyes have some drawbacks [6–9] too as listed below:

1. Requirement of a longer time: Natural dyes require a longer dyeing time for extraction and purification and also for actual dyeing via mordanting in comparison with time required to apply synthetic dyes on same textiles, for additional step of mordanting. Dye extraction steps require additional time and setup. The exhaustion of most of the natural dyes on textile materials is poor in spite of using the mordant which leaves a large quantity of colorant component in the dye bath after dyeing due to poor exhaustion, which increases the cost of dyeing though natural dyes are eco-safe as compared to dyeing with synthetic dyes.

2. Shade range limitations: It is difficult to obtain all desirable shades from natural dyes, i.e. ranges of shade available from natural dyes and pigments are limited. Out of the required three primary colors—red, green and blue—although there are several sources for reddish and greenish dyes, there is only one major source of the blue natural dye, i.e. from natural indigo. As different natural dyes and pigments are differing in their chemistry and application process, only few natural dyes are compatible to be applied together in a binary/ternary mixture and obviously are noncompatible for producing compound shades.

3. Non-reproducible shade: Due to the difference in proportion of constituents, the variation of these agro-products from one crop season to another in terms of location, species maturity period and consistent shade always cannot be obtained. However, after extraction, by UV spectral analysis, if the dye extract can be diluted to bring it to the same concentration level of colored solution, the reproducibility of shades can be better assured even with the extract of varying concentration of colored extract of such selective natural dyes.

4. Fastness properties: Only a few natural dyes and pigments possess a good rating of the overall color fastness satisfying consumers' needs. As mordants with objectionable metal salts such as Cr, Cu, Sn, etc. are not permitted or allowed under eco-norms of different countries (ecomark scheme of India or OTN-100 norms of the UK, etc.), the overall color fastness rating of natural dyed textiles is sometimes poor to medium only. However, recent approaches of a suitable pre-treatment with natural bio-mordants containing tannic acid residues or pre-treatment with natural cationic agents like acid-extracted soya bean seed waste [23] and post-treatment with chitin (natural cationic agent) or post-treatment with natural UV absorber agents (like orange peel, eucalyptus leaves, etc.) are being used for enhancing the color fastness rating for wash and light to an acceptable level.

The present scenario shows that an approximately 1% share of textiles is only being dyed with natural dyes that are used mostly in the cottage sector by traditional artisans and small-scale textile dyers. The reasons for not using natural dyes in the large-scale textile sector were, however, lack of availability of ready extracted and purified dye powder, lack of standardized dyeing processes with assurance of reproducible and uniform shades and non-warranty of the required level of color fastness to achieve. Natural dye application processes cannot be easily implemented by large-scale textile mills as machine dyers. However, recent effort of people, understanding environmental friendliness of natural dyes and finishes with growing demands of customers for such eco-friendly products, ready availability of extracted and purified natural dye powder by certain newer natural dye manufacturers along with their recommendation of standardized mordanting and natural dyeing and finishing processes have partly overcome these issues for machine dyeing viable in large-scale textile sector. The natural bio-indigo dyeing on cotton denim [24] to produce bio-denim using natural reducing agents, as developed by Ama Herbal Lab Pvt Ltd, Lucknow and use of earth colour from plant waste developed by Archoma, Mumbai has become a commercial success in India for large scale industry.

However still natural dyeing industry is a labour-intensive industry concentrating in the small-scale sector of handloom textiles and khadi (hand spun and handloom woven) sector in developing countries like many South-East Asian countries providing livelihoods by creating job opportunities for all those engaged in cultivation, extraction and application of these natural dyes on textiles. Cultivation of natural dye plant is also an alternative cash crop to the cultivators. For the promotion of natural dyeing and natural finishing of textiles including the khadi sector, the availability of extracted and purified natural dye and pigment powder as ready soluble colorants has to be assured to promote such eco-friendly natural dyed textiles in the khadi sector. Some work [19] in this endeavor by Mahatma Gandhi Institute for Rural Industrialization (MGIRI), Wardha, Maharashtra, India, is worth mentioning for the cotton khadi sector. As the small-scale khadi and handloom sectors of textiles lack the resources to install and operate expensive effluent treatment plants needed to bring the synthetic dye's effluent within the eco-limits set by regulatory authorities, they will be more benefitted by adopting standardized dyeing methods with natural dyes and natural finishes on specific natural fiber-based textiles, by mitigating its major problems.

To promote natural dyed textiles, there is an essential need of establishing proper identification methods and test protocols with suitable national and international test standards of identifying natural dyes from the dyed textiles followed by establishing proper certification protocols for natural dyes from the natural dyed textiles in the form of natural dye mark protocol, which would definitely improve consumers' confidence and would benefit both producers and users. In this attempt BIS, India, and also ISO/TC-38 have been working on it since 2015, and till date few such test standards are finalized and published [25, 26] by BIS, India, and ISO/TC-38 has framed a special working group—WG-31—on natural materials for textiles for finalizing such global standards for worldwide consumers and producers benefit, so that both large- and small-scale sectors of textile industries can get benefitted.

2. Results and discussion on a few case studies for the application of natural dyes on textiles

2.1 Characterization and testing of natural dyes

2.1.1 Ultraviolet-Visible UV-VIS spectral analysis of aqueous extracted solution of natural dyes

1% aqueous solution of each purified individual natural dye powder was separately prepared and was subjected to a wavelength scan in a UV-VIS absorbance spectrophotometer for 200–700 or 1100 nm wavelength range to identify the predominant lambda maximum with specific peaks in UV region and specific peaks in visible region in order to understand the UV-absorbing nature of the dye.

Absorbance peaks of UV-VIS spectrum of a dilute aqueous solution of few individual purified natural dyes have been shown in **Table 1**.

In the UV-VIS spectrum of each natural dye, if there are predominating sharp peaks in the UV region, the dye has a reasonable UV-absorbing character, indicating that the said particular dye may have a higher UV absorption rate enhancing the UV-fading rate due to its more UV-absorbing character, and hence specific cares and after-treatments with more stronger UV absorber compounds [8, 13] are necessary for improving light fastness of particular dyed textiles, by additional treatment with more stronger UV absorber compounds having preferential UV absorption, leaving the dye with UV absorber character to absorb less UV light.

It may be observed from the wavelength scan of UV-VIS spectrum of tesu extract (**Figure 1**) in an aqueous medium that in the UV region (190–380 nm), there is hardly any preferential absorbance peak in this UV region, but in the extended UV zone, it shows a small broad hump at 250 nm and a small trough at 385 nm along with showing negative absorbance in a particular wavelength (190–250 nm). Thus, there is almost no preferential UV absorption which occurs in tesu colorant (extracted in the aqueous medium), and hence the light fastness of tesu dye is not to be affected much by UV light exposure, showing predominating hue by large peak at λ_{max}-490 nm for tesu.

UV-VIS spectrum of few selective natural dyes and their peaks is shown in **Figure 2** and **Table 1**, which shows that these selective natural dyes (except tesu) show UV-absorbance peaks either at 242–250 nm or at 370–380 nm in UV region, where it is found to be dominating in 242–250 nm in UV zone for babool and marigold, whereas madder (Rubia/manjistha) has no such preferential UV-absorbance peak in 242–370 nm zone except a small but sharp peak at 380/390 nm, which is possibly due to absorbance of the keto or more specifically for

Purified natural dyes	Wavelengths (nm) for major UV-VIS peaks with λ_{max}
a. Marigold (MG)	250, 598 (λ_{max})
b. Babool (BL)	242, 372, 550, 602 (λ_{max})
c. Jackfruit wood (JFW)	250, 540 (λ_{max})
d. Sappanwood (SW)	242, 373, 610 (λ_{max})
e. Red sandalwood (RSW)	250, 370, 540, 602 (λ_{max})
f. Manjistha/madder (MJ)	380, 540, 610 (λ_{max})
g. Tesu (TS)	250 (very small negligible peak), 385 (small trough), 490 (λ_{max})

Table 1.
UV-VIS absorbance peaks at different wavelength for few purified natural dyes.

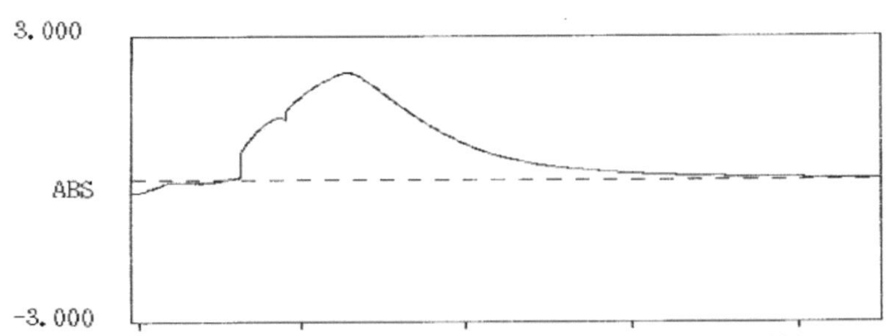

Figure 1.
UV-VIS spectra of the extract of color component of tesu as natural dyes.

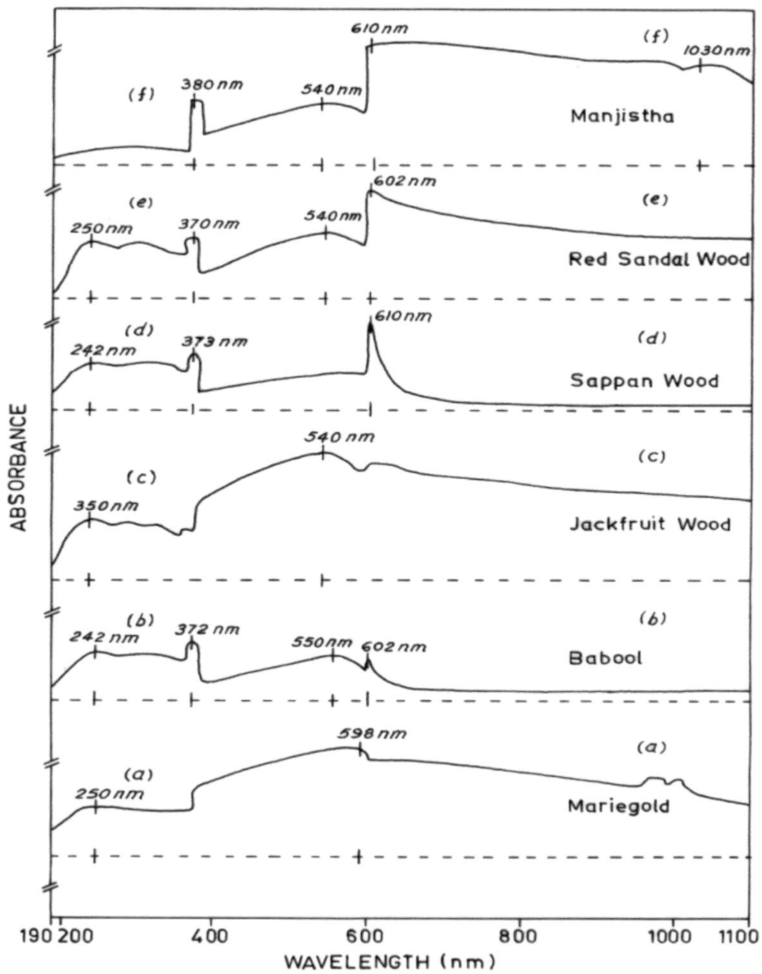

Figure 2.
UV-VIS spectra of the extract of color component of few natural dyes other than tesu (x- and y-axes: Absorbance values in y-axis are −3.0 to +3.0 and wavelength values are 190–1100 nm).

quinone group in anthraquinone structure present in madder/manjistha (Rubia) containing manjisthin, alizarin, etc. [8]. Besides these peaks, the other characteristic peaks in the visible region (400–700 nm), in the UV-VIS spectrum of corresponding solutions of the said natural dyes, are the peaks for the predominating hue of the main color component of the corresponding dye. In some cases (as in madder, red sandalwood, babool, etc.), there are more than one or multiple peaks in the visible region that indicates the presence of more than one color component as a mixture which shows multiple peaks at the visible region closely differing from λ_{max} for predominating hues for each of them [8, 11, 13].

2.1.2 Fourier-transform infrared spectroscopy (FTIR) spectral analysis

The purified dye powder was further washed in distilled water followed by washing in 100% acetone before final drying and was then subjected to FTIR spectroscopy in a double-beam spectrophotometer (in Perkin Elmer spectrum-II FTIR spectrophotometer) using a KBr disc technique. The FTIR spectrum for each purified selective natural dye powder has been shown in **Figure 3** for tesu, **Figure 4** [spectra (a)–(e) for red sandalwood, jackfruit wood and madder/manjistha

(Rubia)] and **Figure 5** [spectra (d)–(f) for marigold, babool and sappanwood]. The characteristic FTIR bands/peaks (absorption band, type and made of vibrations, nature and appearance of bands/peaks along with the description of responsible bonds and associated chemical functionalities) for each of the purified natural dyes observed have been marked with corresponding wave number (cm^{1}) for each spectrum [8, 11] in **Figures 3–5** and are also tabulated in **Tables 2–4** [8, 11].

The FTIR spectral scan analysis is presented in **Tables 2–4**, which are easy to understand for confirming the type of chemical groups indicating types of bonds present therein in the main color components for each purified natural dye mentioned there. The known chemical structures of the major color components of each of these natural dyes have been reported in the earlier literature [7, 8]. The FTIR bands of each purified natural dye are found to be matching with the earlier reported chemical structures of major color components of each individual natural dye mentioned here.

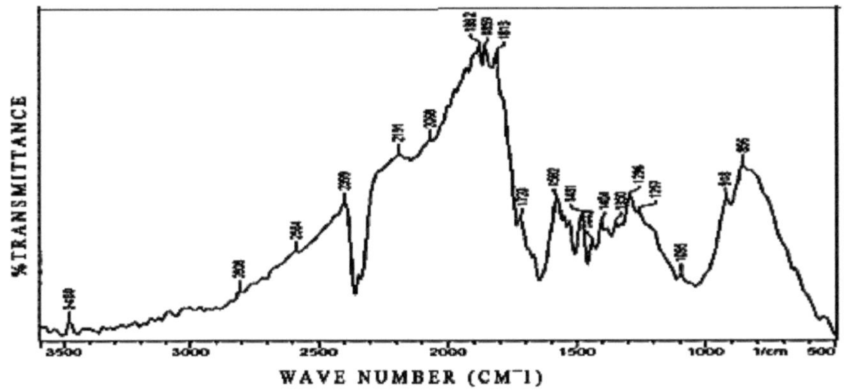

Figure 3.
FTIR spectra of purified tesu as natural dyes (transmittance values are 25–100%).

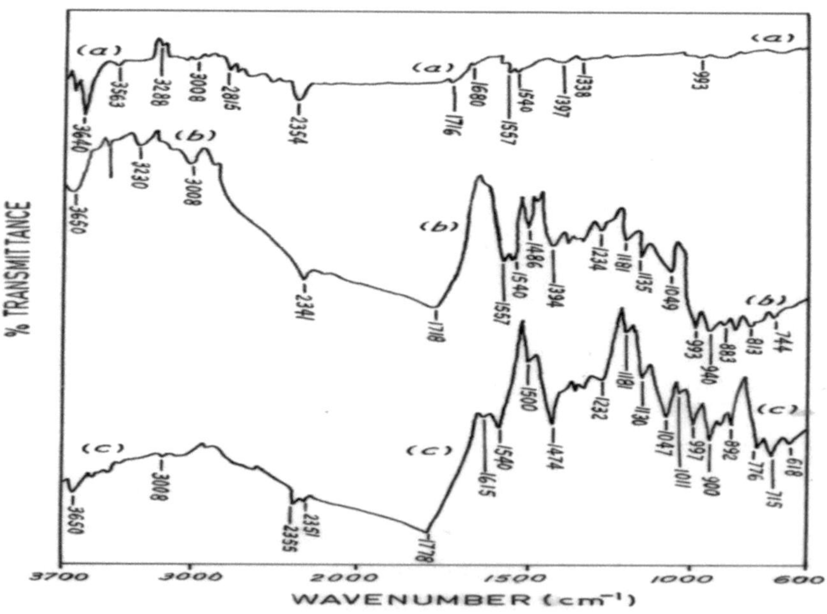

Figure 4.
FTIR spectra of purified natural dyes: (a) red sandalwood, (b) jackfruit wood and (c) manjistha/madder (Rubia) (transmittance values are 25–100%).

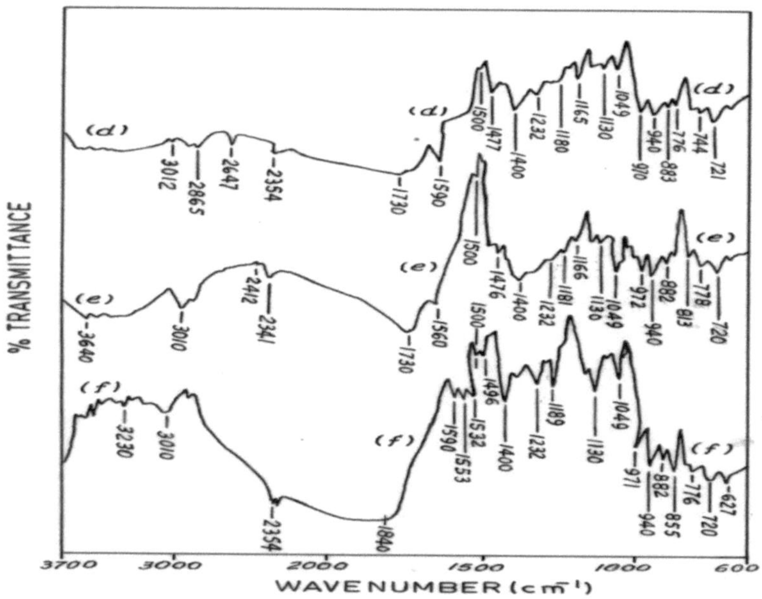

Figure 5.
FTIR spectra of purified natural dyes (d) marigold, (e) babool and (f) sappanwood (transmittance values are 25–100%).

Absorption band, wave number in cm^1	Nature of bands/peaks	Bond and its mode of vibration with associated responsible functional groups
For TS (TESU)		
608–668 (avg. 647) and 1350	Small	−OH out-of-plane bending and −OH in-plane bending
1210–1296 (avg. 1257)	Small	−C−O−C and −C=O stretching combination
1376–1487 cm^{-1} (avg. 1381)	Medium	−CH and $-CH_2$ bending
1540–1557, 1582, 1589 and 1630	Small and doublet	–C=C stretching in nonconjugated –C=C–
1680–1750	Medium sharp	–C=O stretching attached with aromatic ring structure
2359, 2810 and 3008	Weak but intense	–C–H stretching in aromatic ring
3288 and 3563–3650	Small	–O–H stretching in free –OH of phenolic structure

Table 2.
Spectroscopic data for FTIR peaks of a purified extract powder of tesu as natural dyes.

2.1.3 Analysis of DSC thermograms

The purified dye powder for each selective natural dye (except tesu) was further repeatedly washed in distilled water twice followed by washing with ethyl alcohol followed by washing again in 100% acetone and was then subjected to final drying under a vacuum oven. Then 2 mg of each individual dye powder was placed in the aluminium crucible of differential scanning calorimeter (DSC) console (using Shimadzu differential scanning calorimeter Model-DSC-50) and started heating and running it by usual method under flowing nitrogen (N_2 gas flow rate was at

Absorption band, wave number in cm^{1}	Nature of bands/peaks	Bond and its mode of vibration with associated responsible functional groups
a. For RSW		
993	Small	–C–H deformation in benzene ring
1338–1397	Doublet	–C–O–H bending in primary and secondary alcohol
1540–1557	Small and doublet	–C=C stretching in nonconjugated –C=C
1680	Small	–C=C stretching in nonconjugated –C=C
1716	Small	–C=O quinone to aromatic ring or –C=O stretching in ring ketone in aromatic ester and 6C ring ketone
2354 and 3008	Weak but intense	–C–H stretching in aromatic/benzene ring
2815	Small	–C–H stretching in –O–CH_3 group
3288	Small	–O–H stretching in bonded H–bonds
3563–3650	Small	–O–H stretching of phenolic –OH
b. For JFW		
744–883	Multiplet	–C–H out-of-plane deformation in benzene ring
940	Weak but intense	–C–O stretching vibration in higher cyclic ether linkages (as present in morol for jackfruit wood)
993	Weak but intense	–C–H deformation in benzene ring
1049	Medium	–C–O stretching in primary or secondary alcohol
1135–1234	Multiplet and sharp	–C–O stretching and –OH in-plane deformation in phenol
1394	Small	–C–H bending in methyl ketone or methyl groups
1486	Sharp but small	–C=C stretching in aromatic ring
1540–1557	Small doublet	–C=C stretching in aromatic ring
1718	Small	–C=O stretching of quinone to aromatic ring or –C=O stretching of ring ketone in aromatic ester and 6C ring ketone
2341	Small but sharp	–C–H stretching in benzene ring
3008	Small	–C–H stretching in aromatic ring
3230	Medium intense	Hydrogen bond –OH stretching in phenolic structure
3563–3650	Weak but intense	–OH stretching for aromatic phenolic –OH groups

Table 3.
Spectroscopic data for FTIR bands/peaks of other selective purified natural dyes.

50 cm^3/min), and heating rate was maintained at 10 °C/min over a temperature range from ambient (about 28–30°C) to 490–500°C and obtained DSC thermograms are plotted as shown in **Figure 6**.

In each case, the thermal transition temperatures are marked in each thermogram (a) to (f) in **Figure 6**, and the corresponding data on the maximum peak for thermal transition temperatures, nature and appearance of DSC peaks and probable reasons [8, 11] of the observed thermal transitions are tabulated in **Table 5**, except

Absorption band, wave number in cm[1]	Nature of bands/peaks	Bond and its mode of vibration with associated responsible functional groups
c. For madder/MJ, also known as Rubia		
618–900	Multiplet	–C–H out-of-plane deformation in benzene ring
997	Sharp	–C–H deformation in benzene ring
1011–1047	Medium	–C–O stretching in primary or secondary alcohol
1130–1232	Very strong	–C–O stretching and –OH in-plane deformation in phenol
1474	Small	–C=C stretching in aromatic ring
1500–1540	Sharp	–C=C stretching in aromatic ring
1615 and 1718	Small and sharp	–C=O stretching in aromatic ring and ring ketone
2355–2351	Doublet	–C–H stretching in –O–CH_3 group
3008	Small	–C–H stretching in aromatic/benzene ring
3653–3650	Small	–O–H stretching in phenolic –OH groups
d. For MG		
720	Small but intense	–C–H bending in polymethylene (as present in xanthophyll esters present in MG)
744–883	Multiplet and sharp	–C–H out-of-plane deformation in benzene ring
940	Tiny	–C–O stretching vibration in higher cyclic ether linkages
970	Small and sharp	–C–H deformation in aromatic ring
1049	Sharp and intense	–C–O stretching in primary or secondary alcohol
1130–1232	Multiplet	–O–H in-plane deformation in phenols
1390–1477	Sharp	–C=C stretching in aromatic ring
1590	Sharp and intense	–C=C stretching in aromatic ring
1730	Small	–C=C stretching and –C=O stretching in nonconjugated–C=C
2354, 2647	Small	–C–H stretching in –O–CH_3 or methyl ether
3022–3012	Doublet	–C–H stretching in aromatic ring
e. For BL		
720	Small but intense	–C–H bending in polymethylene (as present in xanthophyll)
778–882	Sharp multiplets	–C–H out-of-plane deformation in benzene ring
940	Sharp	–C–O stretching vibration in higher cyclic ether linkages
972	Small and sharp	–C–H deformation in aromatic ring
1049	Sharp and intense	–C–O stretching in secondary or primary alcohol
1130–1232	Sharp multiplets	–OH in-plane deformation for phenolic structure
1400–1500 and 1500	Multiplet or tiny	–C=C stretching associated with aromatic ring structure

Absorption band, wave number in cm[1]	Nature of bands/peaks	Bond and its mode of vibration with associated responsible functional groups
1730	Small	–C=C stretching in nonconjugated structure
2354–2312	Doublet	–C–H stretching in benzene/aromatic ring structure
3640	Small	Vibration of free phenolic –OH groups
f. **SW**		
627	Small	–C–H out-of-plane deformation in benzene ring
720	Small	–C–H bending in $(CH_2)_n$, i.e. polymethylene
776–882	Sharp multiplets	–C–H out-of-plane deformation in benzene ring
940	Sharp	–C–O stretching vibration in higher cyclic ether linkages
971	Sharp	–C–H deformation in aromatic ring structure
1049	Sharp	–C–O stretching in primary or secondary alcohol
1130–1232	Sharp multiplets	–OH in-plane deformation phenols
1400–1500	Multiplet	–C=C stretching in aromatic ring
1532–1590	Multiplet	–C=C stretching in aromatic ring
1840	Very small	
2354	Broad trough	–C–H stretching in benzene ring
3010	Small	–C–H stretching in aromatic ring
3230	Small	Hydrogen bonded –OH stretching in phenol

Table 4.
Spectroscopic data for FTIR bands/peaks of selective other purified natural dyes.

tesu, as tesu is considered in the earlier literature to contain normal heat-resistant butein [29] as the color component (hence, there was no need to study the thermal stability of tesu at dyeing temperature zone of 60–100°C).

The frequently asked question (FAQ) for this study arises: why DSC study of natural dyes is necessary? By DSC thermogram analysis, the thermal degradation temperature of any color components present in any natural dye molecules is indicated for heating under nitrogen at different temperature zones, which very well identify and reveal lower-temperature degradable components, if present in that sample. If any lower-temperature degradable components are there in a specific natural dye extracted sample, then its dyeing should be done below that temperature, i.e. dyeing temperature should not exceed above that temperature at all. Otherwise, if one of the color components degrades thermally at dyeing temperature in dye bath, this will cause some loss of color component and less shade depth. To avoid this, it is also essential to study the DSC parameters for natural dyes also. For example, in DSC thermogram—c for red sandalwood—there is an exothermic sharp thermal degradation peak at 82°C, and hence it should not be dyed above 80°C, i.e. preferably dyed below 80°C or preferably at 70°C. Similarly for DSC thermogram—f for madder—there is also an exothermic broader hump for thermal degradation of one of its component starting at 74°C and continuing up to 125°C, indicating that madder dyeing should be done below 74°C, i.e. preferably around 60–65°C only (not at the water boiling temperature).

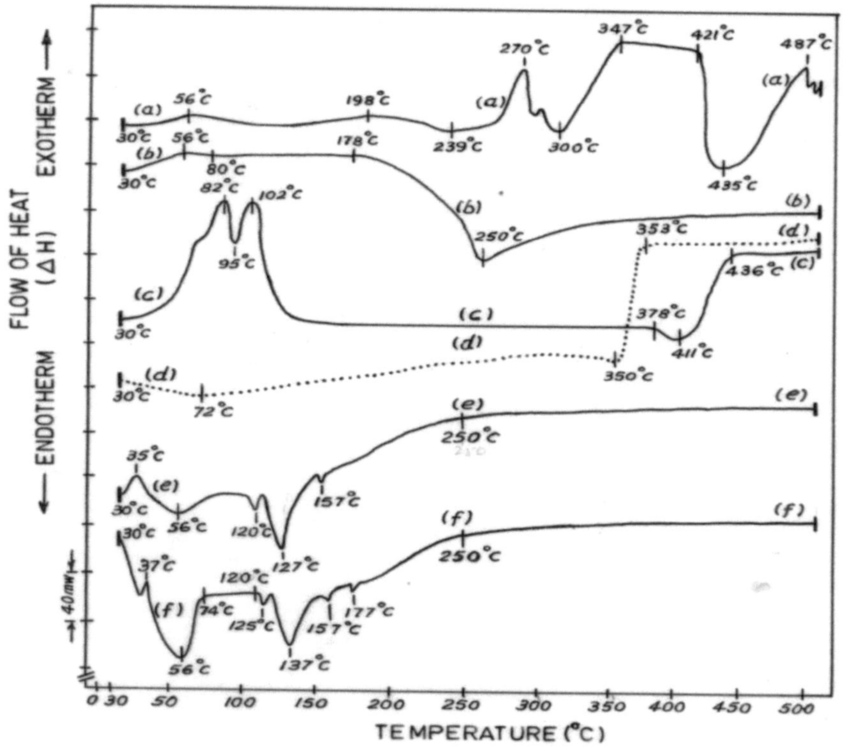

Figure 6.
DSC thermograms of purified natural dyes: (a) babool, (b) jackfruit wood, (c) red sandalwood, (d) sappanwood, (e) marigold and (f) manjistha/madder (Rubia).

2.1.4 Test method for the identification of specific natural dyes from such dyed textiles

A number of test methods developed for different commercially used natural dyes for textiles to test those natural dye powder and natural dyed clothes for the identification of specific natural dyes from such dyed textiles have been adopted by the BIS as IS standards [25, 26], some of which are under favorable consideration of ISO/TC-38 also for global standardization of these test methods as per the ISO format. For example, one such method of test of identification for madder/manjistha (Rubia) as natural dye is discussed here to understand the process.

2.1.4.1 Identification of natural madder (Rubia) as compared to synthetic alizarin by UV-VIS spectral test

Purified natural and synthetic counterparts of similar two red dyes (madder/Rubia known as manjistha containing manjisthin and synthetic alizarin considered being used instead of madder) were taken and weighed separately (0.1 gm) and dissolved in 1000 ml dichloromethane and scanned through UV-VIS spectrophotometer for wavelength scan. For the visible spectrum these two solutions were further diluted five times and were used for UV spectrum study for identification by comparison of the peaks and UV-VIS spectrum with lambda maxima and optical density (OD) values of natural madder (Rubia) and synthetic alizarin red colors, as shown in **Figure 7** and **Table 6**.

It is observed that in the visible region (400–700 nm) of the UV-VIS spectrum, characteristic peaks found are at 426 and 491 nm for synthetic alizarin, and

Compound dyes	Thermal transition temperatures, °C	Nature and appearance of DSC peaks	Probable reasons for observed thermal transitions
a. **BL**	56	Small wide hump (exothermic)	Combined effects of breaking of H-bonds and moisture evaporation
	178	Broader crest (exothermic)	Decomposition of other minor constituents
	239	Broader trough (endothermic)	Decomposition of some other minor constituents
	279	Sharp peak (exothermic)	Thermal decomposition of gallic acid
	300	Deep trough (endothermic)	Decomposition of epicatechin
	347–421	Flat plateau region (exothermic)	No effects
	435	Deep trough (endothermic)	Thermal decomposition of catechin
	487	Small sharp peak (exothermic)	Thermal decomposition of theoflavones or other minor constituents present
b. **JFW**	56	Small hump (Exothermic)	Combined effects of breaking of H-bonds and moisture evaporation
	80–178	Flat plateau region	No effects
	250	Wide trough (endothermic)	Thermal decomposition of morol
c. **RSW**	82	Sharp peak (exothermic)	Breaking of strong H-bonds of deoxysantalin
	95	Sharp trough (endothermic)	Thermal decomposition of deoxysantalin
	102	Sharp peak (exothermic)	Breaking of more strong H-bonds of santalin (A and B)
	129–378	Flat plateau region (endothermic)	No effects
	411	Small trough (endothermic)	Thermal decomposition of santalin (A and B)
	436–500	Flat plateau region (exothermic)	No effects
d. **SW**	72	Small transition (endothermic)	Moisture evaporation
	350	Small trough with sharp transition (endothermic)	Thermal decomposition of brazeline
	353–500	Flat plateau region (exothermic)	No effects
e. **MG**	35	Tiny peak (exothermic)	Breaking of H-bonds
	56	Medium trough (endothermic)	Moisture evaporation [7]
	120	Small trough (endothermic)	Thermal decomposition of xanthophyll ester
	127	Sharp trough (endothermic)	Thermal decomposition of quectrol (flavanol) [7]

Compound dyes	Thermal transition temperatures, °C	Nature and appearance of DSC peaks	Probable reasons for observed thermal transitions
	157	Tiny trough (endothermic)	Decomposition of other minor constituents
	250–500	Flat plateau region (exothermic)	No effects
f. MJ or madder (Rubia)	37	Tiny peak (exothermic)	Breaking of H-bonds
	56	Deep trough (endothermic)	Moisture evaporation
	74–120	Flat plateau region (exothermic)	No effects
	125	Small and sharp trough (endothermic)	Thermal decomposition of purpuroxanthin
	137	Sharp trough (endothermic)	Thermal decomposition of manjistha
	157	Tiny trough (endothermic)	Thermal decomposition of purpurin
	177	Tiny trough (endothermic)	Thermal decomposition of pseudopurpurin
	250–500	Flat plateau region (exothermic)	No effects

Table 5.
DSC thermal transition temperatures of extracted and purified natural dyes.

corresponding peaks for natural madder (Rubia) dye containing manjisthin are at 398 nm (OD 0.801) and 426 nm (OD 0.838) as distinguishing characteristic peaks and their optical densities for differentiating natural madder containing manjisthin and synthetic alizarin by UV-VIS spectral analysis as shown in **Figure 7**.

While in the UV region of the UV-VIS spectrum, optical density at same lambda max. Values for peaks in UV region for natural madder (Rubia) and synthetic alizarin dyes are different. In the UV region (200–380/399 nm) of the UV-VIS spectrum, although both shows the lambda max peaks in UV region at same wavelength, i.e. at 250 nm, but their optical densities values at 250 nm for both the red colors are different (corresponding data are provided in **Table 6**).

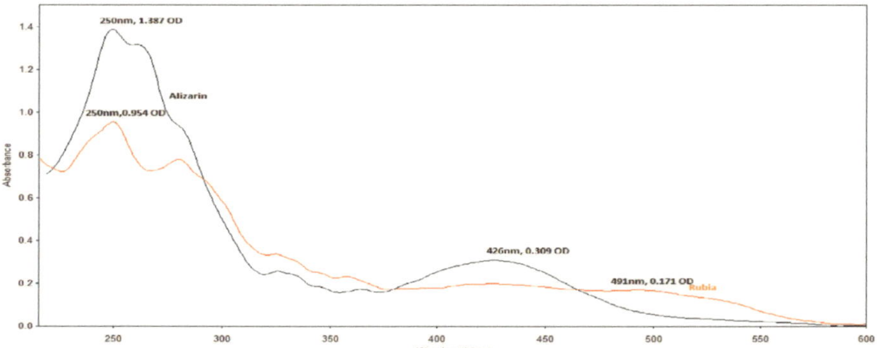

Figure 7.
UV-VIS spectrum of madder (Rubia) and synthetic alizarin dyes.

Peak shown at nm	In synthetic red alizarin	In natural madder (Rubia)	Comments for distinguishing peaks and lambda max values for identification (describing difference with reason)
	250 nm (1.38 OD) and 426 nm (0.309 OD)	250 nm(OD, 0.954) and 491 nm (OD, 0.171)	The pattern of the peaks and optical density at lambda max values in UV and visible region are very different for the two dye samples as per **Figure 7**

Table 6.
Identification of peaks and lambda max values from UV-VIS spectrum of natural madder (Rubia) and synthetic alizarin.

2.1.4.2 Identification of natural madder (Rubia) dye with synthetic alizarin using high-performance liquid chromatography (HPLC) with UV detector

The purified extracted dye powder of natural madder (i.e. Rubia containing manjisthin) and synthetic alizarin as two different red dye samples were analyzed by HPLC method [27]. Both the natural madder and synthetic alizarin purified and washed dye powder were weighed separately (0.1 g) and were dissolved in 1000 ml of methanol. 1 μL of the prepared solution of both one by one was injected to the C-18 reverse phase column of HPLC with UV detector and eluted in HPLC column. The base line showed response within a run of 15 min for natural madder (Rubia) and synthetic alizarin. The parameters of this assay were made to be such that a clean peak of the both the samples are observed. Clear observation was made from these two red dye samples as shown in the two chromatographs given in **Figures 8** and **9**, that for synthetic alizarin, one main peak is at 1.6 min and no peak after 3.5 min, while natural madder (Rubia) has main peak at 1.8 min and another peak at 11.3 min, as per data provided in **Table 7**.

Thus, natural madder (Rubia) and synthetic alizarin show clear differences in their characteristic peaks, retention times and peak heights. Natural madder (Rubia) always shows two equivalent peaks as compared to synthetic alizarin which shows three peaks due to un-purified sample of the latter; this is a mark of identification factor between natural madder (Rubia) and synthetic alizarin by HPLC method with a UV detector.

This method of identifying Rubia (madder) as a natural dye to identify from its dyed textiles has been standardized and adopted as Indian standard—IS-17085-2019, January 2019 [25]—and many more such natural dye test standards for the identification [25, 26] of specific natural dyes from dyed textiles are adopted as Indian standards by the BIS.

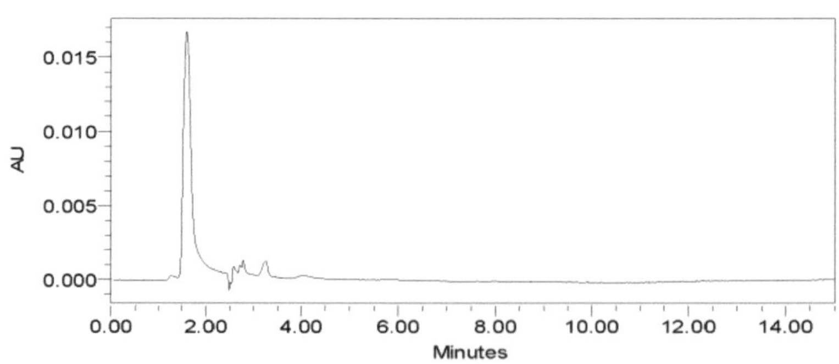

Figure 8.
Chromatogram of synthetic alizarin.

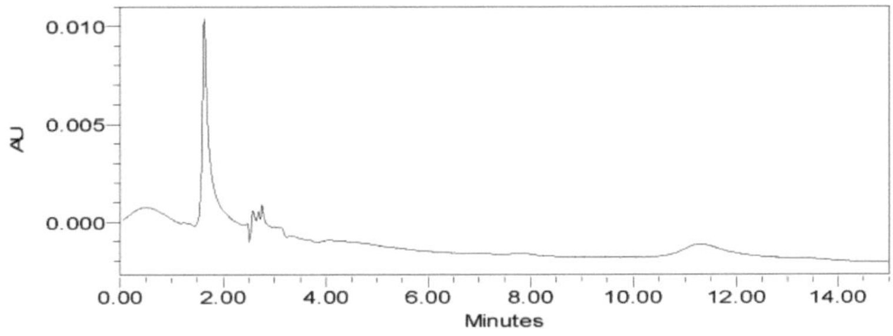

Figure 9.
Chromatogram of natural madder (Rubia) containing natural alizarin.

Peak shown at nm	In synthetic alizarin	In natural Rubia	Comments for distinguishing main peaks and time in min for identification (with reason for differences)
255 nm	One main peak is at 1.6 min and no peak after 3.5 min	1.8 min main peak, small peaks at 2.5 and 11.3 min	One main peak is at 1.6 min and no peak after 3.5 min while natural Rubia has a main peak at 1.8 min and a characteristic peak at 11.3 min

Table 7.
Chromatogram analysis data of natural madder (Rubia) and synthetic alizarin.

2.2 Standardization of extraction, mordanting and dyeing process variables: a specific case study

The study of optimization of extraction conditions, mordanting, and dyeing process variables of few natural dyes on different textiles is available in the literature [6, 10, 28, 29], but the study of dyeing process standardization of madder (*Rubia tinctorum-Indian madder was obtained from M/s AMA Herbals Lab Pvt Ltd, Lucknow, India*) applied on cotton is reported here. In this part of study, initially the optimization of conditions of extraction of madder (manjistha/Rubia) as a natural dye/colorant is reported in **Item 2.2.1**. Later, bleached cotton fabrics were dyed after a sequential double natural pre-mordanting [10% myrobalan (harda) + 10% natural potash alum, i.e. 20% overall application of harda + natural Potash alum in 50:50 ratio], and then dyeing process variables (such as mordant concentration, dye concentration/shade %, pH, dyeing temperature, dyeing time, material-to-liquor ratio (MLR) and electrolyte/Salt concentrations) are reported using madder (Rubia/manjistha) extract as a natural dye to optimize the dyeing process conditions, and the observed results are discussed in **Item 2.2.2**.

*2.2.1 Optimization of aqueous extraction of Indian madder (*Rubia tinctorum*)*

For optimizing the extraction conditions, a colored aqueous extract liquor was obtained from madder (Rubia) dried powder under differently varying conditions of MLR 1:20–1:60, *p*H 4–10, extraction time period 15.0–90.0 min and extraction temperature 50–90°C for optimizing the conditions of extractions of color components of madder/Rubia (manjisthin and purpurin) from its natural source materials. The optical absorbance or OD values of the filtered aqueous extracts of the madder/Rubia were measured at λ_{max} of 540 nm (i.e. maximum absorbance wavelength) using Hitachi-U2000 UV-VIS absorbance spectrophotometer. Results of optical density (color value of dye solution) of aqueous extraction of madder under varying

Extraction variables	OD/absorbance for madder at $\lambda_{max.}$ 540 nm	Extraction variables	OD/absorbance for madder at $\lambda_{max.}$ 540 nm
MLR		**Temperature (°C)**	
1:20	0.238	50	0.198
1:30	**0.279**	**60**	**0.278**
1:40	0.266	70	0.132
1:50	0.252	80	0.164
1:60	0.236	90	0.177
Time (min)		**pH**	
15	0.241	4	0.192
30	**0.273**	5	**0.281**
45	0.188	6	0.188
60	0.173	8	0.193
90	0.182	10	0.205

**Values in bold indicate the optimum values considered while studying the effect of extraction parameters for madder. When one parameter was varied, say for MLR variation of 1:20–1:60, time was 30 min, temp was 60°C, and pH was 5. For time variation of 15–60 min, MLR was 1:30, temp was 60°C, and pH was 5; for temperature variation of 50–90°C, MLR was 1:30, time was 30 min, temp was 60°C, and pH was 5; and for pH variation 4–10, MLR was 1:30, temp was 60°C, and time was 30 min.*

Table 8.
OD or absorbance values of aqueous extract of madder (manjistha/Rubia) (λ_{max} of 540 nm) extracted under varying conditions of aqueous extraction.*

conditions have been shown in **Table 8.** The maximum values of OD [30] at λ_{max} of 540 nm wavelength are identified and marked in bold letters (for corresponding maximum color yield) as shown in **Table 8** and are considered as optimum or best extraction conditions, i.e. *MLR at 1:30, pH at 5, extraction temperature at 60°C and extraction time at 30 min.*

2.2.2 Study of mordanting and dyeing process variables for dyeing cotton with madder (Rubia)

2.2.3 Results of variation of concentrations and type of mordants

Table 9 shows the effect of concentrations and types of mordanting (as single or as double pre-mordanting technique) on surface color depth and other color interaction parameters of cotton fabric dyed with 4% aqueous extract (extract of purified 4% dry powder of madder obtained from M/s AMA Herbals Lab Pvt. Ltd., Lucknow) as colored dye solution used in this natural dyeing. Amongst all the single and double pre-mordanting done (**Table 9**), a combination of 20% overall application of harda and natural potash alum (50:50 ratio), i.e. 10% harda and 10% natural potash alum applied in sequence, satisfies the most desirable required stoichiometric ratio for effective complexing showing maximum surface color strength (K/S value) than that obtained by using either any of the said single pre-mordants or any other double pre-mordants studied for dyeing bleached cotton fabric with madder/Rubia. This may be an important fact that myrobalan (harda) acts as a mordanting assistant or dyeing additive, when used in conjunction with metallic salts. Myrobalan (harda) contain chebulinic acid/tannic acid moieties (having mordantable –COOH/–OH groups) in it, which thus is useful for a higher color yield when used along with the metal salts as second mordant forming insoluble metal

Mordant concentration and dye concentration	K/S at λmx	ΔE*	ΔL*	Δa*	Δb*	ΔC*	ΔH*	BI	MI	CDI	CV % of K/S
(10% harda + 10% potash alum) only mordanted cotton	1.53										
Harda (H) + $KAl_2(SO_4)_2$ (Pa)*, overall 20% application with ratio											
0:100 [**harda (H): $KAl_2(SO_4)_2$**] **+ 4% madder**	1.66	4.54	−3.70	2.12	1.55	2.59	0.45	22.08	0.95	0.86	1.50
25:75 [**harda (H): $KAl_2(SO_4)_2$**] **+ 4% madder**	2.53	2.95	2.55	0.47	1.39	1.31	0.67	19.85	0.36	2.46	2.32
50:50 [**harda (H): $KAl_2(SO_4)_2$**] **+ 4% madder**	**3.67**	1.49	1.23	0.76	0.36	0.32	0.78	15.22	0.33	3.03	2.05
75:25 [**harda (H): $KAl_2(SO_4)_2$**] **+ 4% madder**	3.40	3.49	3.05	0.91	1.43	1.66	0.34	16.34	0.48	4.26	4.15
100:0 [**harda (H): $KAl_2(SO_4)_2$**] **+ 4% madder**	2.05	1.53	1.06	0.13	1.09	1.08	0.19	29.83	0.31	2.78	2.99

**H, harda, and Pa, potash alum.*
Note: K/S, ΔL, Δa*, Δb*, ΔE*, ΔH*, ΔC*, etc. are all color parameters as per the latest CIE formula [30].*

Table 9.
Effect of different single and double natural bio-mordants on color strength and other colour interaction parameters of cotton fabric dyed with 4% madder extract (λ_{max} of 540 nm).

tannates/chebulinate (containing tannic/chebulinic acids) utilising multiple nos of carboxylates (for chebulinic acid)/multiple -OH groups (having high coordinating power) present in harda/myrobalan helping to enhance more fixation of such natural mordantable/anionizable dye molecules forming giant coordinating complex of fibre-harda-metallic mordant-natural dye to fix with –OH groups of cellulosic (cotton) fibers [19]. But it still has enough free adjacent hydroxyl/carboxylic acid groups to form mordant-dye complex to fix the anionizable/mordantable selective natural dye like madder/Rubia on the said harda and alum double pre-mordanted bleached cotton fibers. However, that is why the application of harda alone containing chebulinic acid is not found to provide sufficient color yield and color fastness for these natural dyes, which however when applied with the combination of metallic salts together (here, it is potash alum containing aluminum) give such an encouraging result on the subsequent dyeing with madder as an anionizable/mordantable natural dye. A similar result and higher color yield of anar peel dyeing on cotton were referred in the earlier literature for use of myrobalan/harda and alum double pre-mordanting [19].

Data on color fastness properties of dyed cotton fabric samples pre-mordanted with varying types of single mordant and different ratio of double pre-mordanted samples (with overall 20–40% mordant application combining Harada and potash alum in 50:50 ratio) subsequently dyed with 4% madder extract has been reported in **Table 10**.

It is evident from the color fastness data in **Table 10** that applications of single mordant and also higher mordant concentration (above overall concentration of 20%) do not show better or higher color fastness to wash. However, light fastness appears to be independent on mordant concentration and is steadily always moderate to good to UV light exposure. Fastness to dry rubbing (crocking fastness) of the dyed samples remains relatively higher, whereas fastness to wet rubbing (crocking)

Mordant concn. Used and dyed with madder (manjistha)	Color fastness				
	Washing		Light	Rubbing	
	LOD	Staining		Dry	Wet
10% harda and dyed with 4% madder	3	4	2/3	4	3
10% $KAl_2(SO_4)_2$ and dyed with 4% madder	3	4	3/4	4	3
10% of harda +10% of $KAl_2(SO_4)_2$ and dyed with 4% madder	3–4	4/5	4	4	4
20% of harda +20% of $KAl_2(SO_4)_2$ and dyed with 4% madder	2/3	4	4	3–4	2/3

Table 10.
Effect of different single and double mordants on color fastness properties of madder- (manjistha/Rubia) (λ_{max} of 540 nm) dyed cotton fabric.

becomes a bit lower irrespective of the concentration of harda and natural potash alum (mordant) used.

Moreover, there is little difference in color fastness to washing, light and rubbing for the different single mordants used, but there are some clear differences on color fastness ratings when the double mordanting system with overall 20% application of harda + natural potash alum in 50:50 ratio is used. Amongst the different combinations of double pre-mordanting systems used, 20% overall application of harda and natural potash alum (in 50:50 ratio) applied in sequence shows better color fastness rating amongst all combinations of mordant tried. This may be presumed to be due to some synergistic effects of this particular combination of natural potash alum with harda as a mordanting assistant due to additional coordinating power of chebulinic acid of harda as a mordanting assistant applied, which perhaps facilitates more number of strong and giant bigger complex formation amongst the said fibre (cotton)-mordanting assistants (harda)—metallic mordant (natural alum)—natural dye (madder) system in the presence of both myrobalan (harda) containing chebulinic acid residue and potash alum containing aluminum in combination in equal proportion. This may be noted that the application of 20% harda and 20% potash alum in combination impairs the colour fastness to washing and crocking, and hence 10% harda and 10% potash alum combination applied in sequence as a double mordant is found to be the better option in this case.

2.2.4 Results of variation of different dyeing process variables for uniform dyeing

Results of the effect of different dyeing process variables such as (a) dyeing time in min; (b) dyeing temp in °C; (c) dye bath in pH; (d) dye concentration in %; and (e) salt concentration in % which have been studied to optimize the dyeing conditions for maximum and uniform color yield (in terms of K/S value) for overall application of 20% harda + $KAl_2(SO_4)_2$ in 50:50 ratio pre-mordanted and dyed cotton fabric are shown in **Table 11**. From relevant data in **Table 11**, it is found that when all other variables are kept fixed, with the increase in time of dyeing (15–120 min), the K/S value initially increases up to 60-min dyeing time and then starts decreasing on further increase in dyeing time from 60 to 120 min, whereas there is hardly any change of the K/S value from that obtained in 60-min dyeing time. Thus, the K/S value is maximum for 60-min dyeing time. This may be explained by the possibility of achieving dyeing equilibrium at a medium faster rate for synergistic action of harda and potash alum both as pre-mordanting agents, and hence within 30–60 min, the dye uptake rate may be maximum for a double pre-mordanted cotton with high coordinating capacity of mordantable natural dyes like madder. However, the final dyeing rate depends on the diffusion rate (being the slowest step in dyeing operation) besides transportation, absorption, diffusion and fixation of any dye like madder/Rubia extract. There are also chances of desorption/breaking of dye-fiber-mordant complex at either higher temperature or higher dyeing time which may lead to a dropping trend above the said 60 min of dyeing time.

For variation of dyeing temperature, on increase of the dyeing temperature (40–95°C), keeping other variables constant, the surface color strength (K/S values) is found to show a slow increase to a small extent from 40 to 65°C whereafter it almost remains same or at par up to 80°C, after which there is a further decrease beyond 80°C up to 95°C. An increase in temperature of dyeing inevitably supplies more energy for the transportation of dye molecules, thus facilitating the higher rate of dye sorption and diffusion up to 65°C, and thereafter this dyeing rate does not alter much even after increasing temperature up to 95°C, i.e. the desorption starts at relatively high temperature above 80°C and maybe one of the components of madder starts degrading above 74°C and the color value, i.e. K/S value, decreases noticeably. However, dyeing of cotton with madder extract at warm conditions, i.e. at 50–65°C, is considered to be best suited and cannot be excluded fully for the decentralized sector for energy-saving purpose.

The corresponding data in **Table 11** show almost a similar level of dye uptake in terms of surface color strength (K/S value) with the variation of dye concentration (keeping other variables constant) for purified madder dye concentration from 3 to 7%, showing a maximum K/S value for 4% dye concentration of purified madder extract. For the variation of pH from 6 to 14 (keeping other variables constant), the surface color strength (K/S value) starts increasing slowly showing a maximum K/S value at pH 12, after which color value further reduces. Thus, pH 12 may be considered optimum, though here pH is to be treated as a critical variable for relatively higher differences of color difference index (CDI) values (showing higher differences of maximum CDI and minimum CDI values amongst the results of varying pH of dye bath). However, considering the corresponding color fastness data, as reflected in **Table 12**, it is further clear that pH 12 renders a better overall balance of all types of color fastness data and shows even sometimes better than that obtained at pH 14. So, pH 12 is considered as the optimum value for this dyeing in this case of cotton fiber-(alum + harda) mordant-dye (madder/Rubia) system. It may be fact that at pH 12, as it is found to be alkaline, there are higher chances of ionization of phenoxy hydroxyl groups of color component of madder/Rubia (containing alizarin, manjisthin and purpurin) and hence provide better chances of complex formation with mordant like natural alum and mordanting assistant like harda forming a giant coordinated complex amongst them for better fixation and higher anchoring by formation of the said fibre-mordanting assistant + mordant-dye system in this case.

Keeping other variables constant, with the variation in MLR from 1:10 to 1:50 (**Table 11**), the K/S value is maximum for MLR at 1:10, and then there is a slow decrease after MLR 1:20, after which the K/S value continues decreasing slowly in small quantum with further increase for MLR at 1:50. Though MLR 1:10 shows the highest K/S value as compared to any other MLR used, MLR 1:20 gives the same value of K/S and better wash and other color fastness results. Hence the optimum MLR may considered to be 1:20 instead of 1:10. From color fastness data in **Table 12**, it is indicated that except wash fastness results for MLR 1:50, all other MLR used show almost medium to good overall color fastness data. However, amongst all these, overall color fastness data for all types of color fastness results are found to be quiet acceptable for MLR 1:10 and 1:20 only. Thus, considering both color yield and color fastness data, MLR 1:20 gives a better balance having a more uniform dyeing and hence may be considered as optimum dyeing conditions with respect to MLR.

Dyeing conditions	**K/S at λmax**	**ΔE***	**ΔL***	**Δa***	**Δb***	**ΔC***	**ΔH***	**BI**	**MI**	**CDI**
10% harda and 10% KAl $(SO_4)_2$ pre-mordanted cotton fabric dyed with 4% purified madder dye	1.53	—	—	—	—	—	—	19.85	—	—
Time, min										
15	2.29	2.33	−1.65	0.30	1.62	1.23	1.10	20.34	0.42	2.43
30	2.95	4.46	−3.31	2.94	0.58	2.75	−1.20	18.59	1.04	3.89
60	**3.18**	9.34	−8.55	−3.28	−1.84	−3.73	0.44	15.13	1.29	1.96
90	2.34	2.45	−1.35	1.36	1.53	1.99	0.48	18.40	0.43	1.69
120	2.19	3.24	−2.91	0.21	1.40	1.04	0.95	16.77	0.44	5.35
Temp °C										
40	2.01	3.76	1.42	0.90	0.05	0.71	0.56	18.25	0.71	4.33
50	2.21	5.65	−4.35	3.60	−0.20	3.06	−1.91	17.73	1.39	4.36
65	**3.04**	4.22	−3.38	2.20	−1.22	0.95	−2.33	21.38	0.90	3.79
80	2.10	2.05	0.72	0.49	0.59	0.77	0.02	24.89	0.19	4.81
95	1.81	2.66	−2.05	1.69	0.06	1.16	−1.24	22.22	0.57	5.84
***p*H**										
6	2.14	0.88	−0.74	−0.21	0.43	0.14	0.45	19.04	0.14	9.58
8	2.35	1.51	−1.14	0.56	0.81	0.96	0.21	18.28	0.18	4.81
10	2.58	1.12	−1.34	0.37	0.97	0.24	0.38	16.37	0.15	8.24
12	**2.96**	2.89	−0.30	−0.71	−0.44	−0.81	0.21	18.50	0.27	6.17
14	2.74	2.88	−2.84	0.38	0.36	0.52	0.40	16.16	0.27	4.31
MLR										
1:10	**2.82**	2.27	−1.74	−0.01	1.45	1.09	0.97	22.70	0.32	5.39
1:20	**2.80**	2.05	−0.72	0.49	0.59	0.77	0.02	24.89	0.19	4.81
1:30	2.15	3.59	−2.17	2.67	1.02	2.59	−1.23	21.22	0.87	4.10
1:40	2.09	5.01	−4.21	2.58	0.84	2.40	−1.27	21.06	0.93	2.45
1:50	2.22	1.83	−1.66	0.75	0.17	0.67	−0.38	19.61	0.28	2.39
Dye concn.,%										
3	2.60	1.27	0.20	−0.06	1.25	0.65	1.07	16.96	0.23	3.79
4	**2.82**	5.45	−4.36	3.24	0.38	2.90	−1.50	15.78	1.16	3.44
5	2.57	2.93	−1.81	1.91	1.27	2.30	−0.08	17.55	0.55	2.26
6	2.61	2.19	−1.64	1.33	0.60	1.42	−0.29	17.61	0.52	0.17
7	2.69	4.71	−3.66	2.86	0.77	2.81	−0.95	16.41	0.98	4.23
Salt concn., gpl										
5	3.12	3.31	−2.48	1.94	1.02	2.18	−0.17	13.22	0.70	2.99
10	**3.37**	4.31	−3.69	2.15	0.63	2.14	−0.66	13.36	0.83	3.24
15	3.22	5.45	−5.01	2.13	0.26	1.93	−.093	11.74	0.94	4.69
20	3.13	0.75	−0.72	0.19	0.07	0.20	−0.04	12.45	0.11	2.78

**Values in bold indicate the optimum values considered while studying the effect of process variables of dyeing of madder. When one dyeing parameter was varied, other dyeing parameters were kept constant, say for MLR variation of 1:10–1:50, dye (madder) concn. Was 4%, time was 60 min, temp was 65°C, pH was 12, and salt concn. Was 10 gpl; similarly, for time variation of 15–120 min, MLR was 1:20, dye (madder) concn. Was 4%, time was 60 min, temp was 65°C, pH was 12, and salt concn. Was 10 gpl; for temperature variation of 40–95°C, MLR was 1:20, dye (madder) concn. Was 4%, time was 60 min, salt concn. Was 10 gpl, and pH was 12, and for pH variation of 6–14, MLR was 1:20, dye (madder) concn. Was 4%, time was 60 min, salt concn. Was 10 gpl, and temp was 65°C; for salt concn. Variation of 5–20, MLR was 1:20, dye (madder) concn. Was 4%, time was 60 min, temp was 65°C, and pH was 12. λ_{max} for K/S data of madder was taken as 540 nm from UV-VIS scan in all cases.*
Note: K/S, ΔL, Δa*, Δb*, ΔE*, ΔH*, ΔC*, etc. are all color parameters as per the latest CIE formula [30].*

Table 11.
Color strength, brightness index (BI), metamerism index (MI), CDI value and related color parameter for variation of dyeing process variables for dyeing of 10% Harda and 10% $KAl(SO_4)_2$ pre-mordanted cotton fabric subsequently dyed with 4% madder/Rubia.*

The addition of an electrolyte (common salt) to the dyeing liquor expectedly increases the exhaustion of the dyeing case of most of the anionizable dyes. Common salt (electrolyte) is dissolved completely in the aqueous liquor in the dye bath at a specific temperature of dyeing, thereby positive sodium ion is attracted to –ve cellulosic surface in water and neutralizes the –ve charge of cellulose and thereby anionic natural dye ions are able to be attracted to cellulose increasing exhaustion. But excessive amount of electrolyte/salt above a certain limit, causes a retardation effect in the dye absorption vis-à-vis color yield and renders lower color depth. From the relevant data in **Table 11**, a similar trend is observed that with the increase in salt/electrolyte concentration from 5 to 20 gpl, the color yield in terms of K/S value is increased for the application of 5–10% salt concentration and starts decreasing for use above 15% salt concentration. Moreover comparing overall color fastness properties for the corresponding part, the overall round color fastness properties are found best for 10% salt concentrations, and fastness results for use of 15% salt concentration are a bit inferior. Considering all the above matter, the optimum concentration of common salt for dyeing cotton fabric with madder is selected to be 10 gpl.

Thus, the observed optimum conditions of dyeing of the double pre-mordanted with 10% concentration of harda and 10% concentration of natural potash alum in sequence for bleached finer cotton fabric with aqueous extract of purified madder (manjistha/Rubia) are as follows: dyeing time, 60 min; dyeing temperature, 65°C; MLR, 1:20; pH, at 12; dye concentration, 4% (on weight percentage of dried purified color extract powder); and common salt concn., 10 gpl, considered as optimum.

Table 11 also shows the effects of different process variables on K/S values and other latest CIE color measurement parameters [30], including total color difference (ΔE*), change in hue (ΔH*), change in chroma (ΔC*), general MI, BI and CDI values [13]. It is interesting to observe that amongst the varying dyeing conditions (time, temperature, pH, MLR, mordant, dye concentration and salt concentration), the most important and predominating variables are identified as pH of the dye bath, dyeing time and dye concentration. Therefore, for uniform dyeing using madder (manjistha/Rubia) extract for cotton fabric, special care is to be taken for the control of pH, dyeing time and concentration of dye solution of madder (manjistha/rubia).

The said other color parameters like ΔE*, ΔL*, Δa* and Δb* indicate the variation in color strength and related parameters for varying dyeing conditions in each case, as compared to standard undyed pre-mordanted control cotton fabric. Changes in hue (ΔH*), in most of the cases, are found to be small negative values (**Table 11**), and in very few cases, there are small positive values, indicating that there is a

Varying dyeing conditions 10% harda + 10% potash alum pre-mordanted cotton fabric dyed with 4% purified madder dye	Overall color fastness properties				
	Washing colour fastness at 50°C		Light	Rubbing	
	Loss of depth	Staining		Dry	Wet
Varying parameters of dyeing					
Time, min					
15	3/4	5	3	4/5	4
30	3/4	4	3	4	4
60	**3/4**	**4/5**	**3/4**	**4/5**	**4**
90	2/3	4	4	4/5	4
120	3	4	4	4/5	4
Temp °C					
40	2	4/5	3	4/5	4
50	2/3	4/5	3	4/5	3
65	**3/4**	**5**	**3/4**	**5**	**4/5**
80	2/3	5	3	4/5	4
95	1/2	4/5	3/4	4/5	4
***p*H**					
6	2/3	4	3	4/5	3/4
8	2/3	4	3	4/5	3/4
10	1/2	4/5	3	4/5	3/4
12	**3/4**	**4/5**	**3/4**	**4/5**	**3/4**
14	2/3	5	3	4/5	3/4
MLR					
1:10	**3**	**4/5**	**3/4**	**4/5**	**4**
1:20	**3–4**	**4/5**	**3/4**	**4/5**	**4**
1:30	3	4	3	4/5	4
1:40	3	4	3	4	3/4
1:50	2	4/5	3	4/5	4
Dye concn.,%					
3	2	4	3	4	4
4	**3–4**	**5**	**3/4**	**4/5**	**3/4**
5	3	4/5	3	4/5	3/4
6	2/3	4/5	3	4/5	3/4
7	2/3	4/5	3/4	4/5	3/4
Salt concn., gpl					
5	3	5	3/4	3	4/5
10	**3–4**	**5**	**3/4**	**4/5**	**4**
15	3/4	4/5	3	3/4	4/5
20	2/3	4/5	3	3/4	3

**When one dyeing parameter was varied, other dyeing parameters were kept constant, say for MLR variation of 1:10–1:50, dye (madder) concn. Was 4%, time was 60 min, temp was 65°C, pH was 12, and salt concn. Was 10 gpl; similarly, for time variation of 15–120 min, MLR was 1:20, dye (madder) concn. Was 4%, time was 60 min, temp was 65°C, pH was 12, and salt concn. Was 10 gpl; for temperature variation of 40–95°C, MLR was 1:20, dye (madder) concn. Was 4%, time was 60 min, salt concn. Was 10 gpl, and pH was 12, and for pH variation of 6–14, MLR was 1:20, dye (madder) concn. Was 4%, time was 60 min, salt concn. Was 10 gpl, and temp was 65°C; for salt concn. Variation of 5–20, MLR was 1:20, dye (madder) concn. Was 4%, time was 60 min, temp was 65°C, and pH was 12.*
Bold values are optimum results.

Table 12.
Color fastness data against the variation of dyeing process variables for dyeing 10% Harda +10% potash alum pre-mordanted cotton fabric subsequently dyed with 4% madder/Rubia.*

minor change in the predominating hue in each case. However, the maximum ΔH* value is observed in the case of the variation in temperature from 40 to 95°C which further indicates the high sensitivity of color strength parameter for this particular natural dye for the variation in temperature of dyeing, indicating it as also a critical parameter. The brightness index of dyed products depends on reflectance value of the dye and its orientation along the fiber axis after fixation. However, interestingly it may be noted that at lower pH of 6–10, lower concentration of dye (30–40%) and lower temperature (50–65°C), the reduction in BI values is much lower than that observed in other conditions of dyeing. Expectedly the reduction in brightness index is found to be higher when the application of dye concentration and/or dyeing temperature are higher, due to disorientation of dye molecules during fixation, for either use of higher concentration of dye molecules or higher dyeing temperature. Results of general MI indicates the metameric effect on the madder- (manjistha/rubia) dyed cotton fabric for different conditions of dyeing. In all these cases, the MI varies from 0.11 to 1.39 (**Table 11**) and it is observed that these MI data are not much widely dispersed within a particular condition being varied, but varies to a noticeable degree from one condition to other, indicating its potential metameric nature for varying one condition to the other. Therefore, use of standardization of conditions of dyeing is a must to minimize metamerism for achieving least metameric dyed products like fibre-mordant-dye system for natural dyeing of cotton with madder/rubia.

Moreover, the ecotoxicology property of madder is available in the literature that the hepato-protective [31] activity of an aqueous-methanol extract of *Rubia cordifolia/Rubia tinctorum* was investigated earlier against acetaminophen and CCL_4-induced damage. Acetaminophen produced 100% mortality at a dose of 1 g/K in mice, while pre-treatment of animals with plant extract reduced the death rate to 30%, proving its ecocompatibility. This is just one example by how after a series of lab experiments, dyeing process parameters are standardized to ensure reproducibility, dye uniformity, etc. and critical dyeing process variables are identified, but for the application of a specific natural dye to any specific textiles, this standardized process parameters are to be made available for each dye-fiber combination separately to the dyers as ready-made guide like synthetic dye application manuals supplied by dye manufacturers or suppliers.

Similarly, it has also been revealed from our earlier lab study that the optimized conditions for dyeing of 15% overall harda + alum (50:50) double pre-mordanted cotton fabric with tesu as natural dye are: dye concentration of 30% (aqueous extract based on the weight of dry powder of natural tesu flower petal (*Butea monosperma* or commonly known as palash or flame of the forest)), pH of 12, MLR of 1:30, time of 60 min for dyeing and 90 min for simultaneous dyeing and finishing and dyeing temperature of 90°C with salt concentration of 10 gpl (though the detailed study and results of the optimization of conditions of dyeing process variables for tesu as natural dye applied on the said 15% overall application of alum + harda pre-mordanted cotton are not mentioned here considering its duplication).

2.3 Functional finishing to impart/to improve the antibacterial property and UV protection factor (UPF) using natural resource-based finishing agents for tesu-dyed cotton textiles: a specific case study

Few studies on UV-protective and antibacterial characteristics of some natural materials including few natural dye extracts are sparingly reported [32–41]. Hence, an attempt for such natural dyeing-cum-finishing is studied in the present work as reported below.

2.3.1 Antibacterial finish on natural dyed cotton fabric with natural finishing agents

The antibacterial characteristics of many natural resource materials are explored by many researchers [21, 22, 35–41], which include the antibacterial property of eucalyptus, curcumin, banana peel and few natural dyes like catechu, rheum emodi, etc. All the required natural resource materials and tesu as natural dye were obtained from the M/s AMA Herbals Lab Pvt. Ltd., Lucknow, India. A recent accounting of the present status and advanced studies on natural dyeing and natural finishing of textiles including modern techniques of dyeing processes and analyses in different angles have been compiled comprehensively by Vankar [42]. In this part of report, eucalyptus leaf extract has been used as a natural antibacterial agent applied on tesu-dyed cotton. Effects of few different post-treatments on cotton fabric dyed with aqueous extract of 30% (based on the weight of dry powder of natural solid source) tesu flower petal (*Butea monosperma* or commonly known as flame of the forest) with 10% of both aqueous and MeOH extracts of eucalyptus leaves by a pad-dry-cure technique (by a two-dip-two-nip padding process, followed by drying at 100°C for 15 min and curing at 120°C for 3 min) after dyeing or by simultaneous dyeing and finishing process with or without suitable catalyst system (like citric acid, 5 gpl). mainly for improving the antimicrobial properties of tesu-dyed cotton fabric after pre-mordanting with 15% overall harda + alum (50:50) have been discussed.

For simultaneous dyeing and finishing, predetermined percentages of eucalyptus leaf extract along with citric acid were added to the dye bath, and dyeing-cum-finishing was carried out simultaneously at 90°C for 90 min. Relevant results in the change in surface color strength and other related parameters along with fastness properties are tabulated in **Tables 13** and **14**.

In **Table 13**, relevant data indicates that when a post-treatment of eucalyptus leaf extract is carried out in the presence of a suitable catalyst (citric acid) on the above said double pre-mordanted and 30% tesu-dyed cotton fabric by a pad-dry-cure technique, its color strength value is found to be higher as compared to the simultaneous dyeing and finishing technique with the same agents/compounds. Also the fabric samples post-treated with 10% of both aqueous and MeOH extracts of eucalyptus leaves by a pad-dry-cure technique are found to be much yellower and darker in case of similar post-treated fabric produced by the simultaneous natural dyeing and finishing process.

However, taking the color fastness results into consideration, data in **Table 14** indicate that light fastness rating obtained for simultaneous natural dyeing and finishing process is slightly better than the similar sample produced by natural dyeing followed by separate finishing process by a pad-dry-cure technique of separate post-finishing of tesu-dyed cotton fabric using 10% extract of eucalyptus, as

Treatments	K/S	ΔE	ΔL	Δa	Δb	ΔC	ΔH	BI	MI	CDI
Standard bleached cotton control fabric	0.01							92.11	0.91	—
15% overall application of harda + alum (50:50) (no dyeing)	0.16	6.12	−1.15	−0.91	5.85	−0.044	−5.92	44.65	3.43	—
15% harda + alum (50:50) and 30% tesu aqueous extract dyed	2.65	17.77	−4.58	−1.062	17.13	11.20	−13.01	15.93	7.26	2.84
Post-treatment with 10% extract of eucalyptus leaves on 30% tesu aqueous extract-dyed cotton after pre-mordanted with 15% harda + alum (50:50) pre-mordanted in sequence										
(10%) Eucalyptus leaf aqueous extract + citric acid (5 gpl)	5.49	15.40	3.18	−0.727	4.31	4.37	11.05	18.61	6.44	5.90
10% Eucalyptus leaf MeOH extract + citric acid (5gpl)	7.49	20.40	3.19	−0.824	4.11	5.33	12.00	19.61	6.75	6.80
Post-treatment with 10% extract of eucalyptus leaves on 30% tesu aqueous extract-dyed cotton after pre-mordanted with 15% alum pre-mordanted cotton fabric										
10% Eucalyptus leaf MeOH extract + citric acid (5gpl) (15% alum pre-mordanted)	4.49	14.60	3.00	−0.676	5.36	5.71	11.70	16.33	6.02	4.97
Simultaneous dyeing and finishing with 15% harda + alum (50:50) pre-mordanted cotton fabric with extract of tesu and eucalyptus leaves with citric acid										
Tesu (30%) extract + eucalyptus leaf aqueous extract (10%) + citric acid (5 gpl)	4.85	14.42	2.38	−0.52	3.68	3.72	12.01	17.56	5.61	8.29

Table 13.
Effect of post-treatment with eucalyptus leaf extract by a pad-dry technique and by simultaneous dyeing and finishing technique using a suitable catalyst for natural dyeing and finishing of cotton fabric with tesu extract after pre-mordanting with 15% overall dosage of Harda + alum (50:50).

the natural finishing agent. The washing and rubbing fastness obtained are more or less found to be similar between the two processes. Moderate fastness to washing of only 30% tesu natural dyed cotton can be attributed to the fiber-metal-tannin-dye complex formation between color components of tesu (containing butrin, butein, flavonoids) and alum as a metallic mordant along with harda (containing chebulinic acid) as a mordanting assistant. Darker shades are obtained after 10% eucalyptus extract post-treatment as a finishing treatment on tesu-dyed cotton fabric, which is considered to be due to the addition of active constituents of eucalyptus containing tannins, polyphenols and an active coloring substance called quercetin which has the brightest yellow shade. Hence, in case of post-finishing treatment by a pad-dry-cure technique, the K/S value is found to be higher than that of simultaneous dyeing-finishing treatment owing to the of formation of giant bigger molecule size of complex formed with eucalyptus component in the presence of citric acid used as a catalyst during the finishing process. In case of simultaneous process, a bigger molecular size of citric acid preferably starts crosslinking with cellulose of cotton reducing and preventing an effective fiber-mordant-dye complex formation with the cotton fabric, resulting in a less color yield.

Ten percent eucalyptus post-treatment on 30% aqueous extracted tesu-dyed [after the said double pre-mordanted with 15% overall mordant application with harda and alum (50:50)] and 10% eucalyptus post-treated or finished cotton fabric as well as simultaneous dyeing and finishing (adding 30% tesu and 10% eucalyptus and 5 gpl citric acid and heated at 90°C for 90 min) of the said pre-mordanted

Treatments	Wash fastness		Rubbing fastness	Light fastness
Variation in dyeing and finishing treatments	**LOD**	**Cotton (staining)**	**Dry**	
15% harda + alum (50:50) and 30% tesu aqueous extract dyed	4	3–4	4–5	3
Post-treatment with 10% extract of eucalyptus leaves on 30% tesu aqueous extract-dyed cotton after pre-mordanted with 15% harda + alum (50:50) pre-mordanted in sequence				
(10%) eucalyptus leaf aqueous extract + citric acid (5 gpl)	4–5	4–5	4	4
10% eucalyptus leaf MeOH extract + citric acid (5gpl)	4–5	4	4	4
Post-treatment with 10% extract of eucalyptus leaves on 30% tesu aqueous extract-dyed cotton after pre-mordanted with 15% alum pre-mordanted cotton fabric				
10% eucalyptus leaf MeOH extract + citric acid (5gpl)	3–4	3–4	3	3
Simultaneous dyeing and finishing of cotton fabric dyed with 30% tesu aqueous extract and 10% eucalyptus aqueous extract on 15% harda + alum (50:50) pre-mordanted cotton fabric				
Tesu (30%) extract +10% eucalyptus leaf aqueous extract + citric acid (5 gpl)	4–5	4–5	4	4

Table 14.
Color fastness to washing, rubbing, light after post-treatment of tesu-dyed double pre-mordanted cotton fabric with eucalyptus as well as simultaneous dyeing and finishing.

cotton fabric were subjected to the standard antimicrobial test as per the American Association of Textile Chemists and Colorists (AATCC)-100–2012 method in terms of bacterial reduction (%) as shown in **Table 15.**

As per the results shown in **Table 15**, for both *Klebsiella pneumoniae* and *Staphylococcus aureus* bacteria, there is 99% reduction of bacterial growth on cotton fabric for double pre-mordanted with alum and harda, i.e. after 15% overall application of harda + alum (50:50) only without any dyeing (no dyeing). This can be attributed to the presence of the natural alum (also known as Fitkari) as mordant, which contributes to the prevention of bacterial growth. However, a remarkable increase of bacterium can be observed when mordanted fabrics were subsequently dyed with 30% tesu extract as natural dye, from which it can be concluded that after dyeing with tesu extract, as potash alum is ionized during dyeing and all aluminium present in alum are consumed for complexion of fibre-mordant-dye complex formation and hence, prevention of bacterial growth by alum is reduced and antibacterial property is reduced partly to Gram-positive bacteria (showing results of 68.88% or approx. 69% bacterial reduction only) and reduced fully to Gram-negative bacteria (showing results –No bacterial reduction at all). However, when the said double pre-mordanted and 30% tesu extract-dyed cotton fabric samples is post-treated or finished with 10% aqueous extract of eucalyptus with citric acid catalyst, there is no bacterial reduction against Gram-negative bacteria (showing results—no bacterial reduction at all), but shows 93.01 bacterial reduction against Gram-positive bacteria. While when a similar pre-treated and tesu-dyed sample of cotton fabric is post-treated with 10% MeOH extract (instead of an aqueous extract of eucalyptus leaves as antibacterial natural finishing agents), it shows the highest bacterial reduction results for against both Gram-positive bacteria (showing a bacterial reduction of 99.99%) and Gram-negative bacteria (showing a bacterial reduction of 99.97%).

Simultaneous dyeing and finishing show antibacterial test results in much similar fashion or quite similar to the only pre-mordanted and subsequently 30% tesu-dyed cotton fabric samples, showing there is no reduction in bacterial growth

Treatments	Bacterial reduction (%) as per AATCC-100-2012	
Variation in dyeing and finishing treatments	***Klebsiella pneumoniae*: ATCC 4352 (Gram-negative bacteria)**	***Staphylococcus aureus*: ATCC 6538 (Gram-positive bacteria)**
Standard bleached cotton control fabric	No reduction	No reduction
15% overall application of harda + alum (50:50) (no dyeing)	99.98	99.43
15% [harda + alum (50:50)] and 30% tesu aqueous extract dyed	No reduction	68.88
Post-treatment with 10% extract of eucalyptus leaves on 30% tesu aqueous extract-dyed cotton after pre-mordanted with 15% harda + alum (50:50) pre-mordanted in sequence		
10% eucalyptus leaf aqueous extract + citric acid (5 gpl)	No reduction	93.01
10% eucalyptus leaf MeOH extract + citric acid (5 gpl)	99.97	99.99
Post-treatment with 10% extract of eucalyptus leaves on 30% tesu aqueous extract-dyed cotton after pre-mordanted with 15% alum pre-mordanted cotton fabric		
10% MeOH extract of eucalyptus +citric acid (5 gpl)	99.79	99.52
Simultaneous dyeing and finishing of cotton fabric dyed with 30% tesu aqueous extract and 10% eucalyptus aqueous extract on 15% harda + alum (50:50) pre-mordanted cotton fabric		
Tesu (30%) extract + eucalyptus leaves aqueous extract (10%) + citric acid (5 gpl)	No reduction	54.02

Table 15.
Results of the antimicrobial test as per AATCC-100-2012 for pre-mordanted 30% tesu extract-dyed and eucalyptus extract finished cotton fabric after pre-mordanted with 15% total application of Harda + alum (50:50).

against Gram-negative bacteria (showing results of no bacterial reduction at all) and antibacterial property is reduced partly to Gram-positive bacteria (showing results of 54.02% or approx. 54% bacterial reduction only). But only 15% alum pre-mordanted (without harda) and 30% tesu extract-dyed cotton fabric sample, post-finishing application of 10% MeOH extract of eucalyptus leaves render it highly antibacterial in nature showing almost at par or similar level of bacterial reduction results that obtained for said 15% application of harda + alum (50:50) same fabric samples dyed with tesu and post finished with 10% MeOH extract of eucalyptus, for both against Gram-positive bacteria (showing bacterial reduction of 99.79%) and Gram-negative bacteria (showing bacterial reduction of 99.52%).

All corresponding pictures of petri plates for the said antimicrobial tests as per AATCC-100-2012 against Gram-negative and Gram-positive bacteria are also given in **Figure 10** with petri plate numbers being 1–7, where 1A–7A are samples incubated for 0 h, 1B–7B are samples incubated for 24 h with Gram-negative bacteria and 1C–7C are samples incubated after 24 h with Gram-positive bacteria, against the corresponding samples; the photographs of petri plates are almost matching with bacterial reduction (%) results shown in **Table 15**, which are self-explanatory.

Hence, in case of simultaneous dyeing and finishing, the antibacterial property results obtained are not fully satisfactory as compared to the said double pre-mordanted and tesu dyed and post-treated/finished with 10% aqueous extract of eucalyptus, but that effect of antibacterial property is highly enhanced if, the same

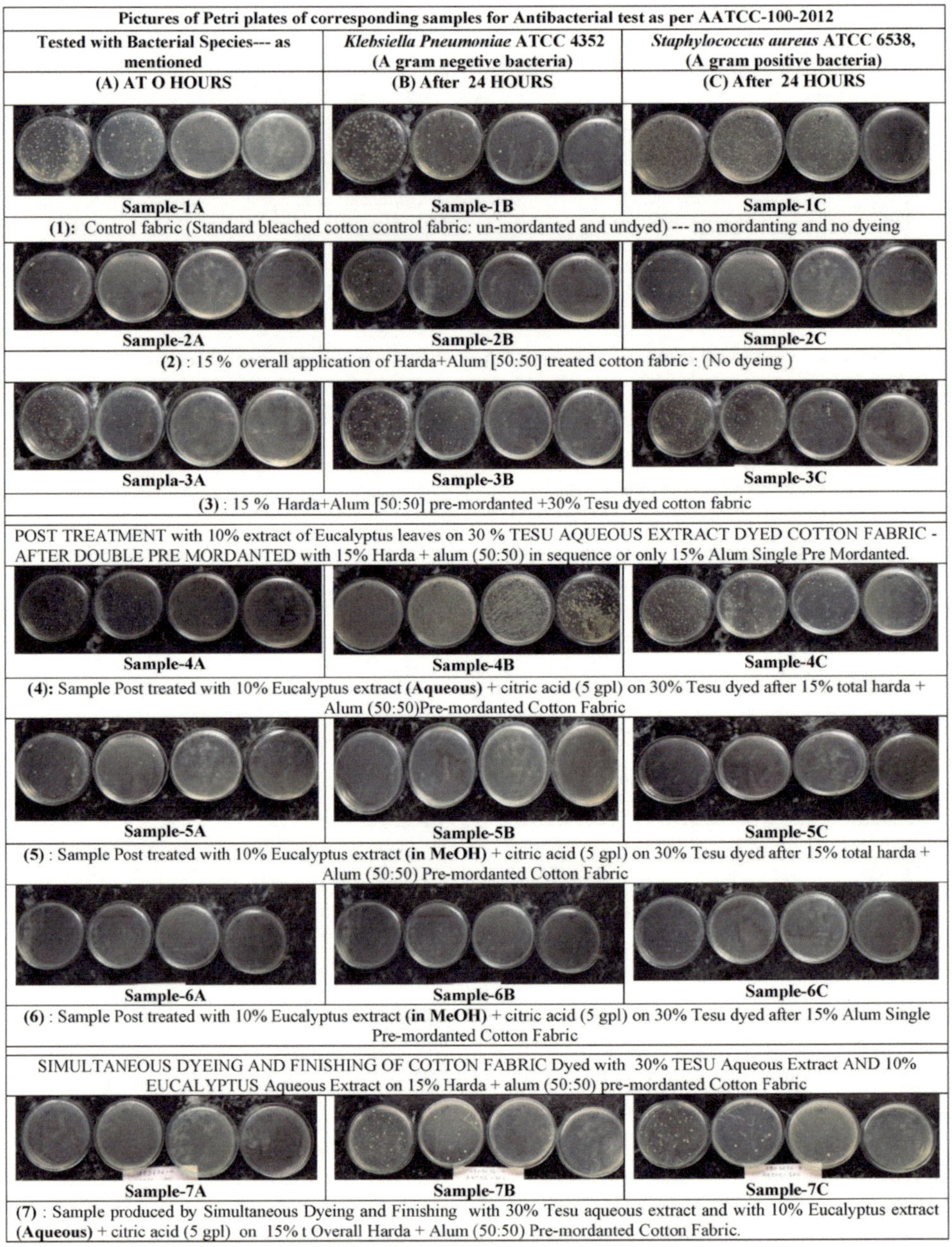

Figure 10.
Pictures of petri plates for antimicrobial test as per AATCC-100-2012 against gram-negative and gram-positive bacteria.

percentage, i.e. 10% MeOH alcohol-extracted eucalyptus leaves, is applied instead of its aqueous extract. So, in order to improve the fullest/highest bacterial reduction finish with a natural agent, the said double pre-mordanted cotton fabric after dyeing with tesu natural dye, the dyed fabric is to be finished separately by pad a dry-cure technique using 10% MeOH alcoholic extracted eucalyptus leaves for good results in the reduction of bacterial growth for both Gram-positive and Gram-negative bacteria. In case of cotton fabrics dyed with methanol (alcohol) aided extracted eucalyptus leaves post-treatment, they give a very darker color depth/ shade as well as excellent antimicrobial finishing properties.

2.3.2 UV-protective finish on natural dyed cotton fabric with natural finishing agent

Few reports are available in the literature on the role of natural dyes and other natural resource materials, when applied on textiles for UV-protective properties [19, 32–34, 43]. Finishing of textiles with vegetable oil is also reported in the literature [44].

But nowadays, UV protection in textiles has become also equally important for the protection of the human skin. Considering this in view, reddening of the skin under any barrier like textiles or other materials can be judged by UPF or sun protection factor (SPF) values. The classification of performance-based UPF value grading/rating is given below.

Performance-related UPF rating and classification	
UP F value	**UV protection rating/category**
15–24	Good protection
25–40	Very good protection
40–50+	Excellent protection

So, as a case study, UV-protective finish on 30% aqueous extract of tesu-dyed cotton fabric (pre-mordanted with 15% overall application of harda + alum (50:50)) was imparted as a finishing post-treatment by a pad-dry-cure technique with aqueous extract of eucalyptus, MeOH extract of eucalyptus and emulsified coconut oil in the presence of citric acid as a catalyst for finishing with these two types of natural UV absorber for imparting UV-protective finish. Relevant data of UV protection factor values before and after the said treatments are measured as per AATCC-183-2010/2014 and are tabulated in **Table 16**. Double pre-mordanted with 15% overall application of harda + alum (50:50) and 30% tesu aqueous extract-dyed cotton fabric samples are when post treated with both aqueous or alcoholic extract of eucalyptus leaves and also with emulsified coconut oil in presence of citric acid as catalyst (in all the cases), UV protection performance are much enhanced as compared to said double pre-mordanted and tesu-dyed cotton fabric (without any post-treatment). Relevant results of UPF values from **Table 16** indicate that best result, i.e. UPF value of 40, is obtained when the said double pre-mordanted cotton fabrics dyed with tesu extract (as natural dye) is finished with MeOH alcoholic 10% extract of eucalyptus leaves in the presence of citric acid (5 gpl) by a pad-dry-cure technique as compared to UPF value obtained as 30 for post-treatment with equal dosages of emulsified 10% coconut oil in the presence of citric acid (5 gpl) under comparable conditions of treatment. However, when cotton fabrics are subjected to simultaneous dyeing and finishing process with the said two types of natural UV-protective agents (emulsified coconut oil and aqueous extract of eucalyptus leaves applied simultaneously in dye bath), the said enhancement of UPF values for simultaneous dyeing and finishing process are much less showing a UPF values 20 for emulsified coconut oil and UPF value is 25 eucalyptus leaves aqueous extract, i.e. UPF values in case of simultaneous dyeing and finishing treatment are not shown to be as high as it shows by sequential post-treatment process by pad-dry-cure process in both the cases respectively. UPF values of 15–24/25 are considered to be moderate to good protection, and hence UPF values 20 and 25 in these two cases of simultaneous dyeing and finishing using either emulsified coconut oil and eucalyptus leaves aqueous extract as natural finishing agents applied simultaneously in the dye bath can be considered to provide a moderate to good UV protection, but not as

Samples/type of post-treatments with UV absorbers	UPF	UV-A (%)	UV-B (%)
Standard bleached cotton control fabric (un-mordanted and undyed cotton)	5	29.67	22.10
15% overall application of harda + alum (50:50) (no dyeing)	10	19.56	11.01
15% [harda + alum (50:50)] and 30% tesu aqueous extract dyed	12	10.15	7.89
Post-treatment with 10% extract of eucalyptus leaves and 10% coconut oil with citric acid as a catalyst on 30% tesu aqueous extract-dyed cotton after pre-mordanted with 15% harda + alum (50:50) pre-mordanted cotton fabric			
10% coconut oil + citric acid (5 gpl) post-treatment	30	2.76	2.02
10% eucalyptus leaves aqueous extract + citric acid (5 gpl)	25	4.16	3.60
10% eucalyptus leaves alcoholic (MeOH) extract + citric acid (5 gpl)	40	2.08	1.81
Post-treatment with 10% extract of eucalyptus leaves on 30% tesu aqueous extract-dyed cotton after pre-mordanted with only 15% alum single pre-mordanted cotton fabric			
10% eucalyptus leaf alcoholic (MeOH) extract +citric acid (5 gpl)	30	2.66	2.89
10% coconut oil + citric acid (5 gpl)	25	5.37	8.22
Simultaneous dyeing and finishing of cotton fabric dyed with 30% tesu aqueous extract and 10% eucalyptus leaf aqueous extract or 10% coconut oil emulsion on 15% harda + alum (50:50) pre-mordanted cotton fabric			
Tesu (30%) extract + eucalyptus aqueous extract (10%) + citric acid (5 gpl)	25	4.49	4.56
Tesu (30%) + coconut oil (10%) + citric acid (5 gpl)	20	3.94	6.08

Table 16.
Result of UPF test as per AATCC-183-2010 for 15% Harda + alum (50:50) pre-mordanted or only 15% alum pre-mordanted cotton fabric dyed with 30% aqueous extract of tesu and post-treatment with natural UV absorbers (aqueous and alcoholic (MeOH) extract of eucalyptus leaves and coconut oil).

good as the UPF value of 40 obtained for post-treatment with MeOH extracted eucalyptus leaves by a pad-dry-cure process on the said double pre-mordanted with 15% overall application of harda + alum (50:50) and 30% tesu aqueous extract-dyed cotton fabric samples. While without any after-treatment or simultaneous treatment with eucalyptus or coconut oil, 15% overall harda + alum (50:50) and 30% tesu aqueous extract-dyed cotton fabric sample shows a UPF value of 12 only as compared to a UPF value of 5 for the standard bleached control cotton fabric (un-mordanted and undyed).

Thus, according to the data in **Table 16**, cotton samples treated with methanol-extracted eucalyptus leaves post-treatment by a pad-dry-cure system provide an excellent UV protection factor against exposure to ultraviolet rays since their UPF value is 40, which lies in the range of UPF category between 40 and 50.

However, when an emulsified coconut oil is treated in the presence of citric acid, it is hydrolysed forming lauric acid (approx. 46–47% content) and α-tocopherol. So during post curing of cotton fabric treated in the presence of citric acid, the hydrolysed emulsified coconut oil (10%), possibly is attached by reaction of –COOH group of lauric acid with hydroxyl group of cellulose forming ester linkage and possible also helps to attach α-tocopherol forming an ether linkage by reaction between hydroxyl group of α-tocopherol and hydroxyl group of cotton cellulose under heat (curing) and acidic catalyst (citric acid). The attachment of eucalyptus to cotton is not yet referenced in any current research, which needs to be further explored by research in this field. However, from compositional analysis, it is known that essence oil parts of eucalyptus contain volatile chemical constituents in

MeOH extract of the eucalyptus leaf having 1,8-cineole, benzene, nerolidol, limonene, alpha-pinene and beta-pinene, which can participate in crosslinking with cellulosic-OH groups in the presence of citric acid and heat showing a better result for protecting bacterial growth and protection against UV rays also, while any type of extract of eucalyptus contain eucalyptol, alpha-pinene, beta-pinene, alpha-phellandrene, gamma-terpinene, caffeic acid, linaool, geraniol and thymol, which are strongly responsible for reducing and blocking the bacterial growth as well. Thus, the eucalyptus leaves MeOH extract can be used as a natural resource-based multiple finishing agent for textiles.

3. Conclusion

The important scientific and technological aspects of natural dyeing on cotton, jute and silk fabrics in terms of extraction and purification, characterization of purified extracted natural dyes by different scientific and instrumental analyses, case studies on the effect of use of different bio-mordanting on color strength, case studies on the standardization of dyeing process variables for optimizing dyeing conditions to get reproducible shades, etc. have led to the generation of scientific ways and means for practising a precise technological control during such processing with variable natural materials, over the traditional and conventional artisan-based natural dyeing processes of textiles.

Similarly the worldwide current research interest on natural resources has improved knowledge of concern textile dyers and other related persons on the use of plant-based bio-mordants, bio-dyes/natural dyes and natural finishing agents, part of which has ultimately come into practice for its commercial utilization and exploitation for eco-friendliness and environmental advantages. Hence, the above said scientific analyses and case studies presented here in brief with natural resource materials for textile dyeing and finishing, may lead to generate further interest of concern readers towards more and more use of these natural resource materials for textile dyeing and finishing like use of blue natural indigo for creating bio-denim, i.e. natural indigo dyeing by natural reduction process applied on cotton has several environmental safety and advantages. A detailed scientific characterization and index-based ingredient identification of such natural dyes and finishes have been also felt essential to formulate test standards for the identification of natural dyes from such natural dyed textile materials for consumers' protection. All effort towards this goal should be encouraged.

Author details

Ashis Kumar Samanta
Department of Jute and Fibre Technology, University of Calcutta, Kolkata, West Bengal, India

*Address all correspondence to: ijtaksamanta@hotmail.com

References

[1] Gulrajani ML, Gupta D. Natural Dye and their Application to Textiles. Delhi: Dept. of Textile Technology, Indian Institute of Technology, Delhi; 1992

[2] Chavan RB. Revival of natural dyes—A word of caution to environmentalists. Colourage. 1995;**42**(4):27

[3] Chavan RB. Revival of natural dyes—A word of caution to environmentalists. Clothesline. 1995; **4**(1):96

[4] Gulrajani ML. Present status of natural dyes. Indian Journal of Fibre & Textile Research. 2001;**26**(3):191-201

[5] Kumar Samanta A, Agarwal P. Application of natural dyes on textiles. Indian Journal of Fibre and Textile Research. 2009;**34**:384-389

[6] Samanta AK, Konar A. In: Kumbassar EPA, editor. Natural Dyes. Croatia/China: In-Tech International Publisher; 2011. pp. 29-56. ISBN: 978-953-307-783-3

[7] Samanta AK, Konar A. Technical Handbook on Natural Dye and Colouration. Kolkata, India: Dept. of Jute and Fibre Technology, IJT, University of Calcutta; 2012. pp. 97-113. Chapter 4

[8] Samanta AK. Chapter 11: Application of natural dyes to cotton and jute textiles: Science and technology and environmental issues. In: Yusuf M, editor. Handbook of Renewable Materials for Coloration and Finishing. New Jersey, USA/Beverly, MA, USA: John Wiley & Sons/SCRIVENER Publishing; 2018. pp. 213-278

[9] Samanta AK, Singhee D, Sengupta A, Rahim SA. Application of selective natural dyes on jute and cotton fabrics by different techniques. Journal of The Institution of Engineers (India). 2003; **83**:22-33

[10] Kumar Samanta A, Agarwal P, Datta S. Dyeing of jute and cotton fabrics using jackfruit wood extract: Part I—Effects of mordanting and dyeing process variables on colour yield and colour fastness properties. Indian Journal of Fibre and Textile Research. 2007;**32**(4):171-180

[11] Samanta AK, Agarwal P, Konar A, Datta S. Characterisation and application of purified selective natural dyes on jute. International Dyer. 2008; **192**(4):25-32

[12] Samanta AK, Agarwal P. Application of mixture of red sandal wood and other natural dyes for dyeing of jute fabric-studies on dye compatibility. International Dyer. 2008; **192**(3):37-41

[13] Samanta AK, Agarwal P, Singhee D, Datta S. Application of single and mixtures of red sandal wood and other natural dyes for dyeing of jute fabric: Studies on colour parameters/colour fastness and compatibility. Journal of the Textile Institute. 2009;**100**:565-587

[14] Kumar Samanta A, Agarwal P, Datta S. Physico-chemical studies on dyeing of jute and cotton fabrics using jackfruit wood extract: Part II—Dyeing kinetics and thermodynamic studies. Indian Journal of Fibre and Textile Research. 2008;**31**(1):66-72

[15] Yusuf M, Mohammad F, Shabbir M, Khan MA. Eco-dyeing of wool with *Rubia cordifolia* root extract: Assessment of the effect of *Acacia catechu* as biomordant on colour and fastness properties. Textiles and Clothing Sustainability. 2016;**2**:10. DOI: 10.1186/s40689-016-0021-6

[16] Vankar PS, Shanker R, Dixit S, Mahanta D, Tiwari SC. Sonicator dyeing of modified cotton, wool and silk with

Mahonia napaulensis DC. and identification of the colorant in Mahonia. Industrial Crops and Products. 2008;**27**:371-379

[17] Kamel MM, El-Shishtawy RM, Yussef BM, Mashaly H. Ultrasonic assisted dyeing III. Dyeing of wool with lac as a natural dye. Dyes and Pigments. 2005;**65**:103-110

[18] Vankar PS, Shanker R, Verma A. Enzymatic natural dyeing of cotton and silk fabrics without metal mordants. Journal of Cleaner Production. 2007;**15**: 1441-1450

[19] Sinnur HD, Samanta AK, Verma DK, Kaware R. Studies on coloration and UV protective action of anar peel (pomegranate rind) as an effective natural colorant for cotton khadi fabric. Journal of The Institution of Engineers (India): Series E. 2018; **99**(1):9-26. DOI: 10.1007/s40034-017-0106-z

[20] Hou X, Chen X, Cheng Y, Xu H, Chen L, Yang Y. Dyeing and UV-protection properties of water extracts from orange peel. Journal of Cleaner Production. 2013;**52**:410-419

[21] Huang S, Wei L, Wang H. Dyeing and antibacterial properties of Liriope Platyphylla fruit extracts on silk fabrics. Fibers and Polymers. 2017;**18**(4): 758-766. DOI: 10.1007/s12221-017-5449-z

[22] Lee Y-H, Lee S-G, Hwang E-K, Baek Y-M, Cho S, Kim H-D. Dyeing properties and deodorizing/antibacterial performance of cotton/silk/wool fabrics dyed with myrrh (*Commiphora myrrha*) extract. Textile Research Journal. July 5, 2016;**87**(8): 973-983

[23] Baite H, Samanta AK, Bhaumik NS, Mallik P, Samanta KK. Study of dyeing process variables on soyabean extract treated cotton and jute fabrics. In: Roy AN, editor. Natural Fibre Resource Management for Sustainable Development. Kolkata, India: Creative Incorporate and TINFS; 2019. pp. 188-197. ISBN: 9788194100904

[24] Chakraborty JN, Chavan RB. Dyeing of denim with indigo. Indian Journal of Fibre and Textile Research. 2004;**29**(1):100-109

[25] Bureau of Indian Standards. BIS-Published-Indian Standards: IS-17084—Textile Dye Stuff-Natural Indigo-Identification and IS-17085—Textile Dye Stuff-Rubia (Madder)-Identification. New Delhi, India: Bureau of Indian Standards; 2019

[26] Bureau of Indian Standards. BIS-Published-Indian Standards: IS-17211—Textile Dye Stuff-Catechu-Identification; IS-17212—Textile Dye Stuff-Lac dye-Identification; IS-17213—Textile Dye Stuff-Punica dye-Identification; and IS-17212—Textile Dye Stuff-Rheum Emodi-Identification. New Delhi, India: Bureau of Indian Standards; 2019

[27] Valianou L, Karapanagiotis I, Chryssoulakis Y. Comparison of extraction methods for the analysis of natural dyes in historical textiles by high-performance liquid chromatography. Analytical and Bioanalytical Chemistry. 2009;**395**: 2175-2189. DOI: 10.1007/s00216-009-3137-6

[28] Singhee D, Samanta P. Studies on dyeing process variables for application of Tesu (*Butea monosperma*) as natural dye on silk fabric. Journal of Natural Fibres. 2019;**16**(8):1098-1112. DOI: 10.1080/15440478.2018.1453430. ISSN: 1544-0478 (Print) and 1544-046X (Online)

[29] Samanta AK, Konar A, Chakraborti S. Dyeing of jute fabric with tesu extract: Part 1—Effects of different mordants and dyeing process

variables. Indian Journal of Fibre and Textile Research. 2011;**36**(1):63-73

[30] Shah HS, Gandhi RS. Instrumental Colour Measurements and Computer Aided Colour Matching for Textiles. Ahmedabad, India: Mahajan Book Distributors; 1990. pp. 76-116

[31] Gilani AH, Janbaz KH. Effect of Rubia cordifolia extract on acetaminophen and CCl_4-induced hepatotoxicity. Phytotherapy Research. 1995;**9**(5):372-375

[32] Mongkholrattanasit R, Kryštůfek J, Wiener J, Viková M. Dyeing, fastness, and UV protection properties of silk and wool fabrics dyed with eucalyptus leaf extract by the exhaustion process. Fibres & Textiles in Eastern Europe. 2011;**19**(86):94-99

[33] Feng XX, Zhang LL, Chen JY, Zhang JC. New insights into solar UV-protective properties of natural dye. Journal of Cleaner Production. 2007;**15**: 366-372

[34] Chattopadhyay SN, Pan NC, Roy AK, Saxena S, Khan A. Development of natural dyed jute fabric with improved colour yield and UV protection characteristics. The Journal of The Textile Institute. 2013;**104**(8): 808-818. DOI: 10.1080/00405000.2012.758352

[35] Ben Fadhel B, Aissi A, Ladhari N, Deghrigue M, Chemli R, Joly J-P. Antibacterial effects of two Tunisian Eucalyptus leaf extracts on wool and cotton fabrics. The Journal of The Textile Institute. 2012;**103**(11): 1197-1204

[36] Singh R, Jain A, Panwar S, Gupta D, Khare SK. Antimicrobial activity of some natural dyes. Dyes and Pigments. 2005;**66**:99-102

[37] Han S, Yang Y. Antimicrobial activity of wool fabric treated with curcumin. Dyes and Pigments. 2005;**64**: 157-161

[38] Khan SA, Ahmad A, Khan MI, Yusuf M, Shahid M, Manzoor N, et al. Antimicrobial activity of wool yarn dyed with *Rheum emodi* L. (Indian rhubarb). Dyes and Pigments. 2012;**95**: 206-214

[39] Khan MI, Ahmad A, Khan SA, Yusuf M, Shahid M, Manzoor N, et al. Assessment of antimicrobial activity of catechu and its dyed substrate. Journal of Cleaner Production. 2011;**19**: 1385-1394

[40] Joshi M, Wazed Ali S, Purwar R. Ecofriendly antimicrobial finishing of textiles using bioactive agents based on natural products. Indian Journal of Fibre and Textile Research. 2009;**34**:295-304

[41] Saleh Mansour S. Antibacterial activity and ultraviolet (UV) protection property of some Egyptian cotton fabrics treated with aqueous extract from banana peel. African Journal of Agricultural Research. 2011;**6**(20): 4746-4752. DOI: 10.5897/AJAR10.845

[42] Vankar PS. Natural Dyes for Textiles (The Textile Institute Book Series) The Textile Institute. UK: Elsevier and Woodhead Publishing; 2017. pp. 17-26; 89-103; 141-166

[43] Grifoni D, Bacci L, Zipoli G, Albanese L, Sabatini F. The role of natural dyes in the UV protection of fabrics made of vegetable fibres. Dyes and Pigments. 2011;**91**:279-285

[44] Stefanovic B, Kostic M, Bacher M, Rosenau T, Potthast A. Vegetable oils in textile finishing applications: The action mode of wrinkle reduction sprays and means for analysing their performance. Textile Research Journal. 2014;**84**(5): 449-460. DOI: 10.1177/0040517513495946

Section 2

Review on Natural Dyes and Pigments and Its Application

Chapter 2

Fundamentals of Natural Dyes and Its Application on Textile Substrates

Virendra Kumar Gupta

Abstract

The meticulous environmental standards in textiles and garments imposed by countries cautious about nature and health protection are reviving interest in the application of natural dyes in dyeing of textile materials. The toxic and allergic reactions of synthetic dyes are compelling the people to think about natural dyes. Natural dyes are renewable source of colouring materials. Besides textiles it has application in colouration of foods, medicine and in handicraft items. Though natural dyes are ecofriendly, protective to skin and pleasing colour to eyes, they are having very poor bonding with textile fibre materials, which necessitate mordanting with metallic mordants, some of which are not eco friendly, for fixation of natural dyes on textile fibres. So the supremacy of natural dyes is somewhat subdued. This necessitates newer research on application of natural dyes on different natural fibres for completely eco friendly textiles. The fundamentals of natural dyes chemistry and some of the important research work are therefore discussed in this review article.

Keywords: colour fastness, dyeing, extraction of natural dyes, natural dyes

1. Introduction

After the advent of mauveine by Henry Perkin in 1856 and subsequent commercialization of synthetic dyes had replaced natural dyes, and since then consumption and application of natural dyes for textiles got reduced substantially. In present scenario environmental consciousness of people about natural products, renewable nature of materials, less environmental damage and sustainability of the natural products has further revived the use of natural dyes in dyeing of textile materials. Natural dyes are having some inherent advantages:

- No health hazard
- Easy extraction and purification
- No effluent generation
- Very high sustainability
- Mild dyeing conditions

- Renewable sources

There are some technical issues and disadvantages related to the application of natural dyes which reduced its applications that are:

- Mostly applicable to natural fibres (cotton, linen, wool and silk)
- Poor colour fastness properties
- Poor reproducibility of shades
- No standard colour recipes and methods available.
- Use of metallic mordants, some of which are not eco friendly.

Hill [1] had given his views that research work with natural dyes is inadequate, and there is need of significant research work to explore the potentials of natural dyes before its important application to textile substrate.

In India initially Alps Industries Ghaziabad (Uttar Pradesh, India) and later Ama Herbals, Lucknow, and Bio Dye Goa done extensive work for industrial research and production of natural dyes and natural dyed textiles. Textile-based handicraft industries in many countries engaged local people to dye textile yarn with natural dyes and weave them to produce specialty fabrics. Printing of textile fabrics with natural dyes in India are specially done in Rajasthan and Madhya Pradesh.

Turkish carpets are recognised for their beauty made with natural dyes. The major importers of natural dyes are the USA and the EU. In the EU the major importers of natural dyes are France, Germany, Italy and the UK. Natural dyes have many advantages [2] like non toxicity, eco friendliness, pleasing shade to eye and having special aroma or freshness of shade [3]; however, natural dyes have some disadvantages to showing poor colour reproducibility, poor or inconsistent composition, average washing fastness [4] and lesser availability in different regions, which are of great concern against its revival. Moreover natural dyes are not having any standard established dyeing [5] method. The final shade depends on the type of mordant used in dyeing. Natural dyes are used in the dyeing of cotton [6, 7], linen [8], wool [9, 10], silk [11, 12], nylon and polyester [13, 14] fabrics. The natural dyes can be classified in different ways such as based on origin/source type, type of hue, chemical structure [15, 16] and colour components. The classification of natural dyes based on origin/source is given below:

- Vegetable origin
- Animal origin
- Mineral origin

For vegetable origin of natural dyes, the best source of natural dyes are the different parts of plants and trees. Most natural dyes are extracted from different parts of plants and trees. Natural dyes and pigments are taken from the following parts of plants/trees:

- Seed
- Root

- Stem
- Barks
- Leaves
- Flowers

Natural dyes are having wide application in the colouration of most of the natural fibres, e.g. cotton, linen, wool and silk fibre, and to some extant for nylon and polyester synthetic fibre. However, the major issues for natural dyed textiles are reproducibility of shade, non availability of well-defined standard procedure for application and poor lasting performance of shade under water and light exposure. To achieve good colour fastness to washing and light are also a challenge to the dyer. Several researchers had proposed different dyeing methods and process parameters, but still these information are inadequate, so this calls for the need of research to develop some standard dye extraction technique and standardisation of whole process of natural dyeing on textiles. Here there are examples of few important natural dyes [17] which are widely used in the dyeing of textile materials, described below.

1.1 Jack fruits (*Artocarpus heterophyllus* Lam)

It is a very popular fruit of south India and other parts of India. The wood of the tree is cut into small chips and crushed into dust powder and then subsequently boiled in water to extract the dye. After mordanting treatment of dyed fabrics, yellow to brown shades are obtained. The cotton and jute fabrics are dyed by this dye. It belongs to the family of Moraceae. The dye consists of morin as colouring molecule (**Figure 1**).

1.2 Turmeric (*Curcuma longa*)

The dye is obtained from the root of the plant. The turmeric root is dried, crushed in powder form and boiled with water to extract the dye. It can be used in the dyeing of cotton, wool, and silk. Proper mordanting treatment improves colour fastness to wash. The brilliant yellow shade is obtained after dyeing with turmeric natural dye. Turmeric is a rich source of phenolic compounds known as curcuminoids. The colouring ingredients in turmeric are called curcumin. Curcumin is diarylheptanoid existing in keto-enol form. Turmeric is a member of *Curcuma* botanical group (**Figure 2**).

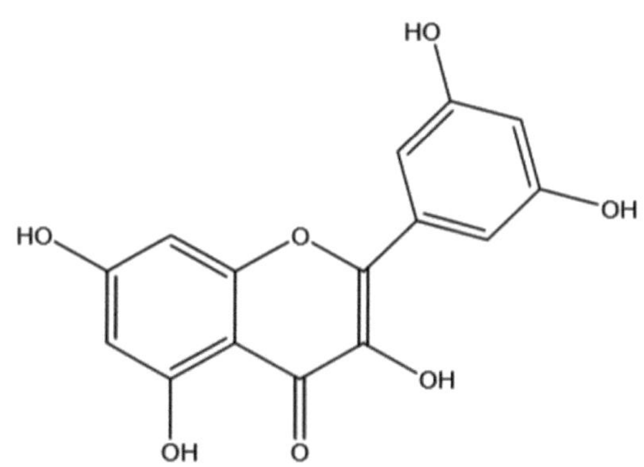

Figure 1.
Molecular structure of morin (3,5,7,2′,4′pentahydroxy-flavone).

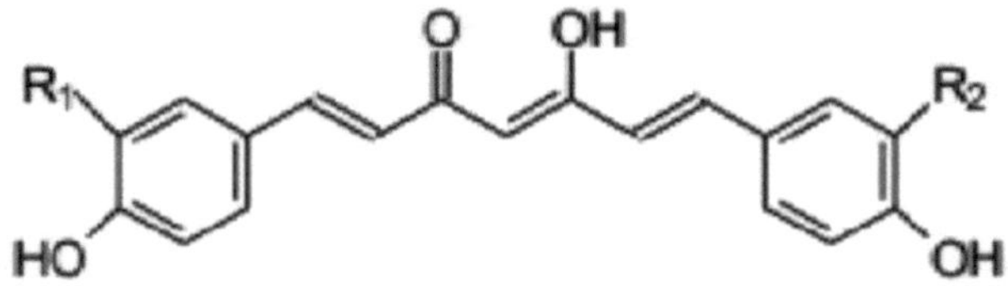

Figure 2.
Molecular structure of curcumin (diarylheptanoid).

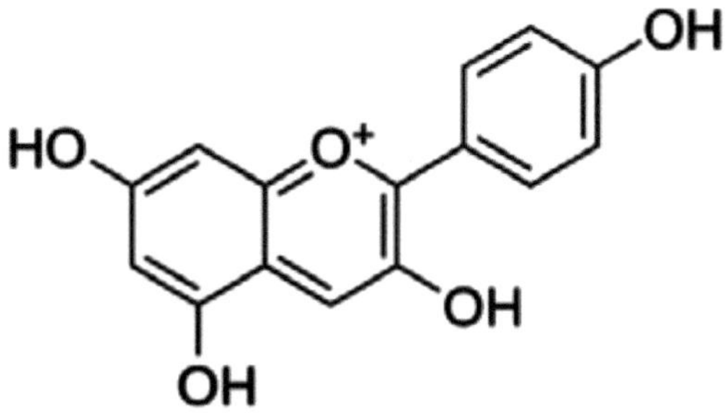

Figure 3.
Molecular structure of pelargonidin (5,5,7,4 tetrahydroxy antocyanidol).

1.3 Onion (*Allium cepa*)

The papery skin of onion is the main source of the dye. Onion skin is boiled to extract the colour and subsequently can be dyed with or without mordanting the fabric. The resulting colour is from orange to brown. It contains colouring pigments called pelargonidin (5,5,7,4 tetrahydroxy antocyanidol). The amount of colouring pigment present varies from 2.0 to 2.25% (**Figure 3**).

1.4 Hina (*Lawsonia inermis* L)

It is the leaf of the plant that is traditionally used in making the coloured design on the hands of women. The leaf of the plant is dried, crushed and subsequently boiled with water to extract the dye from leaf. The mordanted fabric gives colour from brown to mustard yellow. This is the dispersed dye type colour; hence, polyester and nylon can be dyed by hina. However, it stains wool and silk giving a lighter brown colour. Hina is commonly known as lawsone. The chief constituent of hina leaves is hennotannic acid; it is a red orange pigment. Chemically hennotannic acid is 2-hydroxy-1,4-naphthoquinone. The colouring molecules have strong substantivity for protein fibre (**Figure 4**).

1.5 Indigo (*Indigofera tinctoria*)

It is the seed of the plant. The full matured plant has 0.4% colour on weight of the plant. The plants are steeped in the water until the fermentation start. When the hydrolysis of glucoside is completed, the liquor is separated from the plant debris. The extract is aerated which converts indoxyl to indigotin which separates out as a precipitate. The shade of natural indigo is difficult to reproduce exactly. The variety of blue shade on cotton can be obtained by the application of natural Indigo. It is kind of vat dye and hence need reductive vatting with liquid jiggery and citric acid or dithionate.

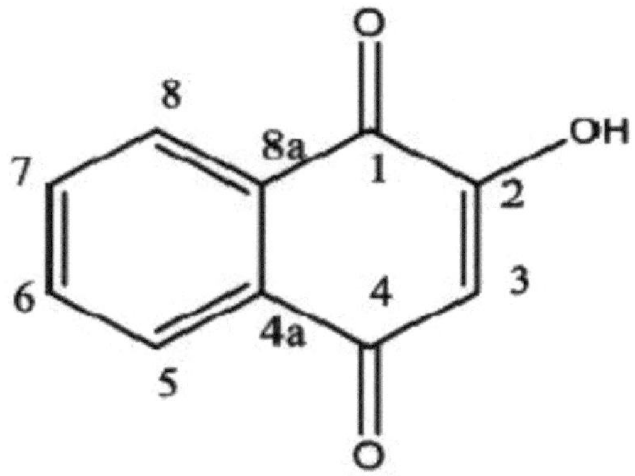

Figure 4.
Molecular structure of lawsone (2-hydroxyl-1,4-naphthoquinone).

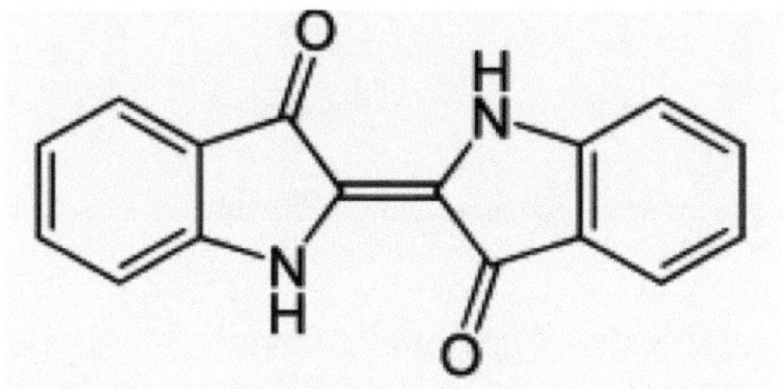

Figure 5.
Molecular structure of natural indigo.

The precursor to indigo is indican which is a colourless water-soluble compound. Indican hydrolyzes in water and releases β-D-glucose and indoxyl. The oxidation of indoxyl resulted in indigotin. The average yield of indican from an indigo plant is 0.2–0.8%. Indigo is also present in molluscs. The molluscs contain mixture of indigo and 6,6′-dibromo indigo (red), which together produce a colour known as Tyrian purple. During dyeing due to air exposure, dibromo indigo is converted into indigo blue, and the mixture produces royal blue colour (**Figure 5**).

1.6 Madder or manjistha or *Rubia* (*Rubia tinctorum*)

The dye is obtained from the root of the plant. The root is scrubbed, dried in sunlight and finally boiled in the water to extract the dye in solution. The dye has red colour. The cotton, silk and wool fibre can be dyed with madder at a temperature of 100°C for time period of 60 min, and subsequently dye solution is cooled. Bright red shade is produced on wool and silk and red violet colour on cotton. This is a mordantable type of acid dye having phenolic (-OH) groups. The colouring matter in madder is alizarin of the antharaquinone group. The root of the plant contains several polyphenolic compounds, which are 1,3-dihydroxyanthraquinone, 1,4-dihydroxyanthraquinone, 1,2,4-trihydroxyanthraquinone and 1,2-dihydroxyanthraquinone (**Figures 6** and 7).

1.7 Tea waste (*Camellia sinensis*)

India is one of the biggest consumer of tea. The left over waste of tea is collectable in large quantity. The extract of tea waste can be used as a natural dye in combination with different mordants, which can produce yellowish brown to brown

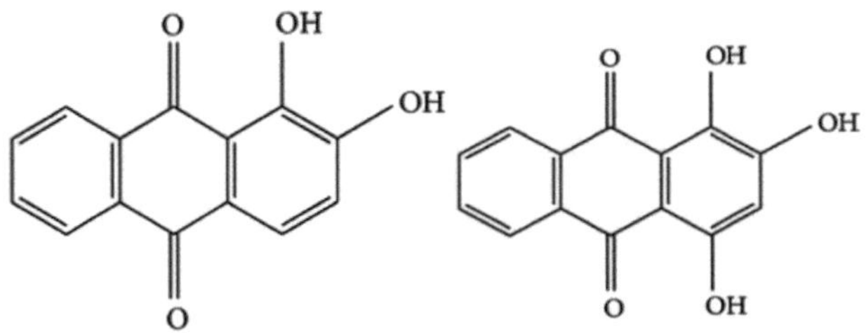

Figure 6.
Molecular structure of alizarin and purpurin.

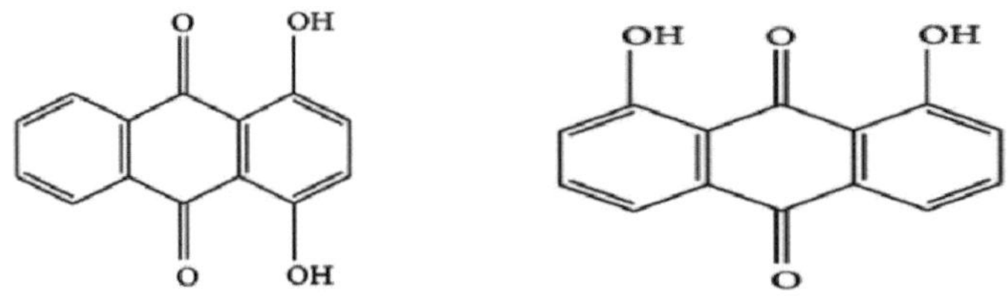

Figure 7.
Molecular structure of 1,4-dihydroxyanthraquinone and 1,8-dihydroxyanthaquinone.

shade. This is a mordantable dye. Flavonoids, flavonols and phenolic acids are the main colouring components in waste of the tea. Polyphenols, which are mostly flavonols, are known as catechins with epicatechin and its derivatives.

1.8 Safflower (*Carthamus tinctorius*)

The safflower petals are soaked in distilled water and subsequently boiled with water for more than 2 h, and it is repeated two times. The solution is filtered and the filtrate is vacuum dried. The obtained powder is having strength of 20–30%. In dyeing it produces cherry red to yellowish red shade. Safflower contains natural pigment called carthamine. The biosynthesis of carthamine takes place by chalcone (2,4,6,4-tetrahydroxy chalcone) with two glucose molecules and that resulted in the formation of safflor A and safflor B (**Figure 8**).

1.9 Sappan wood (*Caesalpinia sappan*)

Aqueous extraction is used to extract the dye from sappan wood. Alkali extraction can also be used. It produces bright red colour. It produces an orange colour in combination with turmeric and maroon shade with catechu. The sappan wood tree is found in India, Malaysia and the Philippines. The colouring pigment is similar to logwood. The same dye is also present in Brazil wood.

1.10 Logwood (*Haematoxylon compechianum*)

The dye is extracted from the stem of the tree. The stems are broken into small pieces and steepened in cold water for several hours followed by boiling.
The extracted dye solution is strained. The logwood natural dye is used to produce black shade on the wool. The logwood trees are found in Mexico, Central America and the Caribbean islands. It is also known as compeachy wood. The colouring matter in logwood natural dye is haematoxylin, which after oxidation forms haematein during isolation (**Figures 9** and **10**).

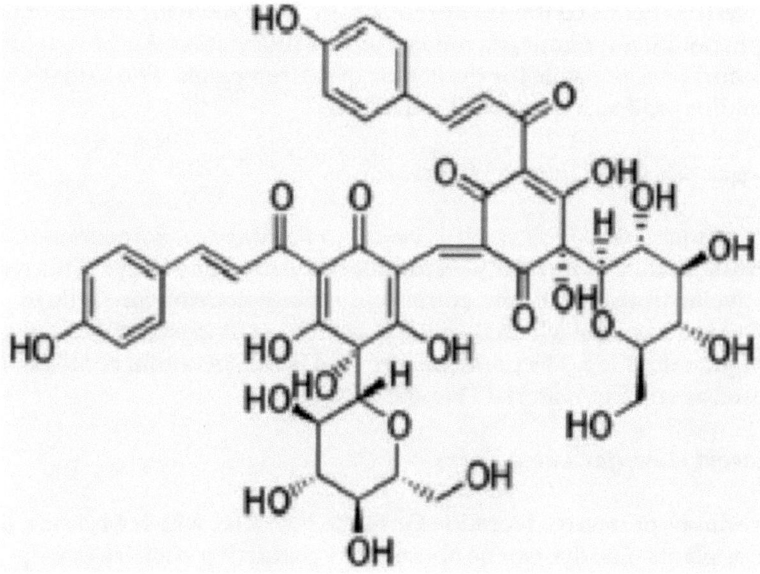

Figure 8.
Molecular structure of carthamine (safflower).

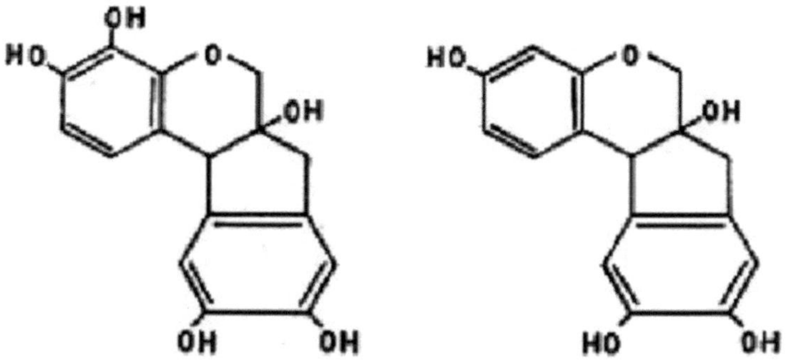

Figure 9.
Molecular structure of haematoxylin and brazilin.

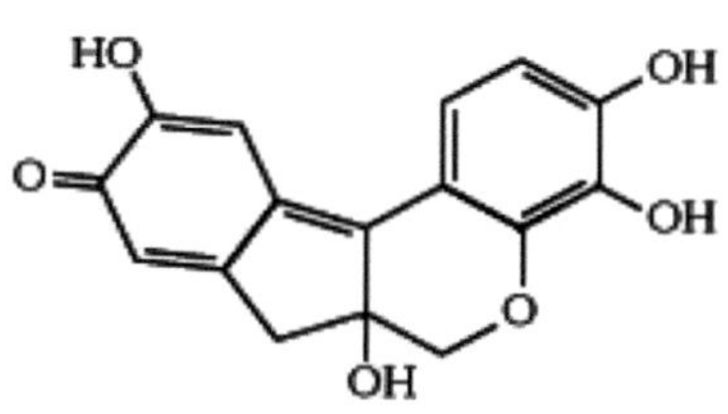

Figure 10.
Molecular structure of haematein.

1.11 Saffron (*Crocus sativus*)

The dye is extracted from the stigma of flower, which is boiled in water, and the colour is extracted. It imparts a bright yellow colour to the textile material. The wool, silk and cotton can be dyed with saffron. Alum mordant produces orange yellow shade which is also called saffron yellow. This is also used as food colouring. Saffron is a perennial plant which belongs to the Iridaceae family. The aqueous extract of saffron petals contains 12% colourant. The colouring matter of saffron contains phenolic compounds, flavonoids and anthocyanins. Anthocyanidins (pelargonidin) is responsible for the colour in saffron petals. The oxidation of anthocyanidins produces flavonol (**Figure 11**).

1.12 Pomegranate rind (*Punica granatum*)

Rind of pomegranate fruit waste is used as a natural dye. Pomegranate fruit is rich in natural tannins. The anar peel produces a yellow colour dye. This natural dye is used in dyeing of wool, silk and cotton fibre. The colouring molecule in pomegranate rind is flavogallol which is called granatonine. It exists in alkaloid form (N-methyl granatonine). The pomegranate rind is rich in tannin content; therefore, it is also used as tanning material (**Figure 12**).

1.13 Lac insect (*Laccifer Lacca* Kerr)

It is a resinous protective secretion from the insect lac which work as a pest on a number of plants. Lac dye can be obtained by extracting stick lac (shellac) with water and sodium carbonate solution and precipitating with lime. Lac contains a water-soluble red dye. It produces scarlet to crimson red shade after dyeing. The lac dye is obtained from an insect named as coccus lacca. Resin which produced by insect is called stick lac. The lac dye contains laccaic acid A and B which are responsible for the colour of the dye. The amount of colouring matter (laccaic acid) is 0.5 to 0.75% on the weight of the resin (**Figures 13** and **14**).

1.14 Cochineal (*Dactylopius coccus*)

Cochineal is obtained from an insect. It produces beautiful crimson, scarlet and pink colour on cotton, wool and silk. After mordanting with alum, chromium, iron and copper; the colour from purple to grey are produced. Cochineal is a scale insect

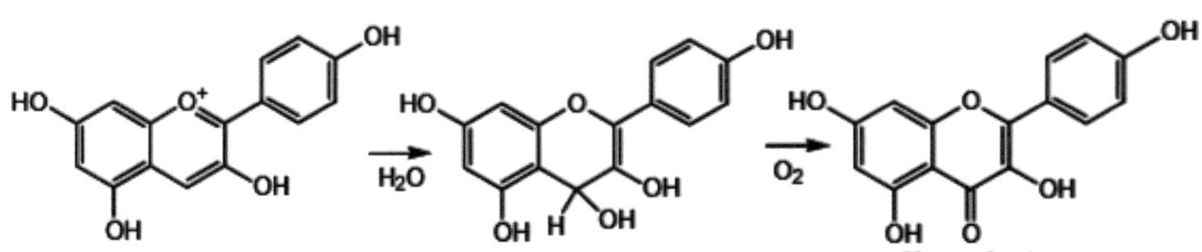

Figure 11.
Molecular structure of pelargonidin (anthocyanidin) purple and kaempferol (flavonol) yellow.

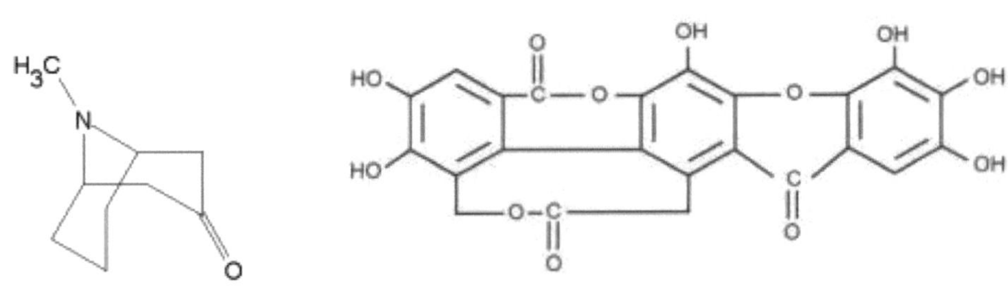

Figure 12.
Chemical structure of granatonine.

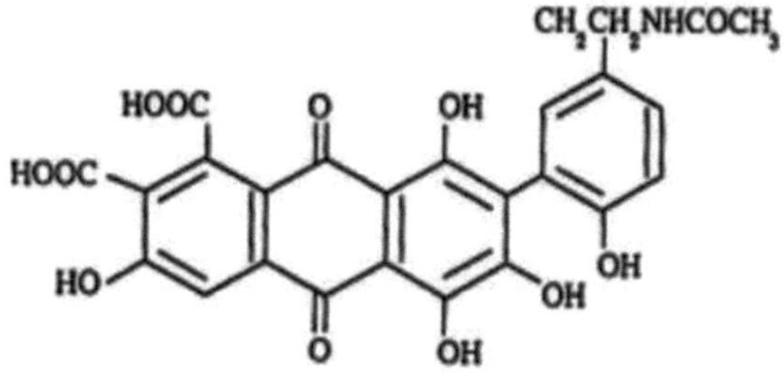

Figure 13.
Chemical structure of laccaic acid A.

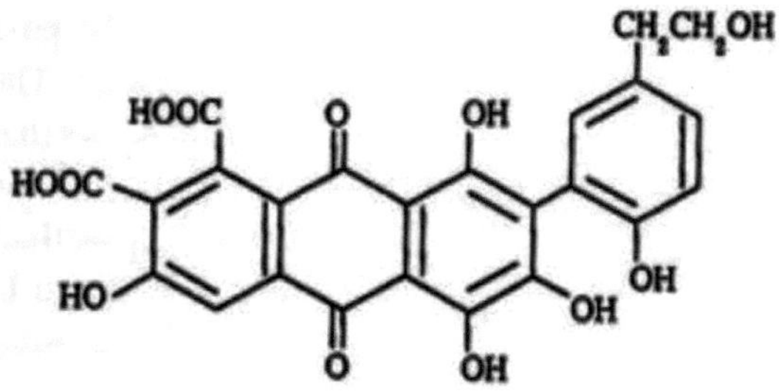

Figure 14.
Chemical structure of laccaic acid B.

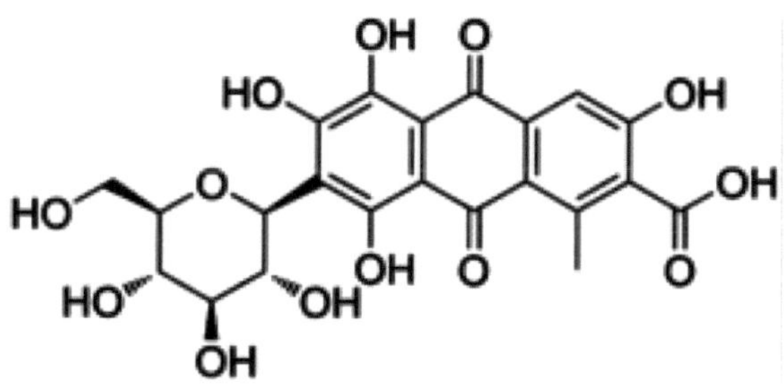

Figure 15.
Chemical structure of carminic acid.

from which natural colourant carmine is derived. Carminic acid is extracted from female cochineal insects. The body of insect is 19–22% carminic acid (**Figure 15**).

1.15 Mineral sources

Some kinds of mineral ores, red clay and ball clay can yield light colours along with mineral salts. But colour composition is not constant and depends on source.

2. Classification of natural dyes

2.1 By chemical constitution

2.1.1 Indigoid class

Two important dyes in this class are indigo blue and Tyrian purple. It occurs as glucoside indicant in the plant. Another blue dye is woad having the same

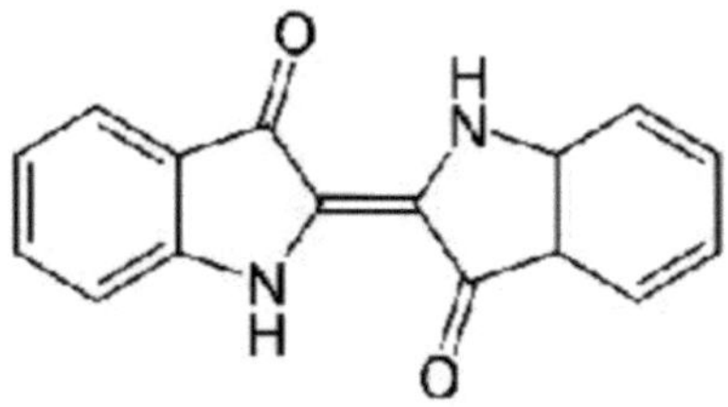

Figure 16.
Indigoid structure.

chemical class. The chemical structure which belongs to indigoid class is shown in **Figure 16**.

2.1.2 Anthraquinone class

Dyes that belong to this class are having anthraquinone structure and obtained from plant and insect. The red shade is specific to this class. Madder, lac, kermes and cochineal are some of the examples. The general chemical structure of this class is shown in **Figure 17**.

2.1.3 Alpha naphthoquinone

The dyes are having alpha naphthoquinone structure such as 2-hydroxy 1-4-naphthoquinone. Hina, lawsone and juglone are examples of this class. The chemical structure of this class is shown in **Figure 18**.

2.1.4 Flavones

The dyes are having yellow shade. The natural dye weld belongs to this category. Most of the dyes are derivatives of hydroxyl and methoxy substituted flavones or isoflavones. The chemical structure of this class of dye is shown in **Figure 19**.

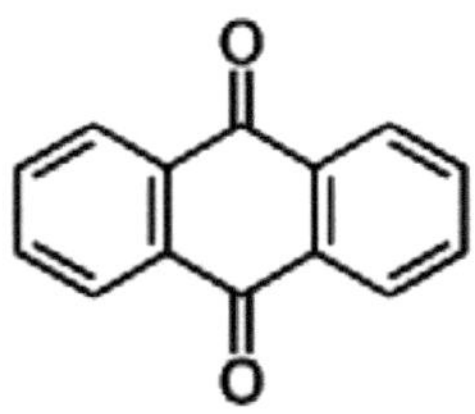

Figure 17.
Anthraquinoid structure.

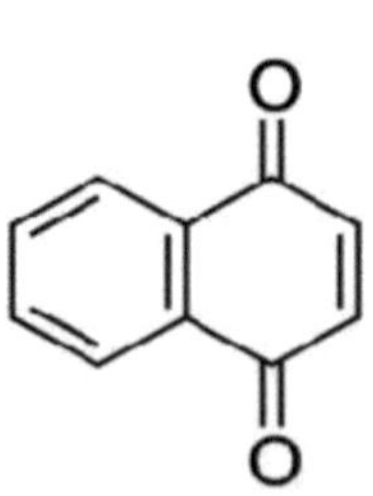

Figure 18.
Naphthoquinone structure.

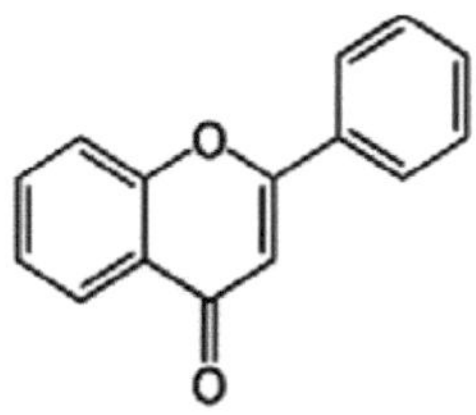

Figure 19.
Flavones structure.

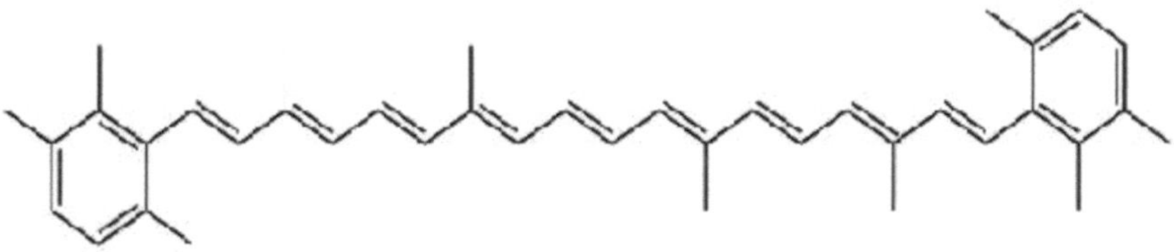

Figure 20.
Carotenoid structure.

2.1.5 Carotenoids

The natural dyes saffron and annatto belong to this class. The dye structure of this class has long-chain conjugated double bonds. The chemical structure of this class is as shown in **Figure 20**.

2.1.6 Dihydropyrans

The dyes which belong to this category are logwood and sappan wood. Logwood, a natural dye, produces dark black shade on silk, wool and cotton.

2.1.7 Anthocyanidins

The natural dye carajurin belongs to this category. The blue and orange shades are obtained from this class.

2.2 Chemistry of natural dyes

Different natural colourants contain different chromophoric and auxochromic groups. Depending on the presence of a particular group in the dye structure, the chemistry of the dyes can be explained in terms of their chromophoric groups. The different dye structures and chromophoric groups are as explained.

2.2.1 Quinoid-based structure

The quinoid-based dye structure can be overviewed as three chemical structures (a) benzoquinone, (b) naphthoquinone and (c) anthraquinone. The natural colourant carthamine belongs to benzoquinone group, and juglone and lawsone are having naphthoquinone structure. Alizarine dye possesses anthraquinone structure.

2.2.1.1 Benzoquinone dyes

In this dye structure the л electron system is small, and the dye contains another unsaturated group in conjugation to л electron system (**Figure 21**). The red colourant

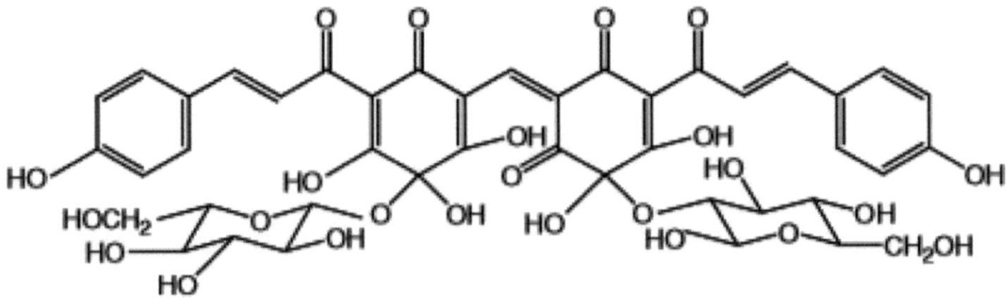

Figure 21.
Structure of carthamine.

carthamine is present in safflower (Natural Red 26). Safflower (*Carthamus tinctorius*) is a subtropical plant and cultivated in India, China, North and South America and Europe. In dyeing, the water-soluble yellow dye (safflor yellow) is extracted [18] by cold water, and then red safflorcamin is extracted by diluted sodium carbonate solution. After the neutralisation of extracted solution, it can be used in dyeing of wool, silk and cotton.

2.2.1.2 Naphthoquinone dyes

Lawsone and juglon natural dye belongs to this category. Lawsone is extracted from hina plant; the leaves also contain flavonoid colourants lutcolin. It is cultivated in countries like India, Africa and Australia. Naphthoquinone is present in glycosidic [19, 20] form named as Hennosid A, B and C. The quantitative analysis of lawsone can be performed by high-performance liquid chromatography on reverse-phase C_{18} column. Chloroform extracted hina leaves were analysed by high-performance thin layer chromatography (**Figures 22** and **23**).

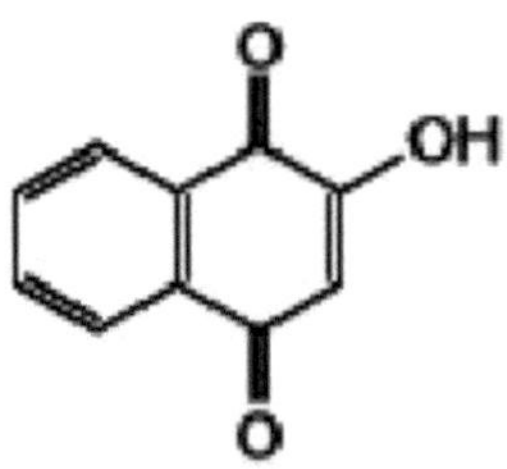

Figure 22.
Lawsone (2-hydroxy, 1,4 naphthalene).

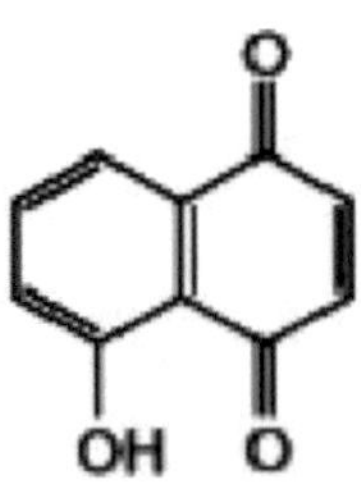

Figure 23.
Juglone (5-hydroxy, 1,4 naphthoquinone).

2.2.1.2.1 Lawsone

Lawsone form 1:2 complex with Fe(II) and Mn (II) and useful in dyeing of wool and silk fibre. The better dye uptake is obtained at pH 3.0. Agarwal et al. [21] studied the effect of different mordants and different mordanting methods to get the different shades. Hina can be used for dyeing of cotton, polyester, polyamide and cellulose triacetate as the structure of dye molecules are similar to disperse dyes [22–24].

2.2.1.2.2 Juglone

Juglone is representative of natural dye with naphthoquinone structure. The dyestuff is extracted from different part of nut trees. Juglone is present as a glycoside form in trees and plants. Wool dyed with juglone are having good resistance with moths and insects. Mordanting treatment further enhances the fastness properties. Dyeing of textile materials with aqueous walnut extract yields brown shade. Wide range of textile fibre, e.g. wool, silk, nylon and polyester, can be dyed with juglone.

2.2.1.3 Antharaquinone

It possess biggest group of anthraquinone dyes. Rhubarb (CI Natural Yellow 23) is extracted from the root of the plant. The extracted dye contains emodin, chrysophenol, aloe emodin and pyscion (**Figure 24**). Rhubarb extract is used in dyeing of wool fibre [25]. It produces yellow to orange shade after mordanting with alum. The mordanting treatment improves light fastness of dyed materials.

Natural dye alizarin, pseudo purpurin and purpurin (**Figure 25**) belongs to plant of Rubiaceae family and has an anthraquinone structure [26]. The dye is obtained from the root of plant.

Madder (C.I Natural Red 8) natural dye produces red colourant; the cultivation of madder is done as a source material for red colour in Europe, Asia and Northern and Southern America. The dyestuff is extracted from the dried roots of the plant. The roots of the plant contain 2–3.0% of di- and tri-hydroxyl anthraquinone glucosides.

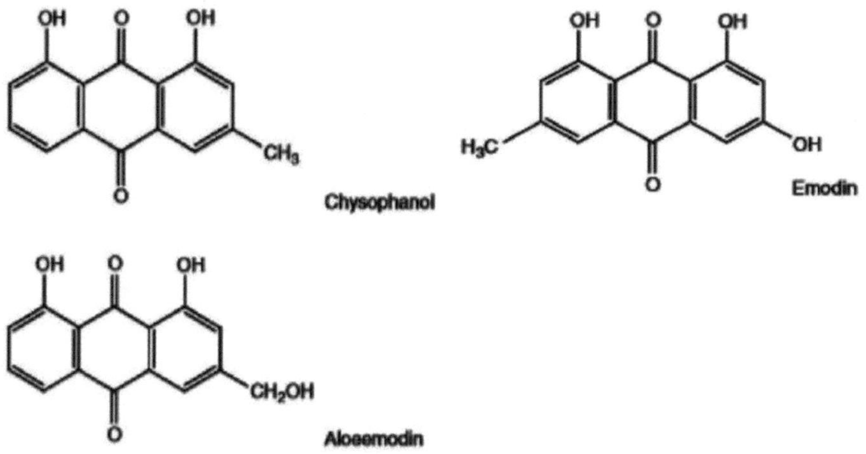

Figure 24.
Different representative structurers of anthraquinone group-based dye molecules.

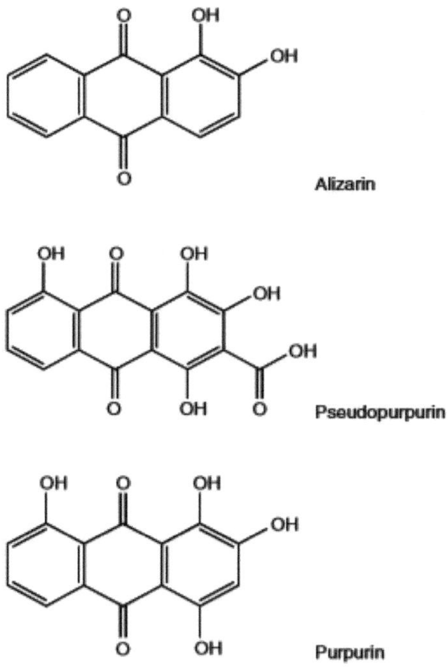

Figure 25.
Structures of alizarin, pseudo purpurin and purpurin.

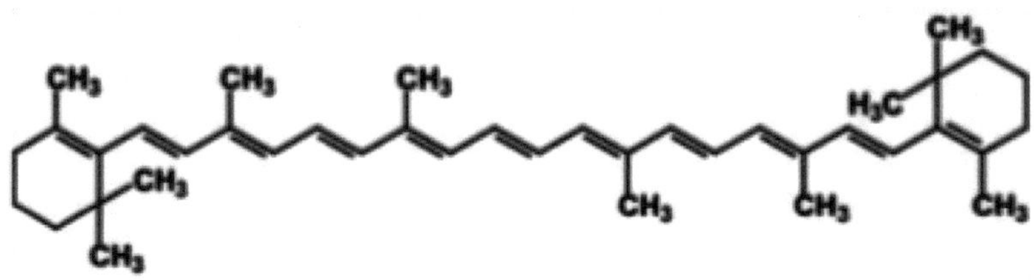

Figure 26.
Structure of β-carotene.

2.2.2 Carotenoids

Carotenoids are red, yellow and orange pigments present in plants and animals [17]. It has a polyisoprenoid structure with a series of centrally located conjugated bonds. The bright colours of many fruits and vegetables are due to carotenoids. Carotenoids are polyisoprenoid structure (**Figure 26**) which contain conjugated double bonds, which acts as chromophore and responsible for characteristic absorption spectra. Carotenoids are divided into two parts:

a. Hydrocarbon carotenoid

b. Oxygen containing called xanthophylls

Structural changes by hydrogenation, double bond migration, isomerization and chain lengthening and shortening resulted in many carotenoid structure. Carotenoids possess strong UV light resistance, and β carotene (**Figure 26**) is a typical structure generally found in natural colourants.

2.2.2.1 Pyron dyes

Pyron dyes contain flavonoids and anthocyanins having structure as shown in **Figures 27** and **28**. The pyron structure is bound to various sugars by glycosidic bonds [17]. Flavonoids are classified as flavonols, flavones, anthocyanidins, isoflavones, flavon-3,4-diols and coumarins. Yellow flavones and flavonols are used as vegetable dyes. The valuable and very popular flavonoid is yellow quercetin which possess several bio effect.

2.2.2.2 Anthocyanins

Anthocyanins are found in fruits and vegetables; some are grape wine, sweet and sour cherries, red cabbage, hibiscus and different varieties of oranges. There are more than 500 varieties of anthocyanins that produces red, pink, violet and orange colours. There are some important anthocyanins which are cyaniding, delphinidin, pelargonidin, malvidin, peonidin and petunidin. Many plants besides anthocyanins also contain quercetin and chlorophylls, and the resulted colour is a mixture of all these.

2.2.3 Dyes from lichens and mushrooms

Violet and purple colours were generally obtained from molluscs and shellfish, and they were source of dyestuff from ancient to the beginning of the Middle Ages. Royale purple and Tyrian purple were the name of the colour obtained originally from molluscs [27]. Lichens and mushrooms are source of natural dyes, and they produce violet and purple colours. Lichens are found in coastal areas and were easier to collect. The dyeing methods with lichens are easy; however, disadvantage associated with lichens is poor light fastness. Therefore, the dyeing of lichens are limited to cheap quality fabrics. Fungi are also used for dyeing of textiles.

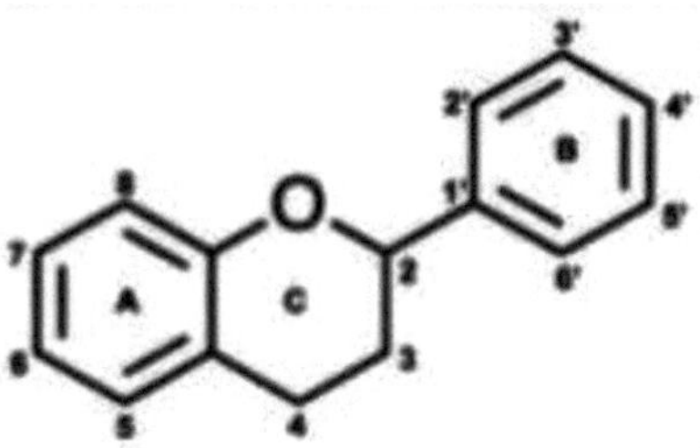

Figure 27.
Structure of anthocyanins.

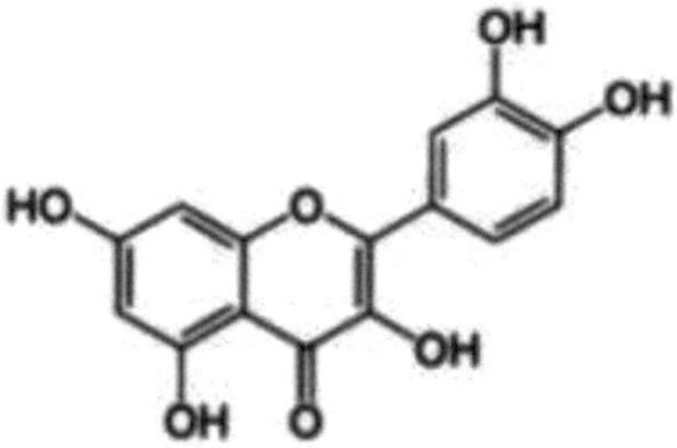

Figure 28.
Structure of quercetin.

In America and India, red colour is obtained from fungus *Echinodontium tinctorium*. In Italy and France, fungi obtained from Polyporales were used to dye the wool.

The colourants in lichens and fungi are benzoquinone derivatives, especially terphenylquinone. Some of these species possess compounds such as *Sarcodon*, *Phellodon*, *Hydnellum* and *Thelephora* [28, 29]. Orchil and litmus are the colourants that are responsible for the colour in lichens. The lichens' colour are produced through pre-compounds of orchil and litmus by consecutive enzymatic, hydrolysation, decarboxylation and oxidation [30] reactions, respectively. Then some pre-compounds are lecanoric acid, atranorin and gyrophoric acid which take part in the formation of orchil and litmus as shown in **Figure 29**.

In the past, the extraction of colourants from lichens were performed by keeping the lichens in water with ammonia for several days. The reaction occurred through enzymatic hydrolysis in which non coloured compounds such as lecanoric acid are converted into orcinol by hydrolysis and decarboxylation. Orcinol after oxidation forms purple orceins or litmus. The colour of both litmus and orchil depend on the pH of the solution [30]. In acidic pH dyestuff forms red cation, and in basic pH, it forms bluish violet anion. The lichens which belong to species *Parmelia*, *Xanthoria parietina*, *Ochrolechia tartarea* and *Lasallia pustulata* are capable to produce yellowish, brownish and reddish brown colours in dyeing of wool with lichens [31]. The dyeing is done by boiling the wool with lichen solution either premordanted or without mordanted wool in presence of ammonia.

The mushrooms which belongs to species *Sarcodon*, *Phellodon* and *Hydlnellum* contain terphenylquinone compounds as a main colourants which produce blue

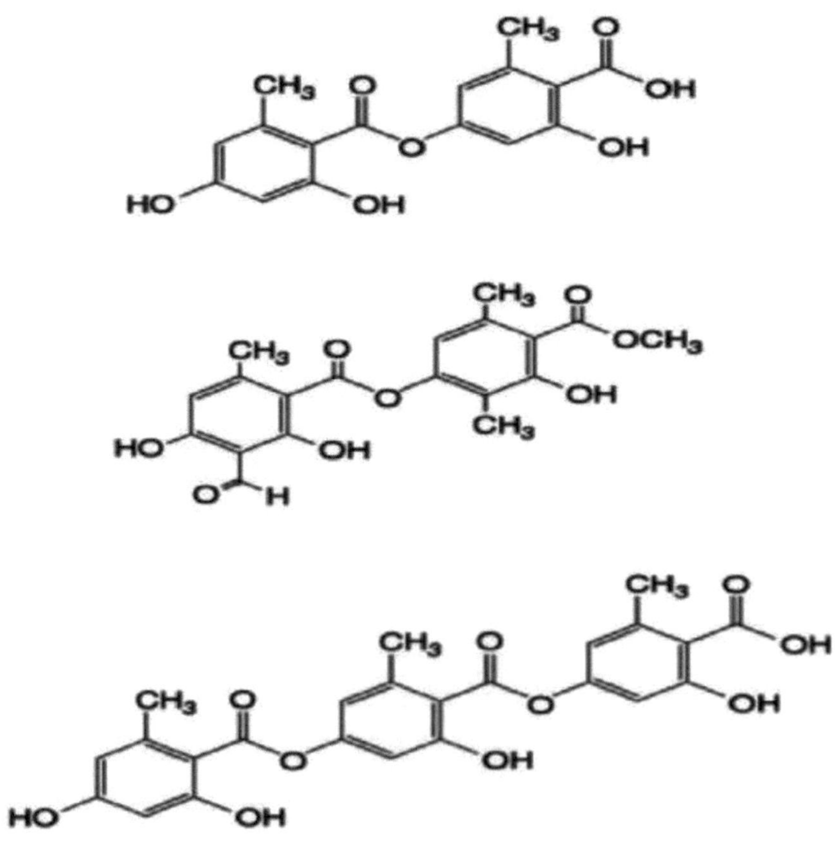

Figure 29.
Structures of different colourants occurring in fungi and lichens.

colour in mushrooms. They are benzoquinone derivatives. The *Cortinarius* species mushrooms are richly coloured in brown, red, olive green and violet. They are anthraquinone derivatives.

2.2.4 Tannins

Tannins are polymeric polyphenols with typical aromatic ring structure with hydroxyl constituents and have relatively high molecular weight. In plants two different groups of tannins are found, (a) hydrolysable tannins and (b) proanthocyanidins (condensed tannin) [32, 33]. Tannins are present in plant cell and are concentrated in epidermal tissues. Tannins are found in wood, leaves, buds, stems, florals and roots [34]. The hydrolysable tannins are concentrated in the roots of several plants. The plants are the source of different variety of tannins. The three major tannins (hydrolysable tannins) are grouped as gallotannins [35] or ellagitannins and which are gallic acid or ellagic acids. The most widespread gallotannins are pentagalloyl glucose. Ellagitannins are esters of hexahydroxydiphenic acids. Gallic acid and hexahydroxydiphenic acid occur together in some hydrolysable tannins [36].

Condensed tannins are polymers of 15-carbon polyhydroxyflavan-3-ol monomer units such as (−) epicatechin or (+) catechin. The complex chemical nature of tannins makes the biosynthesis and polymerisation a difficult task; however, there are some established pathways for bio synthesis. The precursor for biosynthesis of hydrolysable tannins is shikimic acid. The direct aromatization of 3-dehydroshikimic acid produces gallic acid, which upon esterification forms polyol.

The bio synthesis of condensed tannins occurs through two different ways (a) by phenylpropanoid and (b) by polyketide. The polyketide pathway takes malonyl moieties for aromatic ring formation in flavonoid biosynthesis. The phenylpropanoid pathway takes aromatic amino acid, L-phenylalanine, which is non-oxidatively deaminated to E-cinnamate by phenylalanine ammonia-lyase.

2.3 By hue or colour produced

The classification of natural dyes are also done according to the hue of the colour. Some important natural dyes giving primary and secondary colours are:

a. Red: Colour index has 32 red natural dyes. The prominent members are maddar, manjistha, Brazil wood, *Morinda*, cochineal and lac dyes.

b. Blue: There are four natural blue dyes. Some prominent colours are indigo, Kumbh and flowers of Japanese Tsuykusa. Natural indigo blue is known from very ancient time to dye cotton and wool.

c. Yellow: There are 28 yellow natural dyes available which are used in dyeing of wool, silk and cotton. Prominent examples are barberry, tesu flowers, Kamala, turmeric and marigold.

d. Green: Plants that yield green natural colour are very rare; they are made by mixing yellow and blue primary colours. Woad and Indigo produce green colour.

e. Black and brown: There are six black natural dyes. Cutch is used to produce brown shade; for getting black shade lac, carbon and caramel are used.

f. Orange: Natural dyes which produce red and yellow colour are used to produce orange shade. Barbeny and annatto are the examples of orange colour.

2.4 Application based classification

a. Vat dyes: Indigo is a water-insoluble dye, and before application it is solubilised in water. The solubilisation of natural indigo is done with the help of sodium hydrosulphite and sodium hydroxide. After solubilisation, it is applied on cellulosic fibre, and after dyeing the development of colour is done by oxidation with hydrogen peroxide. Indigo dye is the representative of indigoid class of vat dyes

b. Direct dyes: The natural dyes which are water soluble and have a long and planar molecular structure and presence of conjugated (single and double bonds) bonds can be applied by direct dyeing method. The dye molecules may contain amino, hydroxyl and sulphonic groups. Turmeric, Harda, pomegranate rind and annatto can be applied by direct dyeing method. Common salt is used to get better exhaustion of dyes. The dyeing temperature is kept at 100°C

c. Acid dyes: The dye molecules possess sulphonic or carboxylic groups in their structure, which produce affinity for wool and silk fibre. The dyeing is done at acidic pH of 4.5–5.5. After dyeing the fastness improvement is done with tannic acid. The dyeing of wool and silk with saffron is done by acid dyeing method. The presence of common salt in dye bath produces levelling effect

d. Basic dyes: The dye molecules produce coloured cation after dissolution in the water at acidic pH. The dye molecules contain $-NH_2$ groups and react with $-COOH$ groups of wool and silk. The dye bath pH is kept 4–5 by adding acetic acid

3. Extraction of natural dyes

The amount of natural dyes present in natural products are very less [11, 37]. They need specific technique to remove dye from their original source. Here there are some methods which are suitable for extraction of natural dyes from their source materials [28]; the different extraction methods are as follows:

3.1 Aqueous extraction

In this method, the dye containing materials are broken into small pieces or powdered and then soaked in water overnight. It is boiled and filtered to remove non-dye materials. Sometimes trickling filters are also used to remove fine impurities. The disadvantages of this technique are that during boiling, some of the dye decompose. Therefore, those dyes which do not decompose at boiling temperature are suitable by this method. The molecules should be water soluble.

3.2 Acid and alkali extraction

Most of the natural dyes are glycosides; they can be extracted under acidic or alkaline conditions. Acidic hydrolysis method is used in extraction of tesu natural dye from tesu flower. Alkaline solution are suitable for those dyes which contain phenolic groups in their structure. Dyes from annatto seeds can be extracted by this method. The extraction of lac dye from lac insect and red dye from safflower is also done by this method.

3.3 Ultrasonic microwave extraction

Microwave and ultrasonic waves are helpful in extraction of natural dyes. This technique is having several advantages over aqueous extraction. In this technique less quantity of solvent (water) is required in extraction. The treatment is done at lower temperature and less time as compared to aqueous extraction. Ultrasonic and microwaves are sent in aqueous solution of natural dye, which accelerate the extraction process.

3.4 By fermentation

In the presence of bio enzymes the fermentation of natural colour bearing substances becomes faster, and the extraction of natural dyes takes place. Indigo extraction is the best example of fermentation method of extraction. Enzymes break glucoside indican into glucose and indoxyl by the indimulsin enzyme. Amatto natural dye extraction is also done by enzyme method. Cellulose, amylose and pectinase are having application in the natural dye extraction from the bark, stem and roots.

3.5 Solvent extraction

There is use of organic solvents such as acetone, petroleum, ether, chloroform and ethanol in the extraction of natural dyes. It is a very viable technique as compared to aqueous extraction. The yield of dye is good, and the quantity of water requirement is less. The extraction is done at lower temperature.

4. Characterisation of natural dyes

For successful commercial use of natural dyes, there is need of standardized dyeing technique for which characterisation of natural dyes is essential.

4.1 UV-visible spectroscopy

It is useful in characterising the colour in terms of the wavelength of maximum absorption and dominating hue. The application of UV-characterization is to identify the ability of dye molecules to absorb UV wavelength and fading characteristics of dyes. Some researchers [38] had done UV analysis of natural dyes. Mathur et al. [9] studied UV spectra of neem bark, and it has two absorption maxima at 275 and 374 nm. Beat sugar [39] shows their absorption bands at 220, 270 and 530 nm. Gulrajani et al. [40] studied the absorption bands of ratanjot and observed that at acidic pH, the absorption occurs at 520–525 nm, and in alkaline pH, it occur at 610–615 nm. Red sandal [41] wood shows strong absorption peak at 288 nm and maximum absorption at 504 and 474 nm in methanol solution at pH 10. *Gomphrena globosa* flower has peak at 533 nm. The dye does not have difference in peak value at pH 4 and 7 in visible region; however it shifted towards 554 nm [42]. Bhuyan et al. studied the dye absorption extracted from *Mimusops elengi* and *Terminalia arjun* and concluded that dye absorbed by the fibre varies from 21.94 to 27.46% and from 5.18 to 10.78%, respectively, depending on bath concentration [43–45]. He also reported absorption of colour extracted from the roots of *Morinda angustifolia* Roxb using benzene extract. The colour shows absorption at 446, 299, 291, 265.5 and 232 nm.

Name of the dye	Wavelength of maximum absorption
Neem bark extraction	275 and 374 nm
Beet sugar	220, 280 and 530 nm
Ratanjot at acidic pH	520 and 525 nm
Alkaline pH	570, 610 and 615 nm
Red sandal wood	288 nm

The value of the wavelength of the maximum absorption for a particular dye depends on the chemical constitution of the dye molecules which is variable and depends on the growth environment of a particular natural dye. The characterisation of a particular dye is helpful in deciding the hue of the dye.

4.2 Chromatographic technique

Thin layer chromatography is used to identify different colour components in natural dyes. Koren [46] analysed insect dye, madder and indigoid. Guinot [47] analysed plants containing flavonoids colour compounds. Balakina [48] analysed quantitatively and qualitatively red dyes such as alizarin, purpurin and carminic acid by high-performance liquid chromatography. Mc Goven [49] et al. identified the dyes stripped from wool fibre by HPLC with C18 column. Szostek [50] et al. studied the retention of carminic acid, indigotin, corcetin, gambogic acid, alizarin, flavonoid, anthraquinone and purpurin. He studied examination of faded dyes through emission and absorption spectra by non destructive method. Cristea [51] et al. had reported quantitative analysis of weld by HPLC and informed that after 15 min. Extraction in methanol/water mixture, 0.448% luteolin, 0.357% luteolin 7-glucoside and 0.233% luteolin 3′7 diglucoside were obtained. Son et al. [52] reported analysis of longer dyeing time in indigo dyeing and their effect on structural change in dye molecules through HPLC analysis. The derivative spectroscopy and HPLC were used to analyse annatto dyestuff; the sample preparation involved extraction with acetone in the presence of HCl and removal of water by evaporation with ethanol. The residue was dissolved in chloroform and acetic acid mixture for derivatives spectroscopy or with acetone for HPLC.

5. Theory of dyeing

Natural dyes are very suitable for dyeing of protein fibres as compared to cellulosic fibres. Synthetic fibres which contain polar groups such as nylon, acrylic and viscose are also accessible to natural dyes. Natural dyes are thermo unstable and have poor chemical stability, which make the natural dyes unfit for dyeing at high temperature and pressure. The presence of hydrogen bond and Van der Waals force of attraction play important role in the fixation of natural dyes on the fibre. Natural dyes are having poor exhaustion value due to subdued affinity for fibre materials, so to increase the exhaustion of dyes, common salt/Glauber's salt are added in the dye bath. The isotherm of the natural dyes sorption obeys Nernst isotherm [17, 53, 54].

Natural dyes are having poor affinity and substantivity [55, 56] for cellulosic fibres such as cotton and viscose. The absence of reactive groups in fibres and dyes does not allow for bond formation, so they need mordanting treatment to fix the dye on fibre surface. Protein fibres are having bond-forming groups in fibre structure, and the presence of carboxylic groups in natural dyes provides opportunity for

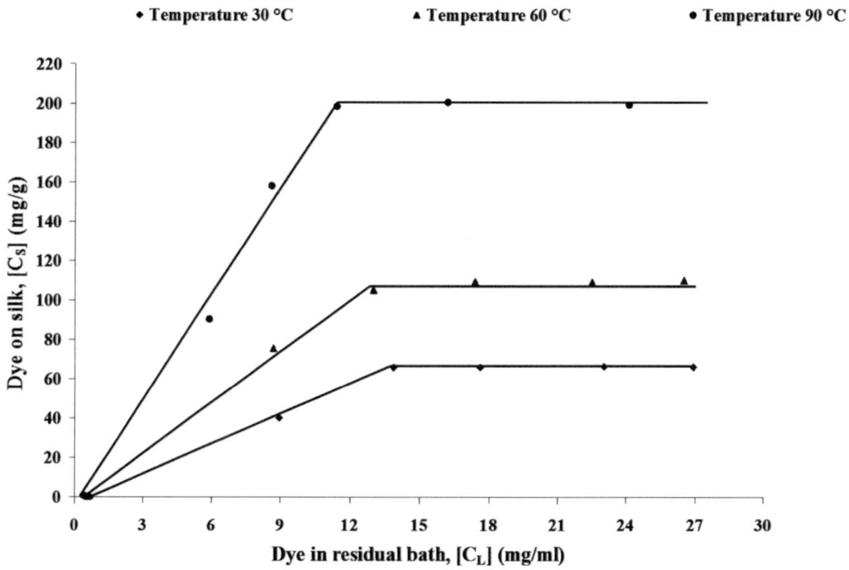

Figure 30.
Sorption isotherm of dyeing of silk fabric (without mordant) with eucalyptus leaves extract at three different temperature 30, 60 and 90°C [17].

bonding and gets bonded with fibre and shows good fastness properties. Natural dyes are having smaller molecular size, and they are not having conjugated linear structure [57]. Therefore, natural dyes are having inferior exhaustion behaviour. Sometimes salt sodium chloride is also used to improve the dye exhaustion % (**Figure 30**).

6. Application of natural dyes

Different researchers had proposed different methods of dyeing of natural and synthetic fibres with natural dyes. The dyeing of textile substrates depends on dyeing parameters which are fibre structure, temperature, time and pH of the dye bath and dye molecule characteristics. The fastness properties of dyes on textile substrates depend on bonding of dyes with fibre. Since natural dyes are lacking in the presence of active groups to make bonds with textile fibres, the fastness properties are not very good. The cellulosic fibres are difficult to dye with natural dyes as they have poor affinity and substantivity. The lack of bonding of natural dyes with cellulosic fibre requires mordanting treatment. Protein fibres have ionic groups and get bonded with natural dyes possessing ionic groups in dye structure.

The dyeing of proteins fibre can be done by exhaust method of dyeing. The dyeing process parameters in wool and silk dying is pH at 4.5–5.5 and dyeing temperature 80–90°C. The exhaustion % of dyes in dyeing is very poor. The longer liquor ratio may be preferred because of poor solubilities of natural dyes in water. Stainless steel-made dyeing machines are suitable in dyeing of wool and silk.

Since natural dyes are having poor affinity for cellulosic fibre and due to poor exhaustion, mordanting treatment [29, 58] is done to fix the dyes on cellulosic fibre. The dyeing of cellulosic fibre can be done at temperature of 80–90°C.The exhaustion of dyes can be increased by adding exhausting agents, sodium chloride or Glauber's salt in dye bath. Most of the dyeing is done at neutral pH. Dyeing of cotton with natural indigo is done at alkaline pH in the presence of sodium hydrosulphite in a container made of stainless steel. The copper container gives deeper shade in dyeing of cellulosic fibre. The mordanting treatment improves the washing fastness of dyed samples. There are three methods of mordanting [44, 45].

6.1 Conventional method of dyeing

In the state of Maharashtra, Gujrat and Rajasthan [59], the people follow conventional method of dyeing of cotton fabric with natural dyes which may be explained with the following process sequences. The fabric is pretreated before dyeing to get the absorbency. The grey fabrics are given dunging treatment followed by washing. The bleaching treatment is given to make the fabric white, after that it is steamed and stepped into alkaline solution, and finally rinsing and washing treatment is given. After thorough pretreatment the fabric is soaked into solution of harda/myrobolan and dried. The dried fabric is premordanted with alum and subsequently dipped into natural dye solution at boiling temperature. After dyeing the fabric is given washing and rinsing treatment and dried in the sun light. Water is sprayed on the fabric to brighten the shade. The process is repeated 2 to 4 days. The dyeing method differs from place to place. Here are some examples:

6.1.1 In Bengal

The commonly used natural dyes are haldi, babul, madder, pomegranate rind and marigold [59]. In the dyeing of fabric with sappan wood, the fabric is dipped in aqueous extract of sappan wood with or without alum solution and boiled for 2–3 hours. In the dyeing of Indian madder, the madder is extracted either from the stem or root and boiled with water to extract the natural colourants. The pretreated fabric is boiled with dye extract solution. Mordanting treatment may be given either before dyeing or after dyeing with alum solution.

6.1.2 In Orissa

The sappan wood chips are boiled with alum and turmeric and after boiling it was cooled. In cooled solution of dye, the fabric materials are kept for 3–4 h. It is a premordanting process. At some places the cold solution of natural dye is taken with sufficient quantity of water, and the fabric is dipped in cold solution for 24 h and finally boiled for 2 h.

6.1.3 In Uttar Pradesh

The application of natural indigo on cotton fabric is done by two methods which are called Khari Mat and Mitha Mat.

6.1.4 Khari Mat

In Khari Mat's process to dissolve natural indigo, 40 gallon of water is taken in an earthen vessel, and in that water there are addition of 2.0 lbs. indigo, 2.0 lbs. of lime, 2 lbs. of sajji mati and 1.0 ounce of gur (molasses). After 24 h of fermentation, the indigo dye became water soluble. The indigo dye solution is ready for dyeing. This technique is successful in hot weather.

6.1.5 Mitha Mat

In this technique, the solubilisation of natural indigo is done by taking 60 gallon of water; in that water there are addition of 4 lb. of lime, and after 1 day again

4 lb. of lime is added. After 4–5 days natural indigo dye became fully soluble. During application this mitha vat is added with old mitha vat with continuous string. The fabric is dyed in the dissolved indigo dye solution at temperature of 50–60°C.

6.2 Dyeing of cotton fabric with natural dyes

There is a standard recipe-based dyeing process for dyeing of cotton fibre/yarn/fabric. The important pretreatments before dyeing are desizing (acid desizing or enzyme desizing), scouring (sodium hydroxide and auxiliaries) and bleaching with hydrogen peroxide (H_2O_2). The fully pretreated fabric free from all impurities and absorbent is premordanted (single or double mordanting, in single either harda or aluminium sulphate in double taking both consecutively) with aluminium sulphate. After mordanting the mordanted fabric is passed through aqueous solution of natural dyes. The dyeing parameters will be:

- Dyeing time = 60–120 min. (depends on depth % of shade)
- Temperature of dyeing = 70–100°C
- M:L ratio of the bath = 1:20–1:30
- Amount of dye in bath = 10–50% (on weight of the material)
- Concentration of common salt = 5–20 g/l
- pH of the dye bath = 10–11

After dyeing, soaping treatment is given to remove any residual/unreacted dyes and auxiliary chemicals from the surface of the fabric. An after treatment with natural dye, fixing agent may be desirable.

6.3 Dyeing of protein fibres

Wool and silk are protein fibre; both fibres have complex chemical structure and susceptible to alkali treatment. Alkaline pH of aqueous solution damage the fibre. At isoelectric pH of 5.0, the wool is neutral and the silk is slightly positive. The wool and silk can be dyed with natural dyes through premordanting or after mordanting. Mordanting is done with tannin-rich natural source chemical like harda or metal salt aluminium sulphate or ferrous sulphate.

In premordanting, the fabric is treated with either harda or metal salt aluminium sulphate (single or double) with 5–20% (on weight of the material) mordant concentration at temperature of 80–90°C for 30–40 min. The M:L ratio is kept 1:5–1:20. After mordanting, drying treatment may be given and subsequently dipped in dye bath containing aqueous natural dye solution. The following dyeing parameters were maintained:

- The pH of the dye bath = 4–5
- Temperature of dyeing = 80–90°C.
- Time of dyeing = 50–60 min.

- M:L ratio of the bath = 1:20–1:30
- Amount of dye in bath = 10–50% (on weight of the material)

After dyeing, soaping treatment is given to remove any residual/unreacted dyes and auxiliary chemicals from the surface of the fabric. An after treatment with natural dye fixing agent may be desirable.

6.4 Dyeing of synthetic fibres

Different synthetic fibres like nylon, polyester and acrylic can be dyed with natural dyes like onion skin extract, babool bark extract and hina. The dyeing can be done either by padding (cold pad batch) method or exhaust method with or without mordanting. Dyeing is carried out at acidic pH. High-temperature high-pressure dyeing gives better results in terms of colour strength than other dyeing methods.

6.5 Fixation of natural dyes

Natural dyes are having poor affinity and substantivity for textile materials. The bonding groups are not present in natural dyes, due to that most of the natural dyes are having poor washing fastness. The fixation of natural dyes on textile materials can be done with the help of mordanting agents. Mordanting agents are dyeing auxiliaries and are salts (chlorides and sulphates) of heavy metals. The heavy metals Al, Cr, Cu and Sn are having vacant d orbitals and easily make coordinate bonds with natural dyes and fibre-active sites. The formed complex has bathochromic and hyperchromic shift. There are different types of mordanting agents such as metallic mordants, tannins and tannic acid and oil mordants. The different heavy metal salts work as complexing agent and chelate with natural dye colourants. Some metallic salts are toxic in nature, but even after that, they are having application in fixation of natural dyes. The different mordanting agents are:

a. Most controversial are lead salts and chromates (potassium, sodium, ammonium dichromate).

b. The salt $SnCl_2$ also works as mordant. It is water soluble, having reducing agent properties. It is toxic in nature.

c. Copper sulphate ($CuSO_45H_2O$) and ferrous sulphate ($FeSO_47\ H_2O$) molecules are also used as a mordant. They are good chelating agents.

d. Tannins are poly phenolic compounds and able to form complexes with metals and bind with organic substances such as proteins, alkaloids and carbohydrates. The tannins are also called bio mordants. Tannins can be used either alone or in association with metal salts. The phenolic groups of tannins can form effective bonds with fibre and natural dye molecules.

6.5.1 Metallic mordants

Metal salts of aluminium, chromium, iron and copper are used as a mordants. The important mordants are potassium dichromate, ferrous sulphate, copper sulphate, stannous chloride and stannic chloride.

6.5.2 Tannins and tannic acid

Tannins are obtained from the excretions of bark and other parts, e.g. leaves and fruits of the plant. Extractions are either used directly or in concentrated form. Large number of tannin containing substances are employed as a mordant in textile fibre dyeing.

6.5.3 Oil mordants

Oil mordants are used in dyeing of madder. Oil mordants make a complex with alum used in mordanting treatment. Metal atom combined with carboxylic groups of oil and bound metal then makes bond with the dye molecules, and in this way, superior wash fastness can be achieved.

6.6 Mordanting process

a. Premordanting: In premordanting process, mordanting is done before dyeing; subsequently the fabric is dyed with natural dye in aqueous media. It is a two-bath process in which the first bath is used for mordanting of fabric and in the second bath, dyeing is done with natural dyes. Dyeing and mordanting are done at the same temperature of 60–70°C. the mordants are complexing agents, and if they are taken in the same bath, they may react to each other, and precipitation of dyes may occur. That deteriorate fastness properties of dyed fabrics

b. Metamordanting: In metamordanting treatment, the mordant chemicals are added with natural dye in the same dye bath; dyeing and mordanting take place simultaneously. The mordanting and dyeing temperature are 80–90°C

c. After mordanting: In after mordanting treatment [53, 54], the dyeing of fabric is done first; after that in the same bath mordanting compounds are added. The temperature of chroming is 80–90°C. after chroming, the temperature is dropped to 60°C, and goods are run for 15 minutes after that liquor is drained

The application of natural dyes on cellulosic materials are done by the pad-dry-washing and pad-dry-steaming-washing method. High-temperature curing is not suggested as dye molecules are susceptible to decompose. Fibre and yarn dyeing can also be done with natural dyes similar to synthetic dye application.

7. Fastness properties of natural dyes

The quality parameters in dyeing is fastness properties. Several test methods are described to access the colour fastness. The fastness properties give idea about the quality of dyeing. In natural dyes, the fastness properties are strongly related to substrate type and mordant used for dyestuff fixation. Besides the dyestuff itself, there are many factors such as water, chemicals, temperature, humidity, light, pretreatments, after treatments, dyestuff distribution in fibre and fixation of dyestuff affect the fastness properties. In natural dyeing the colour and fastness of natural dyes need special attention for careful selection of materials and process. Natural dyes were in use up to end of the nineteenth century. At that time the dyeing with natural dyes were at peak with excellent fastness properties; however, after commercialization of synthetic dyes in the nineteenth century, the proficiency in natural dyeing started to decrease. The different fastness properties of dyes show the resistance of dyes towards different external environment in which fabric containing dyes are exposed. The fastness properties of dyes depend on the structure of dyes, exposure on the environment and fastness improvers and type of mordant used. There is need to explore some natural after treatment agents to improve the light and washing fastness.

7.1 Light fastness

The light fastness of natural dyes is poor to medium. The poor light fastness is due to chromophoric change in dye structure after absorption of light. The chromophoric groups are not very strong to dissipate the energy absorbed through resonance. Cook [60] had reported a comprehensive review on light fastness improvement of dyed textile fibres. He studied the use of tannin related after treatments on mordantable dyes to be used in cotton dyeing for improving light and wash fastness, and his findings were useful in improving fastness properties of natural dyed fabrics. Natural dyes have poor light stability as compared to synthetic dyes. Padfield and Landi [61] observed the light fastness of wool dyed with nine natural dyes such as:

a. Yellow dyes (old fustic and Persian berries), light fastness rating 1–2

b. Reds (cochineal with tin mordent, alizarin with alum mordant, lac with tin mordant), rating 3–4

c. Blue (indigo depends on mordants), rating 4–5 and 5–6

d. Black (logwood), rating 4–5

Mordants highly influence the light fastness of natural dyes. Turmeric, fustic and marigold dyes faded more than any other yellow dyes; however, the application of tin and alum mordants causes more fading than chrome, iron and copper. This shows the dependency of fastness properties of natural dyes on the type of mordants. Samanta et al. [62] reported the light fastness improvement in natural dyes applied on jute fabric by 1% benzotriazole. The biggest challenge in natural dyeing for colour fastness is related with light fastness. The choice of suitable mordent will improve the light stability except some iron salts which lead to shift in the resulting colour. Textile auxiliaries also improve fastness properties. To improve the light stability of natural dyes, Lee [63] commended an UV absorber on protein fibre. Oda [18] suggest singlet oxygen quenchers to improve the light fastness rating. Mussak [64] discussed light-induced photo degradation process of natural dyes. Several attempts were made to improve the light fastness of different textile fabrics dyed with natural dyes out of which some are [65–67]:

a. Effect of various additives on photo fading of carthamin in cellulose acetate film.

b. Critical examination of fading process of natural dyes to reproduce original colour of the fabric after fading.

c. The rate of photo fading effect is effectively suppressed in the presence of nickel hydroxyl-arylsulphonate. The addition of UV absorbers in bath has small effect in reducing photo fading effect.

7.2 Washing fastness

The washing fastness of natural dyes is poor to medium. The bonding of dye with fibre is very poor, and due to that dyes are not very fast with detergent solutions. Duff et al. [29] studied the effect of alkalinity of washing solution in washing of natural dyes dyed fabrics. The alkaline pH of the detergent solution changes the colour value in terms of the hue and value. Logwood and indigo are having good fastness value as compared to others. The mordanting treatment improves the washing fastness of dyes. Samanta et al. [68] reported some improvement in washing fastness by use of fixing agent.

7.3 Rubbing fastness

The rubbing fastness of most of the natural dyes are moderate to good. Samanta et al. [8, 58] reported that jackfruit wood, manjistha, red sandal wood, babool and marigold having good rubbing fastness on jute and cotton fabric.

8. Advantages of natural dyes

8.1 UV-protective fabrics

UV-protected fabrics are required to protect the skin and body of the human being from sunburns, tannings, premature skin burns and skin ageing. Researchers had done the work on to produce fabrics which had sun-protecting effect by the application of natural dyes in dyeing. Sarkar [69] evaluated ultraviolet protection factor (UPF) value of cotton fabric dyed with madder, indigo and cochineal with reference to fabric parameters. Grifani [70, 71] studied the effect of natural dyes on cotton, flax, hemp and ramie and got good results. Metallic mordants [72] have potential to improve the UPF value of wool, silk and cotton. Orange peel extract natural dye applied on wool increased the UPF value of dyed wool fabric considerably.

8.2 Insect proof

Cellulosic materials and woollen are susceptible to moth and fungus attack in humid and warm conditions. Koto et al. [73] studied the effect of natural dyes on wool. The anthraquinone-based natural dyes cochineal, indigo and madder are able to produce insect proof and repellent fabric when used as a dyes in dyeing of wool.

9. Summary and conclusions

- Natural dyes due to its unique character of natural origin are known as eco-friendly dyestuff; however the bonding of dye molecules with fibre-active sites are very poor, and they need some bridging chemicals to anchor the dye molecules with fibre, and mordanting agents are helpful in bridging the dye molecules with fibre. The synthetic mordanting agents are not very eco friendly, and some are toxic which depress the efficacy of natural dyes and sometime become matter of debate.

- Natural dye does not have any shade card to match the samples or reproducing the shade. So there is need of collection of spectral data of natural dyes so that any shade can be reproduced.

- There is need of awareness about natural dyes dyed fabric in people so that it can be popular in big way. and due to that demand and consumption of natural dyed fabric will increase.
- Natural dyes are costly as compared to synthetic dyes. So some research work should be done to reduce the cost of production.
- Big production houses, technical institutions and research houses should organised workshops and symposia to spread the advantages of natural dyes.
- The government should promote the production of natural dyes by giving financial incentives to small manufactures of natural dyes.
- There must be some very strong research and development work to improve the quality of natural dyes in terms of low cost, use of natural mordent and widespread applications.

Acknowledgements

I am very thankful to Prof. A.K. Samanta for inspiring me and giving very excellent suggestions for preparing this review paper. I am very thankful to the editor for his remarkable patience and monitoring.

Author details

Virendra Kumar Gupta
M.L.V. Textile and Engineering College, Bhilwara, Rajasthan, India

*Address all correspondence to: virendra1970@rediffmail.com

References

[1] Hill DJ. Is there a future for natural dyes? Review of Progress in Coloration and Related Topics. 1997;**27**:18

[2] Dedhia EM. Natural dyes. Colourage. 1998;**45**(3):45

[3] Chavan RB. Chemical Processing of Handloom Yarns and Fabric. Delhi: Department of Textile Technology, IIT; 1999. p. 6

[4] Ghosh P, Samanta AK, Das D. Effect of selective pretreatments and different resin post-treatments on jute-viscose upholstery fabric. Indian Journal of Fibre and Textile Research. 1994;**19**:298

[5] Gulrajani ML, Deepti G. Natural Dyes and their Application to Textiles. Delhi: Department of Textile Technology, IIT; 1999. p. 23

[6] Senthil P, Umasankar P, Sujatha B. Ultrasonic dyeing of cotton fabric using with neem leaves. Indian Textile Journal. 2002;**112**(6):14

[7] Saxena S, Iyer V, Shaikh AI, Shenai VA. Dyeing of cotton with lac dye. Colourage. 1997;**44**:23

[8] Samanta AK, Preeti A, Siddhartha D. Dyeing of jute and cotton fabrics using Jackfruit wood extract: Part I—Effects of mordanting and dyeing process variables on colour yield and colour fastness properties. Indian Journal of Fibre and Textile Research. 2007;**32**:466

[9] Mathur P, Metha A, Kanwar R, Bhandari CS. Use of neem bark as wool colourant—Optimum conditions of wool dyeing. Indian Journal of Fibre and Textile Research. 2003;**28**:95

[10] Gulrajani ML, Gupta DB, Agarwal V, Jain V, Jain M. Some studies on natural yellow dyes. Indian Textile Journal. 1992;**102**(4):50

[11] Mahale G, Sakshi, Sunanda RK. Silk dyed with Acalypha (*Acalypha wilkesiana*) and its fastness. Indian Journal of Fibre and Textile Research. 2003;**28**:86

[12] Katti MR, Kaur R, Shrihari N. Dyeing of silk with mixture of natural dyes. Colourage. 1996;**43**(12):37

[13] Lokhande HT, Vishnu A, Dorngade, Nayak SR. Application of natural dyes on polyester. American Dyestuff Reporter. 1998;**40**

[14] Rathi DR, Padhye RN. Studies on application of natural dyes on polyester. Colourage. 1994;**41**(12):25

[15] Paul R, Jayesh M, Naik SR. Natural dyes: Classification, extraction and fastness properties. Textile Dyer & Printer. 1996;**29**(22):16

[16] Teli MD, Paul R, Pardesi PD. Natural dyes, classification, chemistry and extraction methods. Colourage. 2000;**60**:43

[17] Krizova H. Natural dyes. In: Kryštůfek M, Vik W, editors. Chapter 18: Textile Dyeing—Theory and Applications. 1st ed. TUL: Vysokoškolskýpodnik Liberec s.r.o., Studentská 2. Liberec. pp. 317-334

[18] Oda H. Improvement of light fastness of natural dyes. Part 2: Effect of functional phenyl esters on the photo fading of carthamin in polymeric substrate. Coloration Technology. 2001;**117**(5):257

[19] Badan BM, Burkinshans SM. Dyeing of wool and nylon 6.6 with henna and lawsone. Dyes and Pigments. 1993;**221**:15

[20] Kawamura T, Hisata Y, Okuda K, Noro Y, Takeda, Tanka T. Quality evaluation of plant dye henna

with glycosides. Natural Medicine. 2000;**54**(2):86

[21] Agarwal A, Grag A, Gupta KC. Development of suitable dyeing process for dyeing of wool with natural dye henna (*Lawsonia inerma*). Colourage. 1992;**39**(10):43

[22] Gupta DB, Gulrajani ML. Kinetic and thermodynamic studies on 2-hydroxy-1,4-naphthoquinone (lawsone). JSDC. 1994;**110**:112

[23] Singh K, Karr V, Mehra S, Mahajan A. Solvent-assisted dyeing of polyester with henna. Colourage. 2006;**53**(10):60

[24] Bechtold T, Turcann E, Ganglberger, Geisslers. Natural dyes in modern textile dyehouse. Journal of Cleaner Production. 2003;**11**:499

[25] Rita M, Bechtold T. Natural colourants in dyeing. In: Thomas B, Mussak R, editors. Handbook of Natural Colourants. United Kingdom: John Wiley and Sons Ltd.; 2009. p. 316

[26] Bechtold T. Natural colourants. In: Thomas B, Mussak R, editors. Handbook of Natural Colourants. United Kingdom: John Wiley and Sons Ltd.; 2009. p. 154

[27] Cardon D. Natural Dyes, Tradition, Technology and Science. London: Archetype Publications; 2007

[28] Casselman KD. Lichen Dyes and Dyeing: The New Source Book. Mineola, New York: Dover Publications; 2001

[29] Grierson S, Duff DG, Sinclair RS. Natural dyes of the Scottish high lands. Textile History. 1985;**16**:23

[30] Gill M, Steglich W. In: Herz W, Grisebach H, Kirby GW, Ch T, editors. Progress in the Chemistry of Organic Natural Products. Vol. 51. 1987. p. 125

[31] Raisanen R. Dyes from lichens and mushrooms. In: Bechtold T, Mussak R, editors. Handbook of Natural Colourants. United Kingdom: John Wiley and Sons Ltd.; 2009. p. 2003

[32] Porter CJ. Tannins. In: Harborne JB, editor. Methods in Plant Biochemistry. Vol. 1, 389. London: Academic Press; 1989

[33] Seigler DS. Plant Secondary Metabolism. Bostan: Kluwer Academic Publications;

[34] Waterman PG, Mole S. Analysis of Phenolic Plant Metabolites. Oxford: Blackwell Scientific Publications; 1994

[35] Ayres MP, Class JP Jr, Macclean SF, Redman AM, Reichart PB. Diversity of structure and antiherbivore activity in condensed tannins. Ecology. 1989;**78**:1696

[36] Julkunen R, Tiitto, Haggman H. Tannins and tanning agents. In: Bechtold T, Mussak R, editors. Handbook of Natural Colourants. United Kingdom: John Wiley and Sons Ltd.; 2009. p. 2003

[37] Agarwal A, Paul S, Gupta KK. Effects of mordants on natural dyes. Indian Textile Journal. 1993;**1**:110

[38] Erica J, Jiedemann, Yang Y. Fiber-safe extraction of red mordant dyes from hair fibers. Journal of the American Institute for Conservation. 1995;**34**(3):195

[39] Mathur JP, Bhandari CS. Use of beet sugar as wool colourant. Indian Journal of Fibre and Textile Research. 2001;**26**:313

[40] Gulrajani ML, Gupta D, Maulik SR. Bio polishing of tasar silk. Indian Journal of Fibre and Textile Research. 1994;**24**:294

[41] Gulrajani ML, Bhaumiks S, Oppermann W, Handman G. Dyeing

of red sandal wood on wool and nylon. Indian Journal of Fibre and Textile Research. 2003;**28**:221

[42] Sankar R, Vankar PS. Dyeing wool with *Gomphrena globosa* flower. Colourage. 2005;**52**(4):35

[43] Bhuyan R, Sai Kai CN, Das KK. Extraction and identification of colour components from the barks of *Mimusops elengi* and *Terminalia arjuna* and evaluation of their dyeing characteristics on wool. Indian Journal of Fibre and Textile Research. 2004;**29**(12):470

[44] Samanta AK, Priti A. Application of natural dyes on textiles. Indian Journal of Fibre and Textile Research. 2009;**34**:384

[45] Kharbade BV, Agarwal OP. Identification of natural red dyes in old Indian textiles. Journal of Chromatography. 1985;**347**:447

[46] Zvic K. HPLC analysis of the natural scale insect, madder and indigoid dyes. JSDC. 1994;**110**(9):273

[47] Guinot P, Roge A, Aradennec A, Garcia M, Dupont D, Lecoeur E, et al. Dyeing plants screening: An approach to combine past heritage and present development. Coloration Technology. 2006;**122**:93

[48] Balankina GG, Vasiliev VG, Karpova EV. HPLC and molecular spectroscopic investigations of the red dye obtained from an ancient Pazyryk textile. Dyes and Pigments. 2006;**71**:54

[49] Mc Goven PE, Lazar J, Michel RH. The analysis of indigoid dyes by mass spectrometry. JSDC. 1990;**106**(1):22

[50] Szostek S, Grwrys JO, Surowiec I, Trojanowicz M. Investigation of natural dyes occurring in historical coptic textiles by high performance liquid chromatography with UV–Vis and mass spectrometric detection. Journal of Chromatography. 2003;**A1012**:179

[51] Cristea D, Bareau I, Vailarem G. Identification and quantitative HPLC analysis of the main flavonoids present in weld (*Reseda luteola* L.). Dyes and Pigments. 2003;**57**:267

[52] Son A-Y, Heng PJ, Kim KT. Dyes and Pigments. 2007;**61**(3):63

[53] Patel KJ, Patel BH, Naik JA, Bhavana AM. Eco-friendly dyeing with Tulsi leave extract. Man Made Textiles. 2002;**45**(11)

[54] Bhattacharya SD, Shah AK. Metal ion effect on dyeing of wool fabric with catechu. SDC. 2000;**116**(1):10

[55] Maulik SR, Bhowmik KI. Studies on application of some vegetable dyes on cellulosic and lignocellulosic fibre. Man-Made Textiles in India. 2006;**49**(4):142

[56] Siddiqui I, Gous MD, Khaleq MD. Indian Silk. 2006;**145**(4):17

[57] Samanta AK, Priti A. Application of natural dyes on textiles. International Dyer. 2008;**193**(3):37

[58] Samanta AK, Priti A, Siddthartha D. Studies on color interaction parameters and color fastness properties for dyeing of cotton fabrics with binary mixtures of jackfruit wood and other natural dyes. Journal of Natural Fibers. 2009;**6**:171

[59] Mohanty BC, Chandramouli KV, Nail HD. Studies in Contemporary Textiles Crafts of Indian Natural Dyeing Process of India. Calico Museum of Textiles, Ahmedabad: H.N. Patel Publication'; 1987, I and II

[60] Cook CC. Aftertreatments for improving the fastness of dyes on textile fibres. Review of Progress in Coloration and Related Topics. 1982;**12**:78

[61] Pad Field P, Landi S. Natural dyes of the Scottish highlands. Studies in Conservation. 1966;**11**:161

[62] Samanta AK, Konar A, Chakarborty S, Datta S. Dyeing of jute fabric with tesu extract: Part 1—Effects of different mordants and dyeing process variables. Indian Journal of Fibre and Textile Research. 2011;**36**(3):63

[63] Lee JJ, Lee HH, Eom SI, Kim JP. UV absorber aftertreatment to improve light fastness of natural dyes on protein fibres. Coloration Technology. 2001;**117**:134

[64] Mussak R, Bechtold T. Renewable resources for textile dyeing–technology, quality, and environmental aspects. In: Proceedings of the IFATCC International Congress. Barcelona; 2008

[65] Samanta AK, Konar A, Chakarborty S, Datta S. Effect of different mordants, extraction conditions and dyeing process variables on colour interaction parameters and colour fastness properties in dyeing of jute fabric with Manjistha, a natural dye. Journal of the Institute of Engineering. 2010;**91**:7

[66] Micheal MN, NAEl Z. Colourage. 2005; Annual 83

[67] Hofenk JH, Graff D. Conservation restoration of church textiles and painted flags. In: 4th Int Restorer Seminar. Vol. 2. Hungary; 1983. p. 219

[68] Samanta AK, Priti A, Siddhartha D. Application of single and mixtures of red sandalwood and other natural dyes for dyeing of jute fabric: studies on colour parameters/colour fastness and compatibility. Journal of the Textile Institute. 2009;**100**(7):565

[69] Sarkar AK. An evaluation of UV protection imparted by cotton fabrics dyed with natural colourants. BMC Dermatology. 2004;**4**(1):15

[70] Grifani D, Bacci L, Zipoli G, Carreral G, Baronti S, Sabatini F. Laboratory and outdoor assessment of UV protection offered by Flex and Hemp fabrics dyed with natural dyes. Photochemistry and Photobiology. 2002;**85**:313

[71] Grifani D, Bacci L, Zipoli G, Sabatini F, Albanete I. The role of natural dyes in the UV. Protection of fabrics made of vegetable fibres. Dyes and Pigments. 2011;**91**(3):279

[72] Gulrajani ML, Gupta D. Emerging techniques for functional finishing of textiles. Indian Journal of Fibre and Textile Research. 2011;**36**:388

[73] Koto H, Hata T, Tsukada M. Potentialities of natural dyestuff as antifeedants against varied carpet beetle, *Anthrenus verbasci*. Japan Agricultural Research Quarterly. 2004;**38**(4):241

Chapter 3

A Review on Application of Natural Dyes on Textile Fabrics and Its Revival Strategy

Pubalina Samanta

Abstract

A comprehensive review on application of natural dyes on textiles and earlier research findings has been discussed in this chapter. Moreover, recently the consumers have become very much conscious about the environment, renaissance of eco-friendly products and process like dyeing textiles with natural dyes, which has thus become also important now. Thus, revival of natural dye application on textiles and summary of earlier researches on standardization of its method of extraction, mordanting, dyeing process variables and even natural finishing, etc. have been elaborated in this review. Characterization of natural dyes and chemistry of its dyeing, etc. are equally important and hence are discussed here critically. Thus this part has become a unique readymade comprehensive chapter for information on chemistry and application of natural dyes on textiles and its revival strategy.

Keywords: natural dyes, natural pigments, extraction, mordanting, natural dyeing methods, characterization of natural dyes, revival of natural dyeing and textiles

1. Introduction

Natural dyes are known to be used since historic times for coloring food substrate, leather, as well as common textile fibers like cotton, wool and silk. However due to the advent of synthetic dyes and their good fastness properties in comparison to natural dyes, the use of natural dyes have suffered drastically. In the present scenario there has been a rise in concern of eco-friendliness and sustainability of the products used by the consumers for which natural dyes are again starting to experience slight rise in popularity. A study has been conducted by Samanta and Agarwal [1] which reports the characterization as well as chemical/biochemical analysis of various natural dyes available, the different types of mordants as well as different mordanting techniques, the different conventional and non-conventional method of natural dyeing of textiles. The different natural dyes used for the study are madder, henna, held, indigo and others such as annatto pulp, *Rubia tinctorum*. Different methods of extraction are employed such as aqueous extraction, non-aqueous method as well as by acid and alkali. Different types of mordant and method of mordanting significantly affect the rate of fading. For cotton the best mordant combinations used in this study are harda and tartaric acid, followed by tannic acid and harda. Double mordanting is employed by using harda and aluminum sulphate. The various process variables to be considered for dyeing with and extraction of natural dyes are concentrations of dye source material, extraction time, dyeing time, mordant concentration, pH and concentration of salt used.

Another study conducted by Daberao et al. [2] gives us a concept about dyeing with palash or tesu flower petal (*Butea monosperma*) *as* natural dye sources. The dyes were extracted from *Butea monosperma* or in other terms flame of the forest and they were applied on 100% cotton. A different method of extraction by boiling was employed and alum was used as mordant. The fabric was then tested for all color fastness tests. The cotton sample was scoured and bleached for better color uptake. The washing fastness results were observed where the natural dye having not too much affinity with the fiber but by the application of mordant could withstand at least five washes. Also wet rubbing fastness of the dye was found to be poorer in the experimental results. However it was observed that *Butea monosperma* has good perspiration fastness since it is unreactive to acidic and alkaline perspiration.

The use of natural dyes has further started to be increased substantially over the current years for its slow but growing revival phase at present, due to people's concern over reducing environmental pollution and hence to avoid chemically more hazardous synthetic dyes and intermediates. Day by day in export market, demands for natural dyed natural textiles are being increased. Different institutions/organizations and Govt. have started multifold revival strategies for increasing the use of natural dyes as not only as an employment opportunity for several NGOs, weaver and dyers society, designers, industries, small scale cottage industries, etc. but mainly for adopting green technology dyeing. The handicraft industry in India uses local talents to dye yarns and fabrics with natural compounds, where several products are famous worldwide like Kalamkari print. Different countries other than India like Turkey, Korea, Mexico, several countries of Africa have embraced the uses of natural dyes. A study has been conducted by Gulrajni [3] to understand the scope of natural dyes and its present status in the world in addition to different application techniques, extraction of different natural dyes as well as varying mordanting techniques. Different problems associated with such natural dyeing are also highlighted there.

The Tharu tribes of the Devipatan [4] division have found a new source of natural dyeing from the local leaves and stems of *Jatropha curcas L.* The dyes are extracted by simple boiling the leaves in water and then evaporating the extract to dryness. The extract obtained is of yellowish olive syrupy color and when applied to the cotton fabrics the different shades of tan and brown color are obtained.

Another state in our country, Manipur has been considered to be a source of a natural dye namely extracts from *Strobilanthus flaccidifolius* for uses in handicrafts, handlooms, fine arts, etc. Other tribes of Manipur like the Meitei community have been using species like *Parkia javanica*, *Melastoma malabathricum*, *Pasania pachyphylla*, *Solanum incidum*, *Bixa orellana*, *Tectona grandis*, etc. these plants are combined with other plants for extraction and then dye is prepared by indigenous sources. This study is a report by Potsangbam et al. [5] of the dyes extracted from the above sources, the method of extraction as well as their application.

Now from the forest of Chhattisgarh, different dye yielding plants have been identified and collected. A study was conducted by Tiwari and Bharat [6] on the diversity of dye-yielding plants of Chhattisgarh, the indigenous method of dye extraction and ethnic uses of dyes. These colors are being used by tribal folks of this region for different purposes such as ornamentation, cosmetics, decorating houses and coloring home utensils made up of mud.

From the state of Goa [7] natural dye-yielding plants like *Cassia fistula*, *Garcinia indica*, *Tectona grandis* are obtained and studied where Goa is said to house more than 3000 different species of flowering plants. Natural dyes were extracted from various parts of plant like fruits, seeds, bark, flowers, roots, etc. The extraction processes are studied and dyes of different shades are obtained. This study can also encourage the small scale industries to use natural dyes from these sources to be applied on cotton and silk fabrics.

A report by Gaur [8] shows the extensive description of survey, collection of botanical information and review of relevant literature on the vegetable dye yielding resources of Uttarakhand Himalayas. Of which, very little known dye yielding plants are considered like *Acacia nilotica*, *Agrimonia pilosa*, *Careya arboraea*, *Averrhoa carambola*, etc. Different extraction processes for each plant are carried out and subsequently dyeing is done by using different mordants. Some of these plants also have high medicinal values and has no toxicity.

1.1 Sources of different natural dyes and their characterization

Various natural products are being used for dyeing these days in order to fulfill the demand of consumers for sustainable environment. The reviews below are for the various natural dyes used on textile materials.

Bechtold et al. [9] have studied on the quality of Canadian golden rod plant material as a natural dye. Aqueous solutions of the material containing the extracted flavonoid dyes were characterized by means of direct photometry, absorbance after addition of $FeCl_2$ is measured, total phenolics (TPH) in the extract and dyeing on wool yarn are analyzed where only relatively small differences in color depth and shade were noted amongst the major parts of the different materials collected.

A study on natural dye henna was conducted by Rahman Bhuiyan et al. [10]. Henna is a red-orange pigment that has long been used for the coloration of skin and hair as well as textile materials. A large number of studies were carried out on extraction as well as application of henna dye in textile fibers and the standardization and simplification of dyeing techniques were determined. Due to burgeoning environmental conditions and growing awareness on sustainability there has been a renewed interest in expanding the scope and applications in the coloration of textile fibers with some successes and promises. Henna shows an acidic nature due to the presence of polar groups, which promotes its use in the textile dyeing process.

Dyeing of natural dye extracted from *Liriope platyphylla* fruit on silk fabrics have been studied by Huang et al. [11]. From which it has been observed that the total phenolic content (1109.13 ± 69.02 mg), total flavonoid content (530.60 ± 89.44 mg), and total anthocyanin content (492.26 ± 77.79 mg) were measured in 100 g fresh weight of *L. platyphylla* fruits. A broad variation in color shade and color depth has been achieved with mixtures of different combinations of dye extracts and metal mordants. Purple, blue, and pale green were main color shades obtained when dyed with the extracts. The fastness of dyed silk fabrics against light, washing, and rubbing were noted to be acceptable with at least a gray scale rating of 3.

A study has been conducted on orange peel by Hou et al. [12]. Orange peel is an easily available agricultural byproduct and it is cheap as well as abundant. The variation in effects of dyeing methods and conditions, including pH value, temperature, time and concentration of OP extracts on the colors of the dyed wool fabrics, were studied. Eco-friendly mordants of aluminum and iron were used. The optimum dyeing conditions were noted which included dyeing temperature of 100°C, dyeing time of 120 min, pH of 3 for direct dyeing and pH 7–9 for one-bath mordant dyeing. Good colourfastness to washing with soap, good colourfastness to rubbing and acceptable colourfastness to light were displayed by the tested specimen.

Hibiscus is a major source material for natural dyeing. It belongs to the family Malvaceae. Aqueous extracts of these flowers have shown good fastness properties according to the study conducted by Shanker and Vankar [13]. The dye has been found to have a good scope in the commercial dyeing of cotton, silk for garment industry and wool yarn for carpet industry. In the present study dyeing with hibiscus has been shown to give good dyeing results. The material is pretreated with 2–4% metal mordants, keeping M:L ratio as 1:40 on weight of the fabric to plant extract. The dye is cheap and has good commercial value if dyed with cotton, wool and silk.

Another natural material has been found by Vankar et al. [14] to be a good source of natural dyeing which is *Mahonia napaulensis DC.*, common name taming, from the family Berberidaceae. The natural dye is from the stem and has been used by the tribes of Arunachal Pradesh. The fastness properties for dyed cotton, silk fabrics and wool yarn were show to increase substantially when pretreated with metal mordant (2% w/w with respect to the fabric).

An attempt has been made by Kamel et al. [15] in dyeing of wool fabrics using lac as a natural dye in both conventional and ultrasonic techniques. The dye extraction was compared between conventional method and ultrasonic technique and the data were evaluated. Accordingly the effects of dye bath pH, salt concentration, ultrasonic power, dyeing time and temperature were compared. The result of fastness properties obtained was fair to good.

Montazer and Parvinzadeh [16] have dyed wool with marigold as a source of yellow color. At first the wool yarns were premordanted with alum, dyed with marigold and then treated with different percentages of ammonia solutions. After washing with standard soap after color hue alters and there has been no effect of ammonia after treatment on washing fastness however the samples show lower light fastness.

A study on the dyeing properties of woolen yarns using gallnut extract as a natural dye was conducted by Shahid et al. [17]. A conclusion that gallnut extract can be applied on woolen yarn with or without mordants to produce bright ivory to light brownish yellow color with good fastness properties against light, washing and rubbing was obtained from the test.

Natural dyes have been slowly garnering popularity all over the world. So a study was carried out by Mirjalili et al. [18] by extraction of dyes from weld using soxhlet apparatus. The natural dyes were extracted and isolated and the colored substance obtained was used for dyeing of wool fiber. Finally a comparison was made with the synthetic colorants on the color fastness tests. It can be concluded from the study that weld can be used as a non-toxic dye. Good fastness properties were obtained from this natural extract.

An attempt has been made to dye the wool fabric with Limoniastrum monopetalum stems by Bouzidi et al. [19]. Extraction parameters were optimized. The optimization of extraction results obtained were dye concentration of 60 g/l, a temperature of 90°C and time duration of 100 min. The best results were obtained of pH 2, dyeing temperature of 100°C, and time duration of 60 min. Metal mordants were used in this process. The extract has ample natural tannin and polyphenol compounds which are considered as mordants since they have the ability to fix the dyes in bath to the fabric.

Indigo carmine is another renewable resource based blue dye which can be used to color protein fibers. Komboonchoo and Bechtold [20] have worked on use indigo carmine in combination with other natural dyes in a one-bath procedure as a hybrid dyeing concept. Optimum dyeing parameters of pH in the range of 4–5 and temperature between 40 and 60°C were obtained.

A new concept of few natural dyes as dye sensitized solar cell (DSC) was brought forth by Hao et al. [21–24]. Amongst all these photochromatic natural dye-extracts, black rice extract dye shows best results, perhaps due to high interaction between carbonyl [—C=O] and hydroxyl [—OH] groups of anthocyanin present in such dyes. Because of the simple preparation technique, these are considered as widely available and low/cheap cost natural dyes as photo sensitized color of natural dyes, having photo-sensitized solar cell type character. Other materials like achiote seeds, rosella, blue pea flowers, spinach and ipomoea were also reported for such natural dyes having in built photo-sensitized solar cell in it.

Cochineal is an insect species of scientific name *Dactylopius coccus.* Carminic acid is the natural colorant obtained from the dried female body of such insects. This finds application in cosmetics, foods, pharmaceutical sectors as well as textile and plastic industries. The study has been conducted by Borges et al. [25] and the study on the newer process of extraction will be discussed in the extraction methods section.

A study of natural eco-friendly dye extracted from *Plumeria rubra* is carried out by Vettumperumal et al. [26]. Due to the existence of highly delocalized systems absorption spectrum shows a broad absorption in the range of 292–590 nm. This plant also encourages usage of waste lands, afforestation of wasteland and provides consequent additional source of income to rural population.

Rubia tinctorum is commonly known as madder produces anthraquinone pigments in its roots, one of them being alizarin (1,2 dihydroxy anthraquinone) which has been used for dyeing textiles since ancient times. Angelini et al. [27] has evaluated four madder genotypes for their agronomic characteristics as well as for their industrial value and to assess its value as a new industrial dye crop. Industrial assays demonstrated good performance when using dry powder 30% of the weight of material to be dyed for dyeing cotton, wool and silk yarns. Resistance to fading appears to be fairly good for dyed wool when using madder.

Different coloring plants from New Caledonia were considered for research by Toussirot et al. [28] amongst which *Hubera nitidissima*, an Annonaceae, showed an intense yellow color on fibers. Color was extracted from the leaves of the said plant on linen, silk and wool. The color fastness results were obtained where it was concluded that *H. nitidissima* appears as an excellent source of light-fast yellow dye with interesting antioxidant properties. These days natural dye extracted from mangrove bark was also used as a dyeing material.

1.2 Application of natural dyes on different textiles

Punrattanasin et al. [29] applied selectively extracted few natural dyes to a silk fabric by an exhaustion dyeing process where aluminum potassium sulfate, ferrous sulfate, copper sulfate, and stannous chloride were used as mordants. Dyeing was carried out in three different stages of the fabric-premordanted, meta-mordanting and post-mordanting. Color fastness values of each were reported. Dyeing conditions were optimized as dyeing temperature of 90°C, dyeing time—60 min and dye bath pH of 3 was fixed to be optimum. In this work, natural silk textiles were dyed with and without mordants using $SnCl_2$, KAl SO_4, $FeSO_4$ and $CuSO_4$ providing varying degree of color/tone/shade, where $FeSO_4$ produced darker and blackish brown shade, $CuSO_4$ produced lighter to pale reddish brown shade, both showing poorer washing fastness but very good water soaking, perspiration, light and rubbing fastness.

The various physical tests were done and tensile strength, tearing strength and stiffness of the fabrics before and after dyeing were also compared.

Shukla et al. [30] has reported dyeing of woolen textiles with extract of *acacia pennata* plants. The color was extracted from barks of the said plant and applied on wool. *Acacia pennata* is a thorny shrub found throughout India and Burma. Experiments were carried out where acacia pennata was used in conjunction with banana stem. When compared without banana stem it was observed that dye fastness without the banana stem was poorer than when stem was used. It was concluded that banana stem acted as a good mordant thus eliminating the use of metallic, carcinogenic mordants.

An attempt has been made by Gulrajni et al. [31] to dye nylon and polyester with annatto. Annatto also known as *Bixa orellana* has a color component namely carotenoid dye bixin. It has been observed that both of these fibers show good affinity for this dye but moderate fastness to washing and poor light fastness.

An attempt has been carried out by Gulrajni et al. [32] to extract dyes from ratanjot also known as *Arnebia nobilis* for application on cotton, wool, silk, nylon, polyester and acrylic. The process conditions such as pH and temperature were recorded. It has been noted that dye exhibits acute sensitivity to pH in terms of solubility and color and is found to be thermally stable upto 80°C. The different colors shown by various fabrics were noted such as pink color for polyester, blue for nylon and other substrates acquiring a purple hue under similar dyeing conditions.

A study by Gulrajni et al. [33] on the kinetics and thermodynamics of dye extracted from *Arnebia nobilis* on woolen textiles was reported. Physicochemical and dyeing kinetics parameters of this natural dyeing using aqueous extract of *Arnebia nobilis* applied on woolen textiles were reported as compared to the same for other natural colorants like juglone, lawsone and *Rheum emodi*, etc. The results showed here that anthraquinonoid based these natural colors do not form desired coordinated complex with wool and rather are absorbed on wool substrate by partition mechanism following Nernst isotherm like absorption of disperse dye on polyester.

Chakraborty and Chavan [34] reviewed on dyeing of cotton denim with Indigo, which gives the information on the newer application techniques of indigo dyes applicable for natural indigo. Since indigo has negative affinity for cotton conventional methods cannot be applied. The details of indigo reduction, solubilization and dye application has been studied in this reference.

Deo et al. [35] had attempted dyeing of ecru denim with onion extract as natural color using Potash-alum in combination with harda and tartaric acid as mordants. Any of the single mordant did not produced desired shade. Amongst combined mordants used, Potash-alum + harda combination was found to be better than potash-alum + tartaric acid for producing desired depth of shade, but potash alum + tartaric acid (5%:5%, that is, 1:1 combination of each 5% application) post mordanting showed best overall color fastness results.

A study has been carried out by Samanta et al. [36] on standardizing dyeing process variables for its application on bleached jute fabric with aqueous extract of tesu (Palash flower petal). It is observed that higher amount of pre-mordanting with 20% myrobolan (Harda containing chebulinic acid) followed by 20% aluminum sulphate in sequence and dyeing at pH –11.0 produced optimum color yield and all round good color fastness. Improvement in wash and light fastness was also achieved with suitable chemical post-treatment using suitable agents.

Gray jute fabric bleached with hydrogen peroxide in conventional method was mordanted with different concentrations of ferrous sulphate and dyed separately with natural dyes extracted from deodara leaf (*Cedrus deodara* L.), jackfruit leaf (*Artocarpus integrifolia* L.) and eucalyptus leaf (*Eucalyptus globulus* L.). Pan et al. [37] have observed the interdependency of color yield and color fastness properties on dosages, that is, concentrations of mordants ($FeSO_4$) used, higher iron-mordant concentration lead to higher color yield, darker color and better overall good color fastness. But they have not studied loss of strength of due to mordanting and which is essentially needed to be assessed also.

Narayana Swamy et al. [38] have studied the use of *madhuca longfolia* as a dye source. The dried leaves of the said plant are taken as dye source for silk dyeing. The optimum conditions under which the dye has been extracted are pH of 10, time (60 min) and temperature (95°C). Varying range of shades is obtained using different methods with or without using mordants. The dyed samples have been evaluated for color measurements and standard wash, light and rub fastness tests. Eco-friendliness of the dye has been kept into account. The dyed samples are also tested for antimicrobial activity against Gram-positive and Gram-negative bacteria. The dyed silk fabrics show acceptable fastness properties and the results show that *Madhuca longifolia* leaves are promising as a natural colorant, which can thus open new doors towards environment friendly products.

An attempt has been made to color silk using barberry, a cationic type natural dye by Pruthi et al. [39] Barberry bark also known as *Berberis aristata* DC. was used for dyeing of degummed pure silk yarn using four selected mordants; alum, chrome, copper sulphate and ferrous sulphate in different ratio, that is, 1:1, 1:3 and 3:1. Optimized results was obtained for aqueous extraction of barberry as 60 min time, 8% dye source material and optimized dyeing conditions were observed to be pH-of dye bath-4.0, dyeing time-45 min for standard mordanting with chrome + ferrous sulphate (1:3) and chrome + copper sulphate (3:1) produced the higher degree of color fastness properties. Varying shade percentages and color tone were obtained by using varying degree/percentages of different combination of mordants.

Das et al. [40] have worked on the application of *Bixa orellana* on protein textiles viz. on wool and silk. Seeds of annatto have been extracted first and then have been employed on silk and wool in absence and presence of magnesium sulphate, aluminum sulphate and ferrous sulphate. Effective colouration has been achieved at pH 4.5 commonly in the absence and presence of such inorganic salts. Color uptake for wool is found to be more than that for silk under all the conditions studied. When both the substrates are treated with such salt prior to application of annatto there has been significant increase in color uptake. Colored protein fibers, in general, produce light and wash fastness ratings of 2–3. Ferrous sulphate in turn, improves color fastness properties and color retention on washing of wool and silk fibers.

Another study has been conducted by Das et al. [41] on the application of *Punica granatum* on wool and silk textiles. *Punica granatum* also commonly known as pomegranate rind was used on wool and silk fabric in the presence and absence of environment-friendly mordanting agents. Both the dyeing of silk and wool with pomegranate solution is found to be effectively accomplished at pH 4.0. With the application of ferrous sulphate and aluminum sulphate during pre- and post-mordanting has shown improvement in the color uptake, light fastness and color retention on repeated washing. The use of such mordants, however, does not show any improvement in wash fastness property of dyed substrates.

An application of *Terminalia bellerica* fruit extract dyeing on woolen textile under different conditions of pH, concentrations of natural coloring matter, time of extraction/durations and temperatures were studied by El-Zawahry and Kamel [42]. The results evaluated were mainly surface color strength and color depth. The study shows that optimum color yield was obtained using following extraction and dyeing conditions:

Extraction: source coloring matter—5%, temperature of extraction bath—near boil, that is, 90–100°C at pH—7 (neutral).

Mordanting: (i) potassium dichromate + lactic acid-application 0.5 gpl, and (ii) chromic chloride + lactic acid-application −0.5 gpl,

Dyeing: time—60 min and temperature—near boil (95°C),

Shade obtained: moss green with mordanting system (i) as above, mustard yellow with mordant system (ii) as above and muster brown with both copper acetate and ferrous sulphate or ferric chloride as mordant. Thus for such natural dyes-color tone and shade depth are much dependant on type of mordant and its concentration used. Overall color fastness results to washing, to acid or alkaline human perspirations and rubbing/crocking fastness were found to be almost the same for said premordanted and dyed wool fabrics. In case of light fastness, longer the duration of exposure to light, darker the shade and better light fastness were obtained. There has been no change of color strength and fastness properties despite of use of standing dye bath (50 g T.b. fruits/100 ml water) for 8 times.

A new approach of dyeing was undertaken by Naz et al. [43] where Eucalyptus (*Eucalyptus camaldulensis*) bark powder (without any further treatment/irradiation) using gamma ray irradiated natural colorant of dry powder of eucalyptus leaf extract, for producing natural colored textiles of soothing brown color with improved color fastness by required pre and/or post mordanting. Thus, when this fabric was therefore dyed in this case using gamma ray irradiated powder of eucalyptus dry leaf, it showed noticeable improved overall color fastness properties.

A paper was presented by Ferda Eser et al. [44] on dyeing of polyester and polyester/viscose blends dyed with walnut shell extracts. Different extraction conditions were considered such as material—liquor (M:L) ratio, extraction temperature, extraction time and pH in order to obtain highest color depth. Optimal extraction of natural dyes from walnut shells (*Juglans regia*) was obtained at temperature—80°C, time of extraction—75 min using MLR as 1:30 at pH 2. For dyeing polyester and polyester/viscose blends with said extract of wall nut shell using $AlKHSO_4$ or $AlK(SO_4)_2$ or $FeSO_4$ for separate mordanting for 90 min time and subsequent dyeing was studied and found that pre mordanting with $FeSO_4$ offers best dyeing results with good color depth and overall good color fastness, which can be used for future applications for ecofriendly dyeing of polyester and its blended textiles.

A detailed study was carried out by Samanta and Agarwal [45, 46] on dyeing of jute and cotton fabrics with binary mixtures of jackfruit wood along with other natural dyes in combination for producing compound shades after study of their compatibility. Conventionally hydrogen peroxide bleached jute and cotton fabrics were taken and was pre-mordanted with 10–20% harda (myrobolan) followed by 10–20% $Al_2(SO_4)_3$ or $FeSO_4$ salt in sequence as sequential double mordanting as a most prospective mordanting system for subsequent dyeing with aqueous extract of jack fruit wood. Study of dyeing process variables showed that optimum dyeing results were obtained for 90 min dyeing time, 70–90°C dyeing temperature, 11.0 pH, 1:30 material-to-liquor ratio, 20–30% mordants concentration, 30–40% source dye concentration, and 15 gpl common salt. In conventional method, for test of compatibility of these selected binary pairs of natural dyes, in order to obtain progressive depth of shade, two sets of five different samples were produced and tested after dyeing with 1:1 mixture of two dyes at 1% fixed shade depth with varying time and temperature profile in one set as well as by varying total concentrations of the binary pairs of dyes (using varying shade depth with 1:1 equal proportion of mixture of two dyes) keeping time and temperature fixed for second set were obtained and their color parameters of K/S vs. DL and DC Vs. DL were compared to judge compatibility by graphical comparison method. However, in this work, a newer method of compatibility rating procedure with calculation of Color Difference Index data (a newly defined useful color difference parameter) was described and adopted here for easy determination of compatibility rating between two dyes of any binary pairs of selective natural dyes used for applying that binary mixture of natural dyes in the same dye bath for compound shade. Moreover, they have shown methods of improving color fastness to washing by using separate post treatment with cationic agents like CTAB (n-cetyl-N-trimethyl ammonium bromide), or cetrimide, etc. Similarly separate post treatment with 1% benztriozale as an UV absorber had also shown an improvement in light fastness results.

Another attempt of dyeing ratanjot on nylon and polyster was studied by Gulrajni et al. [47] where the observed results indicated that this dye has a good substantivity for both nylon and polyester fibers, probably due to less polar structure of this dye and Nernst partition isotherm of absorption of this dye on these two fibers. However, deep color shade and better fastness to light and washing was obtained.

Sagarika Devi et al. [48] have studied about *Alternaria alternata* for textile dyeing and printing where reddish brown natural pigments which was obtained after extraction of colors from dry mycelium of *Alternaria alternata* in methanol solvent media. At pH −6, this Fungus produce the said extractable colored pigment, which can be applied on cotton for light color with medium grades of color fastness results using pigment dyeing process. This natural color is antibacterial and antifungal as evidenced in this work by AATCC100 test method for both gram positive and gram negative bacteria species for test, showing its antimicrobial nature.

Ke [49] studied the dyeing properties of natural dye extracted from *Rhizoma coptidis* on acrylic fibers. Acrylic fiber was dyed with *Rhizoma coptidis* aqueous solution and its dyeability was studied in terms of the thermodynamic and kinetic properties and dyeing process conditions. This study showed that effect of dyeing temperature is positive, that is, color yield and dye diffusion rate increases with increase of dyeing temperature up to a limit, indicating dyeing temperature and mordant concentration as important critical variables in such dyeing of acrylic with extract of *Rhizoma coptidis*. Color fastness to washing and rubbing are found to be grade—4.

Haque et al. [50] extracted ubiadin dye from *Swietenia mahagoni* and studied its dyeing characteristics onto silk fabric using metallic mordants. Metallic mordants such as $MgCl_2$ and $FeSO_4$ were used and dyeing properties were evaluated. $FeSO_4$ as compared to that of $MgCl_2$ showed good result for color yield and color fastness results.

Mahale et al. [51] studied about natural dyeing of Silk yarn skeins using extract of *Acalypha wilkesiana* leaves using varying concentrations of mordants like potash alum, potassium dichromate, copper sulphate and ferrous sulphate. Potassium dichromate and copper sulphate are not ecofriendly mordant. Potash alum though gives good fastness but considering color yield and fastness both, $FeSO_4$ offers best results of color yield and color fastness.

A study was conducted by Poorniammal et al. [52] for natural dyeing with extracted and purified natural fungal pigment from *Thermomyces* sp. to apply on different textile fabrics to optimize and dyeing process parameters for silk, cotton and woolen fabrics. This extracted pigment color obtained from *Thermomyces* sp. indicated good affinity towards silk fabrics than others, with good light fastness (rating 4), color fastness to washing (rating 4–5) and color fastness to rubbing (rating 3–4). The optimum conditions for dyeing was found to be dyeing temperature—30°C, dyeing pH—3, myrobalan mordant—5%, and dyeing time—20 min duration were suggested. The pigment also gave a reasonable extent of bacteria reduction in such silk dyed sample against *Salmonella typhi* (51.05%).

An attempt is made by Onial et al. [53] on utilization of *Terminalia chebula* Retz. fruits pericarp as a source of natural dye for textile applications. *Terminalia chebula* Retz. of Family-Combretaceae, trade name-Myrobalan fruits pericarp powder was taken for the utilization as a dye. The dried fruits constitute one of the most important vegetable tanning materials and have been used in India for a long time. This fruit pericarp thus can be used as a raw material for natural dyeing.

Das et al. [54] made an attempt to dye wool and silk with *Rheum emodi*. Silk and woolen fabrics, which were dyed with colorant extracted from *Rheum emodi* in the absence and presence of metallic mordants such as magnesium sulphate, aluminum sulphate and ferrous sulphate for producing shades of different colors, ranging from yellow to olive green. Study of dyeing isotherms and kinetics of dyeing process indicated that this dyeing mechanism do not follow coordinated complex formation amongst fiber-mordant-dye, rather follow Nernst type isotherm showing pattern of partition mechanism, for this anthraquinonoid-based colorant where the dye molecules are adsorbed by silk and woolen fabrics as a disperse dye.

However, rate of dyeing is found to be higher for silk than that of wool and that color depth is increased by use of both aluminum sulphate or ferrous sulphate as mordant and considering color fastness test results, the later, that is, ferrous sulphate as mordant is found to be superior (offering wash fastness grade as 3 to 4 or 4) than use of same dosages of aluminum sulphate.

Thus ferrous sulphate is preferred as mordant for obtaining an improvement in the color fastness properties and color retention on washing of both wool and silk fabrics further.

Goodarzian and Ekrami [55] used brown dry rind of pomegranate as dyestuff. Extraction was carried out by solvent extraction method. Woolen fabrics were dyed with both raw and extracted dyestuffs using variations of concentrations. Spectrophotometric evaluations as well as colorimetric studies were carried out to compare the color strength of raw and extracted dye stuff on woolen fabric. It was concluded that color strength of extracted dye from pomegranate rind was more than raw dye stuff.

As mentioned by Garfield and Mauve [56], 'Mauviene' was the first synthetic dye synthesized by William Perkin in 1856. Like every scientific invention with man-made materials, advantages and disadvantages coexist, and the synthesis of synthetic-dyes is no exception. With the present growing global concern for environment protection and use of eco-friendly and bio-degradable materials, the trend of application of natural dyes have once again gained the momentum and for growing concern of consumers on eco friendliness of textile products, application of natural dyes on textiles is slowly being revived again.

Advantages of natural dyes over synthetic are manifolds [57] as they are eco-friendly, safe for body contact and are harmonized as reported by Brian [58]. Many scientists have also suggested and reported the medicinal and antibacterial importance of natural dyes [59, 60]. Yellow dye from rhizome of turmeric has been reported to be traditionally used in medicine as an anti-inflammatory drug [61]. Most of the natural dyes are proved to be non-toxic and eco-friendly, although there are some exceptions.

The natural dyes are the colorants extracted from the vegetables matters, minerals or insects [62]. Although most of the natural dyes have poor to moderate light fastness and the synthetic dyes represent a full range of colors with light fastness properties ranging from moderate to excellent [63], the use of natural dyes on textiles have been reported by many scientists. Dyeing of cotton with leaf-extract of Beilschmiedia fagifolia was reported by Vankar et al. [64], who has used sonicator method to dye cotton with aqueous extracts of *B. fagifolia*. The authors have reported that pre-treatment of cotton with 1–2% metal mordant and dyeing with 5% plant extract produced optimum results with good fastness properties.

In another report, Shah and Datta [65] used floral dye extracted from marigold flower to dye cotton fabrics. Gahlot et al. [66] used colorants extracted from Jatropha integarrima flowers for dyeing of cotton, wool and silk. Dyeing of silk with Onosma echiodes (Goldendrop) was reported by Sidhu and Grewal [67]. Mahale et al. [68] dyed cotton with Arecanut palm extract. Ultrasonic dyeing of cotton and silk with *Nerium oleander* flower has also been carried out [69]. Purohit et al. [70] reported use of natural color from waste leaves of Arotocarpus beterophyllus on different textile substrates like cotton and silk to get standard reproducible shades of golden yellow color.

The application of natural dyes such as turmeric, madder, catechu, Indian rhubarb, henna, and tea and pomegranate rind on manmade fiber nylon has been reported by

Teli et al. [71]. Some studies have also been conducted on application of Lac dyes [72] on different fibers. Application of a natural dye, annatto, on mulberry silk was carried out by Javali et al. [73]. Some studies [74–76] on natural dyeing of silk textiles have been reported in literature for use of Indian madder, *Spathodea campanulata* and lac dyes as natural sources. Patel et al. [77], have reported environmental-friendly and cost-effective method to create various shades on silk with few natural dyes.

There are many historic books documenting the literature on the use of natural dyes or natural dyed materials (textiles, candles, food, furrs, etc.) dating to as far back as the eighteenth century. The significant literary document on natural coloring matter was made available for the first time by Perkin and Everest [78].

Sahid and Mohammad [79], and Mayer and Cook [80] have also reviewed details of chemistry asnd application of natural dyes and more recent report on the structures of quinonoids natural colorants is described in Thomson's book [81]. Recent reviews in this area also include work undertaken by Parris [82] and Hofenk de Graaff [83] the latter includes information on fastness properties and history of use. Studies in the analysis of natural colorants in textiles are a fascinating subject which started as early as 1930s.

Recently, Samanta et al. [84–86] have described thermodynamic analysis of rate of dyeing, half dyeing time, enthalpy, free energy, etc. as a physico-chemical parameter of dyeing jute with red sandal wood, jackfruit wood and tesu as natural dyes.

The analysis of mass spectrometry of textile fibers dyed with indigo has been reported by McGovern [87]. However, Wong [88] was not able to detect 6,6-dibromoindigotin by direct analysis, but only after it had been separated by reductive extraction with sodium hydrosulphite.

Thin layer chromatography (TLC) was used by many workers to identify natural dyes in textiles [82]. Dyes detected were insect dyes and vegetable dyes viz., yellow, red and blue colors. Koren [89] also analyzed the madder and indigoid dyes by HPLC. Guinot et al. [90] also used TLC chromatography analysis to carry out a preliminary evaluation of plants containing flavonoids (flavonols, flavones, flavanones, chalcones/aurones, anthocynanins), hydroxycinnamic acids, tannins and anthraquinones, which are the phylo-compounds (color compounds) found in the plants.

Physicochemical dyeing parameters of red sandal wood as natural dyes and its compatibility with other dyes were analyzed by Samanta and Agarwal et al. [91, 92]. Neem bark [93] colorant showed two absorption maxima at 275 and 374 μm; while beet sugar showed three absorption bands at 220, 280 and 530 μm as per study undertaken by Mathur [93]. The visible spectra of ratanjot [47] in methanol solution was observed at both acidic and alkaline pH by Gulrajani et al. [47]. *Gomphrena globosa* flower colorant showed one major peak at 533 μm. The dye did not show much difference in the visible spectrum at pH 4 and 7; however the peak shifted to 554 μm as reported by Shanker and Vankar [94].

Bhuyan [95] observed the amount of dye absorption for extract of *Mimusops elengi* and *Terminalia arjun* varies from 21.94 to 27.46% and 5.18 to 10.78%, respectively, for the said two dye sources. The color components isolated from most of the barks contain flavonoid moiety. Samanta et al. [96, 97] postulated a new index called color difference index (CDI) value which can be calculated by an empirical formula postulated by them and which made determining dye compatibility easier and simpler for a binary mixture, that is, between a pair of natural dyes.

Identification of dyes in historic textiles though chromatographic and spectrophotometric methods as well as by sensitive color reactions was highlighted by Blanc et al. [98], who studied the retention of carminic acid, indigotin, corcetin, gambogic acid, alizarin flavanoid, anthraquinone and purpurin, etc.

A non-destructive method was reported for identifying faded dyes on textiles fabrics through examination of their emission and excitation spectra. Zin and Moe [99] purified and characterized extracted natural agents and colors from mango bark for application in protein fibers like wool.

Walker and Needles [100] carried out the separation and identification of natural dyes from wool fibers using reverse phase HPLC using a C-18 column. Two quaternary solvent systems and one binary solvent system were reported to be used to obtain chromatograms of by the HPLC analysis of plant and insect based red anthroquinonoid and molluscan type blue and red purple indigoid dyes [89]. This method enables the elution process for the determination of different chemical functionality and class of dyes and significantly shortens the time of test. Son et al. [101] reported HPLC analysis of indigo highlighting the structural changes of indigo component, attributing a decrease/increase in color strength with variation of dyeing time.

Balakina [63] also investigated/analyzed the quantitative and qualitative analysis of red dyes such as alizarin, purpurin, carminic acid, etc. by HPLC. High Performance Liquid Chromatography (HPLC) has been also used by several workers to identify synthetic as well as natural dyes.

Jain and Vashanta [102] characterized antimicrobial activity after eco-friendly dyeing with arcea nut using natural mordant/mordanting additives like myrobolan, lodhra and pomegranate rind and found that pomegranate rind renders best antibacterial activity and Lodhar renders highest color fastness to wash amongst all the moderating additives used.

Mondhe and Rao [103] made an attempt to prepare azo-alkyd dyes by the reduction of nitro alkyds, followed by diazotization of amino alkyds and coupling with different phenol compounds present in *Jatropha curcas* seed oil by using IR spectra.

The toxicity [104, 105] data also provide evidence about the adverse effect to human and environment. Of primary concern are the acute toxicity, irritation effects on the skin and the eye and sensitization potential besides environmental pollution in the society. Furthermore, possible long-term effects such mutagenic, carcinogenic or reproductive toxicity is best judged by LD50 test. The crude methanolic extracts of stem and roots stem, leaves, fruit, seeds of *Artocarpus Hetrophyllus* [106] exhibited good rating of antibacterial activity. The butanol fractions of the same root bark and fruit were also found to be the most active.

Mishra and Patni [107] extracted tannins from gall leaf from oak plant (i.e., oak galls containing gallic acid and tannic acid and helps in better dye fixation) from Himalayan region and dyed cotton, woolen and silk textiles with different metallic mordants and obtained better color fast fabrics, which are skin friendly too. The main reason of revival of natural dyes for textiles are its environmental friendliness and skin friendliness too.

1.3 Natural dyes cum natural antimicrobial finishing agents

A study was conducted by Mari Selvam et al. [108] to investigate the antibacterial and antifungal effects of such dyed textiles dyed with Turmeric, Terminalli, Guava and Henna. The results obtained indicated that at a dose level of 50 μl of Terminalli dye was able to inhibit the growth of all the fungi tested. The absorbance rate of natural dyes was analyzed by UV Spectrophotometer. The absorbance rate obtained were high in Terminalli (2.266) and turmeric (2.255). Hence from this study it was concluded that natural dyes were bound with traditional products to give good color and good antimicrobial activity against isolated fungal pathogens.

Another study carried out by Rajni Singh et al. [109] on antimicrobial activity of some natural dyes like *Acacia catechu*, *Kerria lacca*, *Quercus infectoria*, *Rubia cordifolia* and *Rumex maritimus*, which gives us an idea about to determine their minimum inhibitory concentration (MIC), which was found to be varying from 5 to 40 mg. So, such dyed textile material with these dyes must up take above MIC concentration for effective antimicrobial action in such natural dyed textiles.

Curcumin, a common natural dye used for fabric and food colorations was used by Han and Yang [110] to dye woolen fabric to obtain dyeing and antimicrobial finishing simultaneously showing relation amongst bacterial reduction percentage and dye (curcumin) concentration, and microbial inhibition rate and surface color strength (K/S value). However, durability of antimicrobial action for different nos. of washing cycle after laundering and after exposure to UV light/sun light are also very important criteria, which were also critically discussed in this work.

Shafat Ahmad Khan et al. [111] attempted a work to investigate the antimicrobial action of *Rheum emodi* L. as a potential antibacterial natural dye and they dyed wool yarns with extract of *Rheum emodi* L. as purified dye applying dye concentration of 5–10% with or without mordants like ferrous sulphate, stannous chloride and natural alum for subsequent antimicrobial test against *E. coli* and *S. aureus* following AATCC100 test method. Test results of such *Rheum emodi* natural dyed woolen yarn samples indicated 90% bacterial reduction percentage as well as very high fungal protection showing very effective antimicrobial properties.

Shahid-ul-Islam et al. [112] have studied the use of *Tectona grandis* L. leaves extract plant colorants for dyeing woolen fabrics for simultaneous dyeing and antimicrobial finishing with natural dye cum natural antimicrobial finishing agent. This study indicated that the dyeing woolen yarn with extract of *Tectona grandis* L. is suitable for dyeing cum multifunctional finishing to impart simultaneous dyeing and antioxidant and antibacterial finishing properties to woolen based textile fabrics.

Fatemeh Shahmoradi Ghaheh et al. [113] have found that pre-treatment with aluminum sulphate as pre mordanting and followed by subsequent dyeing with selective natural dyes extracted from green tea leaf, madder route, turmeric route, saffron petals, and henna as natural dye cum natural antimicrobial agents provide moderate to good antibacterial finishing property on woolen fabrics and also led to good durability of the said antimicrobial action even after five cycles of laundering and above 300 min exposure to UV light/sun light.

Mohd Ibrahim Khan et al. [114] have conducted a study on antimicrobial activity of catechu itself and catechu extract dyed woolen yarn. The results indicated to show more than 90% antibacterial reduction as per standard test method. Observed antimicrobial inhibition character indicate that catechu may be a promising natural antimicrobial finishing agent for developing bioactive and antimicrobial dyed textile materials for today's need.

A number of recent studies on simultaneous natural colouration and antimicrobial finishing of different textiles using selective natural dyes/natural agents applied alone or in combination were investigated by several authors as mentioned below for detailed study and further references:

Prusty et al. [115] have studied about simultaneous natural coloring and antibacterial finishing of few natural colorants on silk.

Similarly Gupta and Laha [116] have worked on simultaneous natural dyeing and antimicrobial finishing of cotton fabric using natural tannin-rich extract of *Quercus infectoria* (QI) plant in combination with alum, copper and ferrous sulphate as mordants showing good antimicrobial activity at 12% concentration (owf), inhibiting the bacterial reduction around 45–60% for ferrous sulphate and bacterial reduction increases to 70–90% when mordanted with alum or copper sulphate making it suitable for anti-odor agent for use in medical, sports and home textiles.

Chen and Chang [117] have applied extract of onion skin on plasma pretreated cotton fabric to obtain simultaneous coloring and antimicrobial finishing effect where the plasma-pre-treated cotton samples subsequent dyed/grafted with extract of onion skin showed measurable inhibition zone against *S. aureus* around 1.1–0.8 cm inhibition for 10 min grafting time with onion skin extract and 0.7–0.5 cm inhibition zone for 30 min grafting time of onion skin extract.

Joshi et al. [118] have reported a comprehensive review on natural product based bioactive agents such as chitosan, natural dyes, neem extract and other herbal products for antimicrobial finishing of textile substrates which is useful for further study.

1.4 Natural dyes cum natural UV protective finishing agents

A study has been conducted by Salah [119] about antibacterial and UV property of Egyptian cotton fabrics treated with aqueous extract from waste peel of banana fruits after its extraction in 1% NaOH solution.

Chattopadhyay et al. [120] have worked on developing natural dyed jute fabric with improved color yield and UV protection characteristics using harda (myrobolan) as bio mordant (though it is not truly a mordant, it is rather a mordanting assistant having high coordinating power for promoting fiber-mordant-dye complex formation using several —OH and —COOH groups of chebulinic acid present in it) and pomegranate rind extract as natural dye as well as UV protective agent using ecofriendly ferrous sulphate and potash alum as mordants. Very good ultraviolet (UV) protection ratings were achieved in case of dyeing of jute fabric with pomegranate rind. However, Jute fabric treated with manjistha, annatto, ratanjot and baboolas natural dyes cum natural UV protective finishing agents, applied after pre-mordanting with sequential pretreatment with Harda extract as biomordant and Alum as metallic but natural eco-friendly chemical mordant. Observed results indicated that UV protection properties of the said selective natural dyes cum natural UV protective finishing agents applied on bleached jute fabric follows the following order of UV protective performances: babool > annatto > manjistha > ratanjot.

Hou et al. [12] has used in their study waste orange peel as agricultural bye product for obtaining concurrent natural coloring and UV protective finishing on textiles for potential strong UV absorbance character of orange peel applied on woolen fabrics. The results was encouraging and optimum conditions of this concurrent natural coloring and UV protective finishing of woolen textiles is: optimum temperature of 100°C, optimum time—120 min, dyeing cum finshing bath pH—3 for following dyeing cum finishing without mordant and pH is 7–9 for simultaneous mordanting, dyeing and finishing in one bath using aluminum sulphate or ferrous sulphate, that is, iron as metallic eco-friendly mordant, showing great potential of orange peel extract as useful for this purpose.

Grifoni et al. [121] have shown in their reports that UV protection property not only depends on the surface finishing agents applied whether natural or synthetic, but also depend much on fabric construction, type of fibers, type of natural or synthetic dyes and finishes used (absorption criteria of dyes and finishing agents in UV zone). In this study, they measured UPF value of different types of textile apparels, hats, canopy type shade structure made of textiles with varying fabric construction for vegetable and natural fibers based product and finally dyed with different natural dyes cum natural finishing agents using tannin based natural mordants for obtaining maximum level of safest UV protection from sunlight radiation.

Feng et al. [122] have conducted a study on the UV protective properties of hats and clothing against solar ultraviolet radiation and found that *Rheum emodi* and *L. erythrorhizon* shows equally good and comparable UV-absorption protection character as compared to the common known standard UV-absorber compounds like benzophenone and benztriazole, etc.

Sinnur et al. [123] have reported a study on natural colouration and UV protective finishing using aqueous extract of pomegranate rind, that is, commonly known as anar peel. Besides optimization of conditions of natural color extraction from dried anar peel powder, effect of different single and double mordants in different proportions and concentrations on color yield and optimization of dyeing process variables as well as measurement of UV protective action of such dyed cotton khadi fabric with extract of anar peels (pomegranate rind, i.e., *Punica granatum* L.) as a natural colorant has been reported recently as an encouraging work.

The analysis of mass spectroscopy of cellulosic textile fibers dyed with indigo has been reported as a method of its identification, which may be the basis and can be used as a finger print for identifying natural indigo with TLC and UV VIS spectroscopic results in combination. Similarly for assuring any textiles being only dyed with natural dye, need its identification method. Very recently BIS has published two national IS standards on identifying natural Indigo and madder (IS 17084-2019 for natural indigo and IS 17084-2019 for madder) for test and identification of these two natural dyes from such a natural dyed textiles.

2. Conclusions

Thus, still there are many gaps in standardizing dyeing conditions for specific fiber-mordant-dye combinations and there is still need of required scientific and industrial research on the effects of different ecofriendly chemical mordants and bio mordants (tannin based natural compounds) and mordanting assistants (gallic acid or chebulinic acid based natural compounds) for finally standardization/optimization of dyeing process variables for obtaining uniform and repeatability of shades to produce with natural colors. Another side of utilizing antibacterial/antifungal and UV protective action or deodourizing action of selective natural dyes by detailed scientific study of the effects of different after-treating compounds antibacterial/UV absorbers compounds for improving its uses as high valued textiles. Similarly study of different natural and ecofriendly chemical dye fixatives for improving color fastness to Washing and effects of UV absorber compounds on exposure of such natural dyed textiles to the exposure to sun-light/UV light can be improved by suitable after-treatment with UV absorbers. For improving rubbing fastness of such natural dyed textiles, after treatment with natural binders or different natural reactive thickeners and ecofriendly synthetic binding agents are required, besides approaches to improve antimicrobial and UV protection activity of such natural dyed textiles. It is also important to know and understand well the exact fiber-mordant dye interaction and role of different pre and post treatments on promoting color yield (in terms of K/S values), uniformity of color yield (measurable by CV % of K/S values) as well as rating for antimicrobial and UV protection factor for different fiber-mordant-natural dye combination as applicable particularly to cotton, silk, wool and jute fibers when dyed with aqueous extract of any selected natural dye.

Hence application of natural dyes on high value apparel and functional textiles are gaining worldwide interest for its less toxic nature, better biocompatibility, biodegradability, producing elegant hues and highly functional value-added textiles as environment friendly oeko-tech/ecofriendly textiles for gaining popularity for natural dyed and finished as high valued textiles of tomorrow, if its revival strategies are well created and executed with utmost care with back up of sufficient scientific study with time bound growth plan and correct revival strategy.

Some of the revival strategies include (i) availability of commercially standardized process and standard commercial shade cards for developing desired shades with acceptable repeatability and appreciable color fastness results; (ii) availability of standardized test methods for identifying and assuring customers for proving a dyed textiles is really 100% dyed with natural dye(s) without any synthetic dyes used as adulterant/topping to match shade; (iii) commercial process variables need to be standardized for different dyes for desired shades at economical minimum cost; (iv) to train and educate concern dyers and weavers and any big or large textile industry sector for successful extraction and dyeing with natural dyes and finally (v) future creation of a natural dye mark certification method by suitable national and international bodies for consumer assurance service like khadi mark, silk mark, etc.

Author details

Pubalina Samanta
Department of BFAD, Rani Birla Girls' College, Kolkata, India

*Address all correspondence to: pubalina@gmail.com

References

[1] Samanta AK, Agarwal P. Application of natural dyes on textiles. Indian Journal of Fibre and Textile Research. 2009;**34**:384-389

[2] Daberao AM, Kolte PP, Turukmane RN. Cotton dying with natural dye. International Journal of Research and Scientific Innovation (IJRSI). 2016;**III**(VIII):157-161. ISSN: 2321-705

[3] Gulrajni ML. Present status of natural dyes. Indian Journal of Fibre and Textile Research. 2001;**26**:191-201

[4] Srivastava SK, Tewari JP, Shukla DS. A folk dye from leaves and stem of *Jatropha curcas* L. used by Tharu tribes of Devipatan division. Indian Journal of Traditional Knowledge. 2008;7(1):77-78

[5] Potsangbam L, Ningombam S, Laitonjam WS. Natural dye yielding plants and indigenous knowledge of dyeing in Manipur, Northeast India. Indian Journal of Traditional Knowledge. 2008;**7**(1):141-147

[6] Tiwari SC, Bharat A. Natural dye-yielding plants and indigenous knowledge of dye preparation in Achanakmar-Amarkantak Biosphere Reserve, Central India. Natural Product Radiance. 2008;**7**(1):82-87

[7] Verenkar NGS, Sellappan K. Some potential natural dye yielding plants from the State of Goa, India. Indian Journal of Fibre and Textile Research. 2017;**8**(4):306-315

[8] Gaur RD. Traditional dye yielding plants of Uttarakhand, India. Natural Product Radiance. 2008;**7**(2):154-165

[9] Bechtold T, Mahmud-Ali A, Mussak R. Natural dyes for textile dyeing: A comparison of methods to assess the quality of Canadian golden rod plant material. Dyes and Pigments. 2007;**75**:287-293. DOI: 10.1016/j.dyepig.2006.06.004

[10] Rahman Bhuiyan MA, Islam A, Ali A, Islam MN. Color and chemical constitution of natural dye henna (*Lawsonia inermis* L) and its application in the coloration of textiles. Journal of Cleaner Production. 2017;**167**:14-22. DOI: 10.1016/j.jclepro.2017.08.142

[11] Huang S, Wei L, Wang H. Dyeing and antibacterial properties of *Liriope platyphylla* fruit extracts on silk fabrics. Fibers and Polymers. 2017;**18**(4):758-766. DOI: 10.1007/s12221-017-5449-z. ISSN: 1229-9197 (print version), ISSN: 1875-0052 (electronic version)

[12] Hou X, Chen X, Cheng Y, Xu H, Chen L, Yang Y. Dyeing and UV-protection properties of water extracts from orange peel. Journal of Cleaner Production. 2013;**52**:410-419. DOI: 10.1016/j.jclepro.2013.03.004

[13] Shanker R, Vankar PS. Dyeing cotton, wool and silk with *Hibiscus mutabilis* (Gulzuba). Dyes and Pigments. 2007;**74**:464-469. DOI: 10.1016/j.dyepig.2006.03.007

[14] Vankar PS, Shanker R, Dixit S, Mahanta D, Tiwari SC. Sonicator dyeing of modified cotton, wool and silk with Mahonia napaulensis DC. and identification of the colorant in Mahonia. Industrial Crops and Products. 2008;**27**:371-379

[15] Kamel MM, El-Shishtawy RM, Yussef BM, Mashaly H. Ultrasonic assisted dyeing III. Dyeing of wool with lac as a natural dye. Dyes and Pigments. 2005;**65**:103-110

[16] Montazer M, Parvinzadeh M. Dyeing of wool with marigold and its properties. Fibers and Polymers. 2007;**8**(2):181-185

[17] Shahid M, Ahmad A, Yusuf M, Khan MI, Khan SA, Manzoor N, et al. Dyeing, fastness and antimicrobial properties of woolen yarns dyed with gallnut (*Quercus infectoria* Oliv.) extract. Dyes and Pigments. 2012;**95**:53-61

[18] Mirjalili M, Nazarpoor K, Karimi L. Eco-friendly dyeing of wool using natural dye from weld as co-partner with synthetic dye. Journal of Cleaner Production. 2011;**19**:1045-1051

[19] Bouzidi A, Baaka N, Salem N, Mhenni MF, Mighri Z. *Limoniastrum monopetalum* stems as a new source of natural colorant for dyeing wool fabrics. Fibers and Polymers. 2016;**17**(8):1256-1261. DOI: 10.1007/s12221-016-5664-z

[20] Komboonchoo S, Bechtold T. Natural dyeing of wool and hair with indigo carmine (C.I. Natural Blue 2), a renewable resource based blue dye. Journal of Cleaner Production. 2009;**17**:1487-1493

[21] Hao S, Wu J, Huang Y, Lin J. Natural dyes as photosensitizers for dye-sensitized solar cell. Solar Energy. 2006;**80**:209-214

[22] Gómez-Ortíz NM, Vázquez-Maldonado IA, Pérez-Espadas AR, Mena-Rejón GJ, Azamar-Barrios JA, Oskam G. Dye-sensitized solar cells with natural dyes extracted from achiote seeds. Solar Energy Materials & Solar Cells. 2010;**94**:40-44

[23] Wongcharee K, Meeyoo V, Chavadej S. Dye-sensitized solar cell using natural dyes extracted from rosella and blue pea flowers. Solar Energy Materials & Solar Cells. 2007;**91**:566-571

[24] Chang H, Wu HM, Chen TL, Huang KD, Jwo CS, Lo YJ. Dye-sensitized solar cell using natural dyes extracted from spinach and ipomoea. Journal of Alloys and Compounds. 2010;**495**:606-610

[25] Borges ME, Tejera RL, Díaz L, Esparza P, Ibáñez E. Natural dyes extraction from cochineal (*Dactylopius coccus*). New extraction methods. Food Chemistry. 2012;**132**:1855-1860

[26] Vettumperumal R, Kalyanaraman S, Selvan GT, Selvakumar PM. Fluorescence analysis of natural dyes from *Plumeria rubra* (red and white) flowers. Optik. 2018;**159**:108-114

[27] Angelini LG, Pistelli L, Belloni P, Bertoli A, Panconesi S. Rubia tinctorum a source of natural dyes: Agronomic evaluation, quantitative analysis of alizarin and industrial assays. Industrial Crops and Products. 1997;**6**:303-311

[28] Toussirot M, Nowik W, Hnawia E, Lebouvier N, Hay A-E, de la Sayette A, et al. Dyeing properties, coloring compounds and antioxidant activity of Hubera nitidissima (Dunal) Chaowasku (Annonaceae). Dyes and Pigments. 2014;**102**:278-284

[29] Punrattanasin N, Nakpathom M, SomboonB,NarumolN,RungruangkitkraiN, Mongkholrattanasit R. Silk fabric dyeing with natural dye from mangrove bark (Rhizophoraapiculata Blume) extract. Industrial Crops and Products. 2013;**49**:122-129

[30] Shukla SR, Shinde SC, Banye AS, Patil SM. Dyeing of wool with Acacia pennata. Indian Journal of Fibre and Textile Research. 2004;**29**:350-352

[31] Gulrajani ML, Gupta D, Maulik SR. Studies on dyeing with natural dyes: Part I—Dyeing of annato on nylon and polyester. Indian Journal of Fibre and Textile Research. 1999;**24**:131-135

[32] Arora A, Rastogi D, Gupta D, Gulrajani ML. Dyeing parameters of hydroxynaphthoquinones extracted from Arnebia nobilis Rech.f. Indian Journal of Fibre and Textile Research. 2012;**37**(1):91-97

[33] Arora A, Rastogi D, Gupta D, Gulrajani ML. Kinetics and thermodynamics of dye extracted from Arnebia nobilis Rech. f. on wool. Indian Journal of Fibre and Textile Research. 2012;**37**(2):178-182

[34] Chakraborty JN, Chavan RB. Dyeing of denim with indigo. Indian Journal of Fibre and Textile Research. 2004;**29**(1):100-109

[35] Deo HT, Paul R. Dyeing of ecru denim with onion extract as a natural dye using potassium alum in combination with Harda and tartaric acid. Indian Journal of Fibre and Textile Research. 2000;**25**:217-220

[36] Samanta AK, Konar A, Chakraborti S, Datta S. Dyeing of jute fabric with tesu extract: Part 1—Effects of different mordants and dyeing process variables. Indian Journal of Fibre and Textile Research. 2011;**36**(1):63-73

[37] Pan NC, Chattopadhay SN, Day A. Dyeing of jute with natural dyes. Indian Journal of Fibre and Textile Research. 2003;**28**(3):339-342

[38] Narayana Swamy V, Gowda KNN, Sudhakar R. Dyeing of silk using *Madhuca longifolia* as natural dye source. Indian Journal of Fibre and Textile Research. 2015;**40**(4):419-424

[39] Pruthi N, Chawla GD, Yadav S. Dyeing of silk with barberry bark dye using mordant combination. Natural Product Radiance. 2008;**7**(1):40-44

[40] Das D, Mallik SR, Bhattacharya SC. Dyeing of wool and silk with *Bixa orellana*. Indian Journal of Fibre and Textile Research. 2007;**32**(3): 366-372

[41] Das D, Bhattacharya SC, Mallik SR. Dyeing of wool and silk with *Punica granatum*. Indian Journal of Fibre and Textile Research. 2006;**31**(4):559-564

[42] El-Zawahry MM, Kamel MM. Dyeing of wool cloth with *Terminalia bellerica* fruit extract. Indian Journal of Fibre and Textile Research. 1999;**24**(2):126-130

[43] Naz S, Bhatti IA, Adeel S. Dyeing properties of cotton fabric using un-irradiated and gamma irradiated extracts of *Eucalyptus camaldulensis* bark powder. Indian Journal of Fibre and Textile Research. 2011;**36**(2):132-136

[44] Eser F, Aktas E, Onal A. Dyeing quality of walnut shells on polyester and polyester/viscose blended fabrics. Indian Journal of Fibre and Textile Research. 2016;**41**(2):188-194

[45] Samanta AK, Agarwal P, Datta S. Dyeing of jute and cotton fabrics using jackfruit wood extract: Part I—Effects of mordanting and dyeing process variables on colour yield and colour fastness properties. Indian Journal of Fibre and Textile Research. 2007;**32**(4):171-180

[46] Samanta AK, Agarwal P, Datta S. Physico-chemical studies on dyeing of jute and cotton fabrics using jackfruit wood extract: Part II—Dyeing kinetics and thermodynamic studies. Indian Journal of Fibre and Textile Research. 2008;**31**(1):66-72

[47] Gulrajani ML, Gupta D, Maulik SR. Studies on dyeing with natural dyes: Part III—Dyeing of Ratanjot dye on nylon and polyester. Indian Journal of Fibre and Textile Research. 1999;**24**(4):294-296

[48] Devi S, Karuppan P. Reddish brown pigments from Alternaria alternata for textile dyeing and printing. Indian Journal of Fibre and Textile Research. 2015;**40**(3):315-319

[49] Ke G. Dyeing properties of natural dye extracted from *Rhizoma coptidis* on acrylic fibres. Indian Journal of Fibre and Textile Research. 2014;**39**(1):102-106

[50] Ahsanul Haque M, Khan GMA, Razzaque SMA, Khatun K, Chakraborty AK, Khan MS. Extraction of rubiadin dye from Swietenia mahagoni and its dyeing characteristics onto silk fabric using metallic mordants. Indian Journal of Fibre and Textile Research. 2013;**38**(3):280-284

[51] Mahale G, Sunanda S, Sunanda RK. Silk dyed with Acalypha (*Acalypha wilkesiana*) and its fastness. Indian Journal of Fibre and Textile Research. 2003;**28**(1):86-89

[52] Poorniammal R, Parthiban M, Gunasekaran S, Murugesan R, Thilagavathi G. Natural dye production from Thermomyces sp fungi for textile application. Indian Journal of Fibre and Textile Research. 2013;**38**(3):276-279

[53] Onial P, Dayal R, Rawat MSM, Kumar R. Utilization of Terminalia chebula Retz. Fruits pericarp as a source of natural dye for textile applications. Indian Journal of Natural Products and Resources. 2015;**6**(2):114-121

[54] Das D, Mallik SR, Bhattacharya SC. Colouration of wool and silk with Rheum emodi. Indian Journal of Fibre and Textile Research. 2008;**33**(2):163-170

[55] Goodarzian H, Ekrami E. Wool dyeing with extracted dye from pomegranate (*Punica granatum* L.) peel. World Applied Sciences Journal. 2010;**8**(11):1387-1389

[56] Garifield S, Mauve WW. Natural Dyes for Textiles. New York.: Norton and Company; 2001

[57] Sekar N. Dyeing of jute and jute blends: Recent developments. Colourage. 2000;**47**(8):35

[58] Brian GJ. Dyeing what comes naturally to the dye house. Society of Dyers and Colourists. 1998;**114**(4):240

[59] Badri BM, Burkinshaw SM. Dyeing of wool and nylon 6.6 with henna and lawsone. Dye & Pigments. 1993;**22**(1):15

[60] Alam MM, Rahman ML, Haque MZ. Extraction of henna leaf dye and its dyeing effects on textile. Bangladesh Journal of Scientific and Industrial Research. 2007;**42**(2):217

[61] Shah NC. Traditional use of turmeric (*Curcuma longa*) in India. Journal of Medicinal and Aromatic Plant Sciences. 1997;**19**:948

[62] Indi YM, Chinta SK. Application and properties of natural dye on cotton—phyllanthus reticulatus. Colourage. 2008;**55**(6):52

[63] Yoshizumi K, Crews PC. Characteristics of fading of wool cloth dyed with selected natural dyestuffs on the basis of solar radiant energy. Dyes and Pigments. 2003;**58**:197

[64] Vankar PS, Shanker R, Dixit S, Mahanta D, Tiwari SC. Characterisation of the colorants from leaves of *Bischofia javanica*. International Dyer. 2008;**192**(3):11

[65] Saha P, Datta S. Dyeing of textile fibre using Marigold flower as floral dye. Colourage. 2008;**55**(5):58

[66] Gahlot M, Fatima N, Papnai N. Jatropha flowers; a natural colourant for dyeing of cotton, wool and silk. Colourage. 2008;**52**(2):96-200

[67] Sidhu SP, Grewal JK. Silk dyed with goldendrop (Onosma echiodes) dye. Colourage. 2008;**55**(3):51

[68] Mahale G, Sunanda S, Sunanda RK. Arecanut—A natural colourant for silk. Colourage. 2008;**46**(4):54

[69] Vankar PS, Shanker R. Ultrasonic dyeing of cotton and silk with

Nerium oleander flower. Colourage. 2008;**55**(4):51

[70] Purohit A, Das R, Nayak A, Nanda B, Das NB. Natural dyes from the waste leaves of artocarpus heterophyllus. Colourage. 2009;**56**(5):89

[71] Teli MD, Shah R, Sabale AG. Dyeing of nylon with natural colourants. Colourage. 2010;**57**(3):96

[72] Chairat M, Rattanaphani V, Bremner JB, Rattanaphani S, Parkins DF. An absorption investigation of the interaction of lac dye with metal ions. Dyes and Pigments. 2004;**63**(2):141-150

[73] Javali UC, Sreenivasa, Radhalakshmi YC. Application of natural dye-annatto on mulberry silk. Colourage. 2009;**56**(2):50-52

[74] Sakata K, Katayama A. Dyeing of silk fabrics with the pigments separated from a powdered dry Indian madder. Journal of Sericultural Science of Japan. 1996;**63**(3):170

[75] Sudhakar R, Gowda KNN. Dyeing of silk with flower extract of *Spathodea campanulata*. Man-Made Textiles in India. 2005;**48**(7):255

[76] Rattanaphani SS, Chairat M, Rattanaphani JB. An adsorption and kinetic study of lac dyeing on silk. Dyes and Pigments. 2007;**72**(1):88

[77] Patel BH, Bhatia KB, Parekh UD. Environmental-friendly & cost-effective method to create various shades on silk. Indian Silk. 2005;**44**(6):24

[78] Perkin AG, Everest AE. The Natural Organic Colouring Matter. London: Longmans Green; 1918

[79] Sahid M, Mohammad F. Recent advancements in natural dye applications: A review. Journal of Cleaner Production. 2013;**53**:310-331

[80] Mayer F, Cook AH. Chemistry of Natural Colouring Matters. 3rd ed. Newyork: Reinhold; 1943

[81] Thomson RH. Naturally-occurring Quinones. 2nd ed. USA: Academic press Inc; 1971. 3rd ed. Netherlands: Springer; 1987

[82] Parris RE. Natural dyes. In: Othmer K, editor. Encyclopedia of Chemical Technology. 3rd and 5th eds. 1979. USA: Wiley; 2007

[83] Hofenk de Graaff JH. Chapter 7: Dyeing black in 17th century. In: Eri I, editor. Conserving Textiles. Rome Italy/Holland: ICCROM; 2004. pp. 54-60

[84] Samanta AK, Agarwal P, Konar A, Datta S. Characterisation and application of purified selective natural dyes on jute. International Dyer. 2008;**192**(4):25-32

[85] Samanta AK, Agarwal P, Datta S. Physico-chemical studies on dyeing of jute and cotton fabrics using jackfruit wood extract as natural dyes: Part-II: Kinetic and thermodynamic studies. Indian Journal of Fibre & Textile Research. 2008;**33**:66-72

[86] Samanta AK, Konar A, Datta S. Dyeing of jute fabric with Tesu extract: Part-II: Thermodynamic parameters and kinetics of dyeing. Indian Journal of Fibre & Textile Research. 2012;**37**:172-177

[87] McGovern PE, Lazar J, Michel RH. The analysis of indigoid dyes by mass spectrometry. Journal of the Society of Dyers and Colourists. 1990;**106**(1):22

[88] Wong ON. The identification of natural dyes in pre-Columbian Andean textiles by mass spectrometry [MA thesis]. Cambridge, USA: Massachusetts Institute of Technology; 1977. p. 20

[89] Koren ZC. HPLC analysis of the natural scale insect, madder and

indigoid dyes. Journal of the Society of Dyers and Colourists. 1994;**110**(9):273

[90] Guinot P, Roge A, et al. Dyeing plants screening approach to combine past heritage and present development. Colouration Technology. 2006;**122**(2):93-101

[91] Samanta AK, Agarwal P, Datta S. Physico-chemical studies on dyeing of jute textiles with natural dyes extracted from red sandal wood. Journal of Institution of Engineers (India)—Textile Engineering. 2006;**87**:16-26

[92] Samanta AK, Agarwal P. Application of mixture of red sandal wood and other natural dyes for dyeing of jute fabric-studies on dye compatibility. International Dyer. 2008;**192**(3):37-41

[93] Mathur JP, Metha A, Kanawar R, Bhandari CS. Use of neem bark as wool colourant—Optimum conditions of wool dyeing. Indian Journal of Fibre & Textile Research. 2003;**28**(1):94

[94] Sankar R, Vankar PS. Dyeing wool with gomphrena globosa flowers. Colourage. 2005;**52**(4):35

[95] Bhuyan R, Saikai CN, Das KK. Extraction and identification of colour components from barks of musops elengi and terminilia Arjuna and evaluation of their dyeing charecterestics on wool. Indian Journal of Fibre & Textile Research. 2004;**29**(12):470

[96] Samanta AK, Agarwal P, Datta S. Dyeing of jute with binary mixtures of jackfruit wood and other natural dyes: Study on colour performance and dye compatibility. Indian Journal of Fibre & Textile Research. 2008;**33**:171-180

[97] Samanta AK, Agarwal P, Singhee D, Datta S. Application of single and mixtures of red sandal wood and other natural dyes for dyeing of jute fabric: Studies on colour parameters/colour fastness and compatibility. Journal of the Textille Institute. 2009;**100**:565-587

[98] Blanc R, Espejo T, Montes AL, Toress D, Crovetto G, Navalon A, et al. Sampling and identification of natural dyes in historical maps and drawings by liquid chromatography with diode-array detection. Journal of Chromatography. 2006;**1122**(1-2):105

[99] Zin NW, Moe MS. Purification of natural dyestaff extracted from mango bark for application in protein fibres. Proceedings of World Academy of Science, Engineering and Technology. 2008;**36**:540-544

[100] Walker C, Needles HL. Historic Teles and paper materials. Advances in Chemical Series, Engineering. 1986;**212**:175

[101] Song JY, Jing L, Rui D. Preparative isolation and purification of indigo and indirubin from *folium isatidis* by high-speed counter-current liquid chromatography. Photochemical Analysis. 2012;**23**(6):637-641. https://onlinelibrary.wiley.com/toc/10991565/2012/23/6

[102] Jain H, Vashantha M. Ecofriendly dyeing with natural dye -Arcea nut; enhancing colour fastness with natural mordants (Myrobolan, Lodhra and Pomogrenate) and increasing the antibacterial activity. Archieves of Appliaed Science Research. 2016;**8**(8):1-7

[103] Mondhe PS, Rao JT. Studies on *Jatropha curcas* seed oil incorporated alkyl dyes part I. Colourage. 1993;**43**(5):43

[104] Zippel E. Analysis of dye colourants and coloured textiles. Review of Progress in Coloration and Related Topics. 2004;**34**(1):1-11

[105] Joshi M, Purwar R. Developments in new processes for colour removal from effluent. Review of Progress

in Coloration and Related Topics. 2004;**34**(2):58

[106] Khan MA, Khan M, Srivastav PK, Mohammad F. Extraction of natural dyes from cutch, ratanjot and madder. Colourage. 2006;**56**(1):61

[107] Mishra P, Patni V. Extraction and application of dye extracted from eriphyid leaf galls of Quercu Leucotrichophora _A Himalayan blue jack oak. African Journal of Biochemistry Research. 2011;**5**(3):90-94

[108] Mari Selvam R, Ranjit Singhand AJA, Kalirajan K. Antifungal activity of different natural dyes against traditional products affected fungal pathogens. Asian Pacific Journal of Tropical Biomedicine. 2012;**2**(3):S1461-S1465

[109] Singh R, Jain A, Panwar S, Gupta D, Khare SK. Antimicrobial activity of some natural dyes. Dyes and Pigments. 2005;**66**:99102

[110] Han S, Yang Y. Antimicrobial activity of wool fabric treated with curcumin. Dyes and Pigments. 2005;**64**:157-161

[111] Khan SA, Ahmad A, Khan MI, Yusuf M, Shahid M, Manzoor N, et al. Antimicrobial activity of wool yarn dyed with *Rheum emodi* L. (Indian rhubarb). Dyes and Pigments. 2012;**95**:206-214

[112] Shahid-ul-Islam SAW, Mohammad F. Imparting functionality viz color, antioxidant and antibacterial properties to develop multifunctional wool with *Tectona grandis* leaves extract using reflectance spectroscopy. International Journal of Biological Macromolecules. 2018;**109**:907-913

[113] Shahmoradi Ghaheh F, Mortazavi SM, Alihosseini F, Fassihi A, Shams Nateri A, Abedi D. Assessment of antibacterial activity of wool fabrics dyed with natural dyes. Journal of Cleaner Production. 2014;**72**:132-145

[114] Khan MI, Ahmad A, Khan SA, Yusuf M, Shahid M, Manzoor N, et al. Assessment of antimicrobial activity of catechu and its dyed substrate. Journal of Cleaner Production. 2011;**19**:1385-1394

[115] Prusty AK, Das T, Nayak A, Das NB. Colourimetric analysis and antimicrobial study of natural dyes and dyed silk. Journal of Cleaner Production. 2010;**18**:1750-1756

[116] Gupta D, Laha A. Antimicrobial activity of cotton fabric treated with *Quercus infectoria* extract. Indian Journal of Fibre and Textile Research. 2007;**32**:88-92

[117] Chen C, Chang WY. Antimicrobial activity of cotton fabric pretreated by microwave plasma and dyed with onion skin and onion pulp extractions. Indian Journal of Fibre and Textile Research. 2007;**32**:122-125

[118] Joshi M, Wazed Ali S, Purwar R. Ecofriendly antimicrobial finishing of textiles using bioactive agents based on natural products. Indian Journal of Fibre and Textile Research. 2009;**34**:295-304

[119] Salah SM. Antibacterial activity and ultraviolet (UV) protection property of some Egyptian cotton fabrics treated with aqeous extract from banana peel. African Journal of Agricultural Research. 2011;**6**(20):4746-4752. DOI: 10.5897/AJAR10.845

[120] Chattopadhyay SN, Pan NC, Roy AK, Saxena S, Khan A. Development of natural dyed jute fabric with improved colour yield and UV protection characteristics. The Journal of the Textile Institute. 2013;**104**(8):808-818. DOI: 10.1080/00405000.2012.758352

[121] Grifoni D, Bacci L, Zipoli G, Albanese L, Sabatini F. The role of natural dyes in the UV protection of

fabrics made of vegetable fibres. Dyes and Pigments. 2011;**91**:279-285

[122] Feng XX, Zhang LL, Chen JY, Zhang JC. New insights into solar UV-protective properties of natural dye. Journal of Cleaner Production. 2007;**15**:366-372

[123] Sinnur HD, Samanta AK, Verma DK, Kaware R. Studies on coloration and UV protective action of Anar Peel (pomegranate rind) as an effective natural colorant for cotton Khadi fabric. Journal of the Institution of Engineers (India): Series E. 2018;**99**(1):9-26. DOI: 10.1007/s40034-017-0106-z

Chapter 4

Review on Natural Dyes for Textiles from Wastes

Deepali Singhee

Abstract

Agriculture and food processing industries generate a large amount of organic waste that still contains colouring pigments. Their sustainable use in dyeing textiles will expectedly solve the problem of their disposal. Some studies involving the use of agro and industrial waste have been documented in this chapter to provide a guideline for further research. Information on some selected wastes that have been used in dyeing of textiles by several authors have been listed, their composition and production details highlighted and their use is explained systematically. The documented studies have been placed in the form of case studies to highlight the different approaches of the authors for explaining the effectiveness of such wastes as a source for textile colourants. As reported in some studies, the extracted dye from the waste doubled up as a mordant itself. Most studies also indicate good dye-ability with appreciable fastness with respect to the textiles dyed with such wastes. Some wastes have also reportedly rendered antibacterial and sun/light protective properties to the dyed fabric.

Keywords: natural dyes from agro-waste, natural dyes from industrial waste, natural dyes from flower waste, natural dyes from fruit waste, sustainable natural dyes, waste utilisation

1. Introduction

Natural dyes have the advantage of having a renewable source and are bio-gradable in nature having low environmental impact [1], but are still associated with problems of poor to moderate colour fastness [2], absence of standardised procedure for extraction and application [3], non-reproducibility of shades [2], pollution caused by use of metallic mordants [4], high energy consumption during extraction and exhaust dyeing [5] and, high cost [6]. Natural dyes also have a significantly lower affinity for fibres resulting in lower dye-exhaustion from the dye-bath on to the fibre surface [7]. Several efforts have been undertaken all over the world to address these shortcomings of natural dyes and find suitable alternative sources in view of the tremendous environmental advantages that they offer. Moreover, the content of the colour component in most natural dyes is a tiny percentage of its total solid weight and large quantities of the dye source are needed to colour small quantities of the textile material. For this, enormous amounts of the dye source has to be procured, which may lead to overexploitation of natural resources; specifically, if they are from the vegetable or animal origin. This would also threaten some endangered vegetal and animal species [8]. In this respect, use of a waste material that are available at no or little cost to dye textiles would expectedly make the process of natural dyeing cost-effective and at the same time bring about increased sustainability in the textile dyeing operations.

Natural dyers have long used numerous parts of trees for dyeing textiles due to their rich tannin content, which can act as a dye or mordant [9, 10]. The use of different parts of a plant/tree to get dyestuff leads to the question of sustainability particularly if the dyestuff-source needs to be renewed and made viable for recurring usage. Production of the plant material for extraction of natural colourants must not compete with farming of crops for food production. Also, at present the production of natural dyes by direct farming, results in substantially high specific cost per kilogram of plant materials and correspondingly per kilogram of the dyed material [11]. The cost can be lowered by the use of by-products from agriculture and food industry, and wastes from forestry. Fruit of trees could be considered the most sustainable in this respect, with new produce grown each year and processed for value-addition and uneatable parts of the fruits like peels and seeds thrown away as waste. Agro-waste such as peels, shells, seeds, etc., are rich sources of pigments and natural dyes can be obtained from them. Sources such as pomegranate peels [12, 13], onion peels [14], leaves of eucalyptus [15], walnut husks [12, 16], walnut shells [6, 17, 18], almond shells [19], peanut shells [20], beetle nut [21], indigo seeds [22], mango seeds [23], tamarind seeds [24], safflower seeds [25], strawberries [26], beetroot [27], discarded/used marigold flowers [28], etc., have been explored for their use as natural colourants. Such studies are however limited and inadequate.

Use of fallen leaves and branches having a rich dye content is another sustainable approach. While it's possible to collect roots and bark from trees without harming the plant, it requires skill and detailed knowledge of a tree's structural and nutritional needs and thus, is not recommended. A healthier practice would be to collect these items from trees that are already harvested for manufacturing or for timber control.

At the same time, disposal of waste generated as by-products by the agriculture, forestry and industries pose a serious threat to the environment. With the environmental awareness and scarcity of space for landfills, wastes or by-product utilisation becomes an attractive alternative to direct disposal.

Thus, a promising concept for production of natural dyes with lowered cost could involve the use of plant materials discarded as waste/by-products from different sources.

1.1 Waste from agriculture

Agricultural wastes are non-product outputs resulting from various *agricultural* operations. Most of these wastes are discarded by burning and dumping or piling into landfills. This on one hand leads to wastage of resources, and on the other hand their disposal poses a serious threat to the environment. Effective utilisation of this agro-waste utilisation would not only address disposal issues, but also provide additional income to the farmers or processing industries that generate the waste [29].

Agro-wastes may include, among other materials, harvest-*wastes* including barks, flowers, fruits, leaves, roots, woods and seeds yield dyes.

Some studies in this area using agro-waste as natural dyes have already been undertaken. Wool yarns have been dyed with sawmill waste of regional dyewoods like Kansas black walnut (*Juglans nigra*), osage orange (*Maclura pomifera*) and eastern red cedar (*Juniperus virginiana*) [30]. Extracts from plants such as *Albizia coriaria*, *Butea monosperma*, *Callistemon citrinus*, *Tagetes erecta* L. and *Camellia sinensis* (oolong tea) have been studied and recommended as sources of natural dye for dyeing cotton fibres [7, 30–35]. Agricultural waste such as eggplant and mangosteen have also been reported as effective raw materials for the production of natural dyes [16, 36]. A wide range of shades with good fastness have been obtained using natural dye extracted from pomegranate peel extracts [4, 12–14, 36, 37]. Such peels have also been reported to exhibit antimicrobial properties [38, 39] and good UPF properties [40]. Onion peels/scales have been used for dyeing [11, 41–45]. Dyeing textiles with turmeric rhizome is well known, but use of the turmeric leaves discarded as waste to dye silk and cotton has also been explored in search of newer alternative and novel source for natural dyes [46]. Banana peel, an inexpensive by-product/waste has been used to dye cotton with excellent antibacterial activity and high UV protection properties [47, 48]. Banana floral stem sap has been used for dyeing cotton knitted fabric with good colour fastness properties barring light fastness [49]. Cotton has also been dyed with banana leaves [47]. Extracts from banana fruit peel has been used to dye mordanted cotton and silk [48]. Natural dyes from orange peel extract applied to wool fibres have a UV protective effect that is six times higher than that achieved with synthetic dyes [50]. Pashmina shawl with very good washing and light fastness has been dyed with petals of the saffron flower, which is thrown as a waste after harvesting [18].

1.2 Waste from forestry

Forestry operations can generate large quantities of brash wood chip, bark, fallen branches and leaves due to deforestation, woodland clearing and firebreaks from plants such as deodar, cedar, jackfruit, walnut and eucalyptus, etc., that can be utilised for textile colouration. Waste is also generated from timber industry in the form of saw dust, timber slash, extracted liquor from timber control and mill scrap. These residues are mostly used as a renewable energy source and in the making of bio-mass fuel and contain natural colouring pigments. Fallen leaves enrich soil, they can pose a threat to the environment as they can decay into harmful methane gas. Many of these leaves contain pigments that can dye textiles.

Some studies have been undertaken to dye textiles with forest wastes. Eucalyptus liquid waste from lumber steaming has been used by to dye cotton [51]. Sawmill waste from Kansas black walnut, osage orange, and eastern red cedar timber have been used as dyestuffs for textiles [29]. Waste bark extracts from Turkish red pine (*Pinus brutia* Ten) have been applied on cotton, flax, wool, tencel, polyamide and acrylic [46]. Silk and wool yarns have been dyed with extract from mango tree bark [52]. Cotton was dyed with colour extracted from eucalyptus bark [53, 54]. Silk and wool fibres have been dyed with using eucalyptus leaves [15].

1.3 Waste from industry

The food industry produces large quantities of liquid and solid wastes that are potential pollutants with associated disposal problems. These residues are used either as animal feed or are disposed or composted. However, wastes from pressed berries, grapes, distillation residues, wastes and peels from vegetable processing containing dyestuff are available at almost free cost.

Some studies on the use of plant waste from the food and beverage industry could be suitable as a source for the natural dyeing of wool fibres [11, 55] have been reported.

There is a need to explore the new sources of natural dyes. Although literature has reported use of agro, forest and industry wastes in dyeing of textile, research in this field is limited, scanty and sporadic. Also a common database on wastes used as natural dyes in dyeing textiles is not available. This chapter thus deals with few wastes that can be used as potential sources for textile colourants. Many of them are associated with multi-functional properties. Research already undertaken by people have been discussed in the form of case studies to provide a clear understanding of the waste source, its related composition and chemistry and effectiveness in imparting colour to textiles, sometimes with additional functional features.

2. Case studies for specific natural dyes from waste

2.1 Grape

Kingdom: Plantae; Family: Vitaceae; Botanical Name: *Vitis vinifera* L.; Common Name: Grape or *angoor* (Hindi); **Part used for dyeing**: Grape pomace, comprising of seeds, stalks, grape marcs, lees and stems.

2.1.1 Occurrence

Grapes belong to Vitaceae family and is cultivated mainly in the Mediterranean region, central Europe, south-western Asia, north and south Germany and eastern and northern Iran. About 5000–10,000 varieties of grapes are known of which only a few are of commercial significance and are used for eating (as a fruit) and for wine making. The fruit is classified as a berry and in the wild variety, are 6 mm diameter that ripen to give dark-purple to blackish colour. In cultivated plants, the berries are usually much larger, up to 3 cm long, and can be green, red, or blackish-purple.

2.1.2 Chemical constitution

Grape pomace potentially constitutes an abundant and relatively inexpensive source of polyphenols like phenolic acids, flavonols and anthocyanins and the colouring component in it is tannins (procyanidin, prodelphinidin, glucosylated procyanidin and gallated procyanidin) [56, 57].

2.1.3 Waste generated for grapes

Grape is one of the world's largest fruit crops and the world wine manufacturing converts currently 10–25% of raw grapes into residues, known as pomace, mostly comprising of seeds, stalks, grape marcs, lees and stems [58]. Because of a low extraction during wine making, the solid residues still retain high levels of condensed tannins [57]. Though rich in polyphenolic compounds; the use of grape pomace or waste has been restricted to animal feed and fertilisers and in some cases they are simply dumped.

Case study 1

Objective of the study: Dyeing of wool, silk, acrylic and cationized cotton and polyamid fibres with aqueous-extract of grape pomace [58].

Baaka et al. [58] have explored the use of a by-product, pomace, abundantly available at relatively inexpensive costs, to dye natural fibres like cotton, wool and silk; and synthetic fibres like polyamide and acrylic using sulphate salts of alum and ferrous as mordants. Since cotton and acrylic are associated with poor affinity to dyes, they were subjected to additional pre-cationization process prior to mordanting for improving dyeability. Surface colour strength of the dyed fabric was found to improve with use of metallic mordants, and good fastness properties to rubbing (4–5), wash (4–5) and light (4) were obtained. Authors have concluded remarkable improvements in the UPF of the fabric dyed with grape pomace in the presence and absence of metallic mordants (**Table 1**).

Case study 2

Objective of the study: Dyeing of silk with anthocyanins extracted from grape pomace [59].

Mansour et al. [59] have optimised the method for extracting natural dye from grape pomace, a winery waste and exhaust-dyeing silk with it using tannic acid ($C_{76}H_{32}O_{46}$, Aldrich) as a natural mordant using the meta-mordanting process.

Reporting that hydration of plant materials is important to enhance the overall extraction rate, the authors have used ethanol containing a weak acid like acetic acid to extract the more labile acylated anthocyanin pigment from *Vitis vinifera* L., leaves and have reported 60:40 as the optimal ratio of water-acidified ethanol mixture. Noting that anthocyanins are sensitive to temperature, a liner increase in yield was seen with rise in temperature and 32°C was found to be optimal for extraction. A time of 38 h was sufficient to swell and hydrate grape pomace permitting diffusion of the solvent into the plant matrix to leach out the pigment.

Mordanting with tannic acid at 95°C showed an increase in K/S value, but it lowered the L^* value. Positive a^* and negative b^* coordinate values of silk dyed with pomace extract indicates redness and blueness respectively and thus the extract rendered a purple and violet shade to the dyed silk fabric. The use of tannic acid at 85°C however increased the b^* coordinate values giving a bronze shade. Due to the ionic character of the –OH groups in anthocyanine, the dye from pomace extract probably reacted ionically with the pronated terminal amino groups of the silk fibre in an acidic pH, giving high surface colour strength. Thus, dyeing at pH 2.5 was considered optimum. The authors have also found that dyeing without any salt gives highest surface colour strength.

Through the use of Box–Behnken model of experimental design in response surface methodology (RSM), the authors have concluded that the optimal depth of shade on silk can be obtained at (a) 2.62 (pH), (b) 83.6°C (dyeing temperature), and (c) 0 gL^{-1} (salt concentration).

The dyed silk showed good light, wash and rubbing fastness that improved through the use of a mordant. The use of tannic acid as a mordant specifically improved the light fastness of silk fabrics dyed with grape pomace and gave darker shades probably due to the effective cross-links formed between the silk fibres and anthocyanin of the dye facilitated by the high molecular weight phenolic hydroxyl group contained in the tannin (mordant).

Fibres	Control (with dyeing and mordanting)	Fabric dyed with grape pomace extract without the use of any mordants	Fabric simultaneously dyed with grape pomace extract and mordanted with aluminium sulphate	Fabric simultaneously dyed with grape pomace extract and mordanted with ferrous sulphate
Cotton	5	25	40	35
Wool	5	30	50+	50+
Silk	2	18	24	40
Polyamide	20	26	40	35
Acrylic	25	30	30	50+

Table 1.
UPF rating of fabric composed of different fibres before and after dyeing with grape pomace extract with and without metallic mordants [58].

2.2 Pomegranate rind/peels

Kingdom: Plantae; Family: Lythraceae; Botanical Name: *Punica granatum* L.; Common Name: Pomegranate or *anar* (Hindi); **Part used for dyeing**: Outer rind/peels of the fruits.

2.2.1 Occurrence

Pomegranate belongs to Punica Genus which has two species, the most important being *Punica granatum* L. and is found in the region extending from Iran to northern India. It is a fruit bearing deciduous and thorny plant which bears cone-shaped red flowers singly, or in clusters.

The edible part of the pomegranate fruit (50%) consists of arils (40%) and seeds (10%). The remaining 50% of the total fruit weight comprises of tough, leathery and yellow or deep pink/red coloured peels.

2.2.2 Chemical constitution

The colouring matter from pomegranate peels consists of flavonoids (mostly flavan-3-ols or granatonine) and condensed tannins (26% of the total chemical composition) [60]. It also contains complex polysaccharides and many more compounds such as phenols, proanthocyanidin and ellagitannin (punicalagins and its isomers), low amounts of punicalin, gallic acid, ellagic acid, and ellagic acid glycosides [61].

2.2.3 Waste generated for pomegranate

The world pomegranate production amounts to approximately 1,500,000 tons [62]. The peel of the fruit consisting of pericarp/rind/hull comprises approximately of 60% of the weight of the fruit and is considered as an agro-waste though it is

Case study 3

Objective of the study: Dyeing of khadi cotton with aqueous extract of pomegranate rinds [40].

Sinnur et al. [40] have used pomegranate rinds as a natural colourant to dye cotton khadi fabric using single and double mordants. They have found that 15% myrobolan (*harda*) along with potash aluminium sulphate when used as mordants in 50:50 ratio offers maximum K/S value and renders good overall good colour fastness than any other combination of the two mordants. The authors have also reported the standardised conditions of dyeing khadi cotton fabric with aqueous extract of pomegranate rinds to be (a) 60 min (dyeing time), (b) 80°C (dyeing temperature), (c) 1:30 (dye-bath MLR), (d) 9.0 (dye-bath pH), (e) 20% (dye concentration) and (f) 3% (common salt).

Through the results of the FTIR and AAS studies, the authors have postulated the possible corresponding reaction mechanism resulting in the formation of large complexes between the mordant, fibre and the dye. The FTIR studies have indicated the presence of aluminium in dyed cotton khadi fabric, thereby confirming the formation of a complex between the cotton fibre, aluminium sulphate mordant and dye from pomegranate rinds together with the presence of some zinc and sulphur.

The dyed cotton khadi fabric shows medium to good level of ultraviolet protection as the UPF rating increases from 5 for the control untreated sample, to 15–20 for the cotton dyed with pomegranate rinds.

Case study 4

Objective of the study: Precipitate of dye extract from pomegranate peels has been used to dye wool yarn, cotton jersey and flax fabric [14].

For ease in use, transportation and storage, extracts can be concentrated either by evaporation of the aqueous extract or by membrane concentration techniques. The former evaporation technique uses considerable amounts of energy and the latter membrane concentration technique calls for a huge capital investment.

To overcome both the disadvantages of source-substitution and poor dye-concentration, Ali et al. [14] have used agro-waste like pomegranate peels, onion peel and Canadian goldenrod in the form of concentrates prepared by energy-friendly precipitation method to dye wool yarn, cotton jersey and flax fabric. Extracts of these dyes from agro-waste have been precipitated as Al-based dye lakes/concentrates using alumnium sulphate. On an average, the precipitate yielded 5% of the original plant weight as dry lake, the concentration of total phenolics (TPH) in this dry solid mass ranged between 20 and 40% and aluminium content varied between 3 and 5% (w/w).

The authors have shown how these Al-based dye lakes that are soluble in oxalic acid, can be used as direct dyes in the presence of aluminium-salt or iron-salt as mordants. Though the associated colour strength/depth of colour of the lakes was found to be lower, but the chroma of the dyeings (chromaticity) was higher. The authors have assigned the reason for the production of darker and more brilliant dyeing (higher b* coordinate values) by the direct use of precipitated extracts to the use of purified condition of the extract that contains more colour in the same amount of plant material.

a rich source of antioxidants, phenols, flavanoids [63]. Reportedly the peels also possess antibacterial, antifungal, anti-cancer and anti-inflammatory medicinal properties and have mostly been used as cattle feed.

2.3 Onion Peel

Kingdom: Plantae; Family: Amaryllidaceae; Botanical Name: *Allium cepa* L.; Common Name: Onion or *pyaz* (Hindi); **Part used for dyeing**: Outer papery skin of the fruit.

2.3.1 Occurrence

Onion, botanically known as *Allium cepa* L., onion belongs to the Liliaceceae family. The plant originated in the middle-east and in India it is cultivated extensively in South India and Bengal. The onion plant is a perennial herb growing to about 1.2 m in height, with 4–6 hollow cylindrical leaves. The underground bulb of onion plant comprises of fleshy leaf sheaths forming a thin-skinned capsule, and varies greatly in size (2–20 cm), shape (flattened, spherical, or pear-shaped) and colour depending on the variety.

2.3.2 Chemical constitution

The main colouring component found in the outermost dry skin of onion is quecertin ($C_{15}H_{10}O_7$), a flavonoid, proto-catechuic acid, kaempferol, anthocyanadine and some tannins [64].

2.3.3 Waste generated for onion

The outer layer of onion, i.e., the onion peel accounting for 10–25% (w/w) of the total weight are essentially removed before usage and thus forms a natural by-product of the food industry. With 22.43 million tons of onion produced by India as the second largest producer in the world [26], significant quantities of onion peel is generated as a waste.

Case study 5

Objective of the study: Dyeing of silk with a combination of a purified soxhlet extract of onion peels and pomegranate rinds [65].

Singh and co-workers [64] have used extract of peels of fruits and vegetables discarded as wastes (outer scaly peels of onion and rinds of the pomegranate fruit) to impart antibacterial functionality to degummed silk fabric dyed with the extract. Aqueous extraction of colour, from both the sources was done separately, and each solution was further purified through soxhlet distillation method. About 3 and 5 g/l of the both the extracts were applied on silk fabric as a mixture in a combination of 50:50, 25:75 and 75:25.

About 5 g/l concentration in the combination of 50:50 of the both the colourants gave the most optimised results and rendered a reduction of 98.1% in the bacterial count as evident from **Table 2**. Higher concentration of the dye when applied singly increases the antibacterial effect and higher proportion of pomegranate peel in the mixture has a slight improved edge over onion peel in rendering this effect.

2.4 Walnut shells

Kingdom: Plantae; Family: Juglandaceae; Botanical Name: *Juglans regia* L. (Persian Walnut) & *Juglans nigra* L. (American Black Walnut); Common Name: Walnut or *akhrot* (Hindi); **Part used for dyeing**: Outer hard shells enclosing the fruit.

Conditions of dyeing of silk	Concentration of the extracts	
	3 g/l	5 g/l
Only pomegranate peel extract	2.4×10^6 (96.1%)	2.1×10^6 (96.6%)
Only onion peel extract	2.8×10^6 (95.5%)	2.5×10^6 (96.0%)
Combination of pomegranate peel and onion peel extracts in the following ratios		
25:75	2.0×10^6 (97.1%)	1.9×10^6 (97.0%)
50:50	1.4×10^6 (97.7%)	1.2×10^6 (98.1%)
75:25	1.7×10^6 (97.3%)	1.6×10^6 (97.4%)

Data in parenthesis corresponds to the percentage in reduction.

Table 2.
Bacterial count (CFU/ml) and percentage reduction of E. coli *on silk fabric dyed with a single and combination of purified extracts of pomegranate peel and onion peel [65].*

Case study 6

Objective of the study: Dyeing of silk with onion peels discarded as agro-waste using unconventional energy effective methods of pad-dry-cure and pad-batch-dry techniques [44].

Natural dyeing is generally done at higher temperature that adds to the cost of the final dyed product. Singhee and Dhanania [44] have explored the challenge and provided solutions.

The authors have optimised conditions of aqueous extraction of colour from onion skin at (a) 5 (pH), (b) 1:20 (MLR) and (c) 80 min (time) and (d) 90°C (temperature). Use of 10% (owf) alumnium sulphate (mordant) by the pre-mordanting process was found optimum with respect to minimum strength loss maximum surface colour strength of the treated silk fabric. 80% (owf) dye concentration at pH 4 using MLR 1:40 at 60°C for 40 min gave optimum results with respect of surface colour strength and fastness properties of the dyed silk fabric. Strict control of pH, temperature of the dye-bath and dye concentration would result in uniform dyeing of silk with onion skins extract as indicated by high dispersion in the CDI values of silk dyed under variable conditions. These three have been identified as the predominating dyeing process parameters for dyeing silk fabric with onion peel.

To bring down energy costs, dyeing was carried out are room temperatures by the pad-dry-cure and pad-batch-dry methods. In the pad-dry-cure method, silk was treated with purified onion peel extract (extracted in ethyl alcohol and toluene solvents using soxhlet apparatus) at room temperature and then padded on a two-bowl padding mangle at 80% pick-up by 2 dip-2 nip process followed by drying at 90°C for 5 min. For the pad-batch-dry method, the padded silk fabric was rolled on a glass rod, covered with a plastic sheet and kept at room temperature for 24 h. Among both these two simpler dyeing processes that uses less water and fuel, the pad-batch-dry method gave highest K/S values which was higher than even that obtained when aqueous extract of the dye is used (**Table 3**). Colour fastness properties to wash and rubbing though were found be lower for pad-batch-dry method.

Samples dyed with aqueous extract of onion peel were subsequently treated with UV-absorbers (benztriazole, benzophenone and MEK) for improving light fastness, and dye fixing agents (Tinifix WS Conc., cetrimide and CTAB) for improving wash fastness properties by both exhaust process and pad-dry-cure method. UV-absorbers like benztriazole and benzophenone gives ½ to 1 grade improvement in the light fastness; while among the different dye fixing agents used, CTAB (cetyl trimethyl-ammonium-bromide) shows higher degree of improvement in the wash fastness when applied by the pad-dry-cure process on silk fabric dyed with aqueous extract of onion peel than by the exhaust method.

2.4.1 Occurrence

The walnut is edible fruit and belongs to the Juglandaceae family. The two most commercially important species are *J. regia* L. for timber and nuts and *J. nigra* L. for timber. The deciduous tall trees (10–40 m) are native to central Asia and the western Himalayas and have been cultivated in Europe as early as 1000 BC. Walnut is single-seeded stone fruit. The kernel or meat, which is usually made up of two halves separated by a partition and is encased in ligno-cellulosic shell that forms the thin endocarp or husk of the fruit.

Dye Concentration	K/S at λ_{max}	LF	Wash Fastness (ISO-II)		Rubbing Fastness	
			LoD	ST	Dry	Wet
Exhaust method (using 80 gpl of aqueous extract at standardise condition of dyeing)	8.2	3–4	1–2	4	3	4
Pad-Dry-Cure (using 5 gpl of purified dye)	2.7	2–3	1	4	4	4
Pad-Batch-Dry (using 5 gpl of purified dye)10.9	2–3	2	3	3	2–3	

LF—light fastness, LoD—loss in dept. of shade, ST—extent of staining on cotton

Table 3.
Surface color strength and color fastness properties of pre-mordanted silk fabric exhaust-dyed, dyed by the pad-dry-cure and pad-batch-dry method with purified extract of onion peel.

2.4.2 Chemical constitution

The colouring component present in walnut shell is juglone (CI 75500), a napthoquinone (5hydroxy-1, 4-napthoquinone) which imparts brown shade to textile substrate [66].

2.4.3 Waste generated for walnut

The walnut shells comprises 67% of the total weight of the fruit and as of 2009, the worldwide production of walnut was approximately 2.2 million tons. With this,

Case study 7

Objective of the study: Wool has been dyed with tannin-rich extract of walnut shells and pomegranate peel (agro-wastes) in combination with metallic mordants [66].

Shahmoradi et al. [65] have exhaust-dyed wool fabric pre-mordanted with metallic mordants with aqueous extracted walnut shells and pomegranate peels. The effect of the mordants, alumnium sulphate, copper (II) sulphate, iron (II) sulphate, iron (III) sulphate, potassium dichromate, tin (II) chloride and tannic acid, on colorimetric and antibacterial properties has been studied. Assigning the occurrence of –OH and C=O in the tannins present in the dye, the authors have reported improved colour strength in the dyed and mordanted wool fibre due to formation of a complex between the metal salt of the mordant and the dye.

K/S values appeared to be dependent on the concentration of the colourants in the solution from the two sources, and lower dye concentration though resulted in brighter shades (high L* value), but the shades were less red (lower a* coordinates). Among the various mordants studied, iron (II) and (III) sulphate and copper sulphate significantly improved K/S values of the dyed fabric.

Washing fastness of all samples treated with both the colourants was high, regardless of the type of mordant used, though light fastness ranged from excellent to poor. The authors have assigned the cause of poor light fastness to oxidation reaction of tannins in the dyes in the presence of light. If the dye-fibre bond promotes transfer of excitation energy from dye to fibre macromolecule, it makes it stable to light thereby improving light fastness. The authors have used this explanation to elucidate the role of chrome and copper salts as mordants in imparting dyeings with higher light fastness for both the dyes.

Pomegranate peel rendered better antibacterial activity against gram-negative bacterium (*E. coli* and *P. aeruginosa*) to the dyed wool irrespective of the mordant used compared to walnut shell extract. Increase in dye concentration improved the activity significantly and 5% concentration of dye in the presence of mordants dramatically enhanced the antibacterial properties. The antibacterial properties against *S. aureus*, *E. coli* and *P. aeruginosa* reduced significantly after washing and exposure to light due to breaking of the metallic complexes as a result of oxidation of tannins in the dye to hydroxyls and ketones in the presence of light. One cycle in a Launder-O-Metre was taken as equivalent to five home washing. Copper, aluminium and tin salts were found to be the most effective mordants for improved antibacterial activity that was durable to washing and light (could sustain five home washing). Interestingly the authors have also reported no significant antibacterial effect on the dyed fabric when chrome was used as a mordant.

Case study 8

Objective of the study: Dyeing of polyamide fibres using purified dye from walnut shells [16].

Mirjalili et al. [16] have used soxhlet extract of walnut shells to dye polyamide fabrics using 3% (owf) of different mordants (potassium aluminium sulphate, cupric sulphate, and ferric sulphate) by simultaneous mordanting and dyeing process. They have reported considerable increased in dye absorption leading to higher *K*/*S* values in case of mordanted samples compared with unmordanted ones, with ferric sulphate giving the highest colour strength.

Use of mordant also imparted higher antibacterial activity and here again ferric sulphate was more effective against *S. aureus* than other mordants; while cupric sulphate was better against *E. coli*. An increase in the dye concentration resulted in a tangible effect on antibacterial activity of dyed fabrics indicating that increased presence of phenolic and naphthoquinone compounds in the higher concentration of the dye extract is responsible for the improved antimicrobial activity. Use of metallic mordants also improved the wash durability of antimicrobial finish.

around 1.5 million tons of walnut-shells is generated as agro-waste [67]. Farmers mostly dispose the agro-waste residues including the shells, husk, stalks, etc., by burning them. But this is associated with serious environmental problems.

2.5 Almond shell

Kingdom: Plantae; Family: Rosaceae; Botanical Name: *Prunus dulcis* L.; Common Name: Almond or *badam* (Hindi); **Part used for dyeing**: Outer green hull and the woody hard shells enclosing the fruit.

2.5.1 Occurrence

Almond, a shrub or tree of the *Rosaceae* family belongs to the same genus, *Prunus* as the apricot, cherry, etc., but it differs from them in having a leathery fruit, which can only be eaten when immature, and has a comparatively large stone and kernel. This deciduous tree grows best in warm temperature where there is no spring frosts or tropical humidity and thus it is common to areas extending from West Asia to the West Mediterranean region. When the fruit ripens, its husk or flesh splits open, exposing the drupes (nut) which is rugged and pitted with irregular holes. The almond fruit consist of four parts, the kernel or meat covered by thin leathery brown seed-coat and enclosed in wrinkly, hard and woody endocarp shell. The shell is encased in the outermost green shell cover called the hull that becomes brittle as the fruit ripens.

2.5.2 Chemical constitution

Almond shells are rich in lignin and other phenolic compounds and the major flavonoids present are (+)-catechin, (−)-epicatechin, kaempferol, and isohamnetin. HPLC analysis has also revealed presence of quercetin, isorhamnetin and morin [30].

2.5.3 Waste generated for almond

Almonds are consumed as a valuable food and its production generates millions of tons of residues (shells, hulls, pruning, leaves, skin and inedible kernel disposition), the bulk of which are the almond shells (35–75% of the total fruit weight). Three million tons of almonds was globally produced annually as of 2014 and with this about 10.5–22.5 million tons of shells are generated and generally discarded [68].

Case study 9

Objective of the study: Wool has been dyed with extract of the outer green hull of the almond fruit [17].

Use of natural renewable and sustainable sources to dye textiles is one of the solutions as an environmentally safe technology that will expectedly conform to ecological legislations and limits. Keeping this in view, Ismal and co-workers [17] have dyed wool with extract of the outer green hulls of the almond fruit using metallic mordants as well as natural bio-mordants. Valex (acorn of *Quercus ithaburensis* ssp. *macrolepis*), pomegranate rind (*Punica granatum* L.), rosemary (*Rosmarinus officinalis*) and thuja leaves (*Thuja orientalis*) have been used as bio-mordants; while alum, iron (II) sulphate, copper (II) sulphate and potassium dichromate have been used as metallic mordants. Wool was dyed simultaneously with 2 g/l of aqueous extract of dried almond hull and varying amount of the bio-mordant (3–20 g/l) at 100°C using 1:50 MLR for 1 h. In case of the metallic mordants, pre and post mordanting techniques at 100°C using 1:50 for 1 h was also used. Unique and rare colour gamut of rose, brown, cinnamon and burgundy/reddish hues were obtained with the metallic mordants and the bio-mordants yielded completely different hues.

Among the metallic mordants, copper (II) sulphate gave highest K/S in case of all of mordanting process types (pre, post and simultaneous). Also, simultaneous-mordanting with any of the metallic mordant studied followed by dyeing with almond hull extract rendered lowest K/S value to the wool fibre. Baring valex, all bio-mordants increased the corresponding K/S values of the wool simultaneously mordanted with the bio-mordant and dyed with almond hull extract in contrast to when they were used individually in comparable concentrations. Valex when used alone, gave the highest colour yields among the bio-mordants and its K/S value for all concentrations was higher than that of the control sample. Relatively higher b^* value for valex resulted in a significant increase in yellow nuance of simultaneous bio-mordanted and dyed wool. The authors concluded that colour yields with bio-mordants can compete with metallic mordants barring copper sulphate when used through one-bath simultaneous dyeing and mordanting process.

Most bio-mordants can themselves be a source of colour and can dye fibres. The authors have also tried to study the synergistic effect on shades developed by natural bio-mordants when used in conjunction, at various concentration levels, with the dye from almond hull. Rosemary, valex and pomegranate bio-mordants show a synergistic effect with respect to the redness of the combined shade; their individually low redness value expectedly reduces the redness of the combined shade. Contrarily in case of thuja, the redness value of the combined shade is higher in spite of the lower redness value of thuja itself. Synergistic effect with respect to the blueness value can also be observed when valex, pomegranate and thuja are used in combination with almond hull extract; the blueness of the combined shade is higher than that of the control sample dyed only with almond hull extract or with the individual bio-mordant. L values of combination dyeing are always higher as compared to individual dyeing with almond hull extract for any of the bio-mordants under study, barring rosemary.

Wash fastness properties of wool simultaneously dyed with almond hull extract and bio-mordants improves slightly, but the corresponding values are lower when wool is mordanted with only with the bio-mordant without dyeing. Among the bio-mordants, valex and pomegranate rinds improve the light fastness of the wool when applied simultaneously with the dye from almond hulls.

Case study 10

Objective of the study: Silk has been dyed with an aqueous extract of almond shells [19].

Deepali and Shraddha [19] have used almond shells waste that is available abundantly and at almost no cost, to dye crepe silk fabric and have optimised process condition parameters (time, temperature, pH, material to liquor ratio and dye concentration) for pre-mordanting with 25% (owf) aluminium sulphate and exhaust-dyeing degummed crepe silk fabric with aqueous extract of the almond shell. The optimised extraction condition with respect to the highest optical density at maximum wavelength for onion peel has been established at (a) 25% owf (mordant concentration), (b) 80°C (extraction temperature), (c) 15 min (extraction time), (d) 1:30 (MLR), (e) 11 (pH).

The optimised dyeing condition with respect to both surface colour strength and fastness properties for dyeing silk crepe fabric using almond shell extract was established at (a) 100°C (dyeing temperature), (b) 60min (dyeing time), (c) 1:40 (MLR), (d) 2 (pH) and (e) 800% (dye concentration). CDI values of the dyed samples are widely dispersed for variations in temperature and pH indicating that special care should be taken with respect to the control these dyeing process parameters while dyeing protein fibres like silk with almond shell extract (**Table 4**). Dyeing under acidic pH gives darker shades and the fastness (wash and rubbing) properties of the dyed samples range from moderate to good.

Varying parameters	*K/S* at λ_{max}	**CDI**	**RCR** (CDI_{max} - CDI_{min})
Degummed and alum pre-mordanted silk (**CONTROL**)	1.00	—	—
Variation in **Time**			
15 min	1.8	55.8	49.44
30 min	1.6	11.3	
45 min	1.8	26.2	
60 min	1.8	17.5	
75 min	1.8	23.1	
90 min	1.8	6.3	
Variation in **Temperature**			
RT °C	1.2	67.6	66.37
40 °C	1.5	3.2	
60 °C	1.5	1.3	
80 °C	1.6	3.6	
100°C	2.0	4.4	
Variation in ***p*H**			
2	5.9	9.6	109.01
4	5.2	110.1	
7	2.0	101.9	
9	1.9	3.2	
11	1.3	1.1	
Variation in **MLR**			
1:10	1.2	0.0	5.53
1:20	1.3	0.3	
1:30	1.5	3.3	
1:40	1.9	1.9	
1:50	1.7	5.6	
Variation in **DYE CONCENTRATION**			
25%	1.2	0.1	4.02
50%	1.2	0.8	
100%	1.6	4.1	
200%	1.8	0.0	
400%	2.16	0.1	
800%	2.39	0.5	

CDI—colour difference index postulated by Samanta et al. [69].

Table 4.
Color strength and color difference index of crepe silk fabric pre-mordanted with 25% (owf) aluminium sulphate and dyed with aqueous extract of almond shell using variable conditions of dyeing [19].

2.6 Groundnut/Peanut skin

Kingdom: Plantae; Family: Fabaceae; Botanical Name: *Arachis hypogaea* L.; Common Name: Peanut or *mungphali* or *chinia badam* (Hindi); **Part used for dyeing**: Outer hard shells enclosing the fruit and the thin papery skin (seed-coat) covering the fruit.

2.6.1 Occurrence

Peanut is a legume and oil crop grown worldwide and classified as *Arachis hypogaea* L. and belongs to the Fabaceae family. Originating in South America, it is grown also in China, Africa, Indian, Japan and United States of America. The fruit is enclosed in a pod that consists of outer covering (shell/hull). The veined brown shell or pod of the peanut contains two or three peanut kernels. Each oval-shaped kernel or seed is comprised of two off-white lobes/cotyledons that are covered by a brownish-red paper-like covering called the seed-coat.

2.6.2 Chemical constitution

Peanut skin have high content of phenolic compounds and contains eight different types of flavonoids along with and two alkanoids [70], phenolic acids, procyanidins dimmers, oligomers and some tannins. The main colouring components in peanut skin are vanillin, catechin and epicatechin (polyphenolic compounds) [71].

2.6.3 Waste generated for peanut

Substantial amounts of by-products (peanut meal, skin, hull and vines) are generated in the process of peanut harvest, peanut oil extraction and making of peanut butter, peanut oil, peanut confections, roasted snack peanuts, extenders in meat product formulations, soups and desserts [72]. Though they contain appreciable amount of polyphenols. Only a portion of these by-products are used as animal feed or as fertilisers. A large portion of the waste is discarded, incinerated or lands up into landfills.

The peanut shells are obtained as agricultural by-product when graded peanuts are passed through shelling machines to obtain the kernels. The production of peanut shells has been estimated to be 230–300 g of peanut shells per kg of peanut [73].

About 35–45 g of peanut skin being generated per kg of shelled peanut kernel and the annually worldwide production of peanut skin, as a by-product of the peanut processing industry is over 0.74 million metric tons [74]. In India, 3.0 lakh kg is available annually and can approximately dye 4.0 lakh kg of the fabric. Thus, a good source of dye from agro-waste associated with good sustainability is available at lost cost.

Case study 11

Objective of the study: Microwave energy has been used to dye cotton, silk and wool with aqueous extract of peanut skin (seed-coat) [20].

Peanut skin discarded as waste from food processing industry contains tannins and thus Pandey et al. [20] have dyed cotton, silk and wool fabrics with aqueous extract of peanut skin without using any mordants by both the conventional exhaust method and new method using microwave energy. Peach shades were obtained on all fibres dyed with peanut skin extract and the colour yield on silk was higher than wool; it was least on cotton. Increase in temperature of dyeing yielded better K/S values and dyeing at 90°C gave highest results on all fibres. Microwave dyeing gave good results, but the surface colour strength of the dyed fabric was less than exhaust dyeing at 90°C except in case of wool. The tenacity of the fabric decreased notably when dyeing was carried out using microwave energy.

Good light (6) and wash (4) fastness, excellent rubbing fastness (4–5) and good perspiration fastness (4) was observed on most dyed fibres. Peanut skin extract rendered excellent sun-protective function (UPF) to the dyed silk and wool fabric.

2.7 Tamarind seeds

Kingdom: Plantae; Family: Fabaceae; Botanical Name: *Tamarindus indica* L.; Common Name: Tamarind or *imli* (Hindi); **Part used for dyeing**: Shells and seed-coat of the fruit.

Case study 12

Objective of the study: Cotton has been dyed with extract from peanut pods (including shells and edible seeds) using metallic mordants [75].

Chhipa et al. [73] have used peanut pods to dye cotton using 10% (owf) metallic mordants (alum, copper sulphate and ferrous sulphate) using pre, post and simultaneously mordanting and dyeing processes. Authors have reported higher surface colour strength with increase in dye concentration irrespective of the process of mordanting used except in case of the pre-mordanting process, which indicates a decrease in the K/S values. In all cases, alum as a mordant gave better colour yield (K/S) and ferrous sulphate the least. Simultaneous mordanting and dyeing gave higher colour strength on cotton in all cases except when the cotton fabric was dyed by the post-mordanting process using copper sulphate.

Irrespective of the sequential process of dyeing and mordanting used, light, wash and rubbing fastness was reportedly good for cotton dyed with aqueous extract of peanut pods.

2.7.1 Occurrence

Tamarindus indica L. belongs to Fabaceae family, and is one of the most important multipurpose evergreen tree cultivated throughout the Indian sub-continent except in the Himalayas and western dry regions. The fruit of this tree is enclosed in a pod that comprises of pulp (55%), seed (34%) and hard brown shell (11%). The seed comprises of a brown coloured hard coat/testa (20–30%) and a kernel (70%). The kernels are separated from the seed-coat either by roasting or by soaking the seeds in water.

The tamarind fruits are bean-like, irregularly curve and bulged pods that usually vary from 2 to 7 cm in length. The pods are initially green and turn brown or greyish-brown as the fruit ripens. The pod-skin becomes brittle and the pulp inside dehydrates naturally to a sticky paste that encloses a few coarse strands of fibre extending along the length of the stalk.

2.7.2 Chemical constitution

The tamarind seed-coat contains 38–40% water soluble matter, 80% of which is a mixture of tannins, polyphenolic compounds [76] as condensed tannins and colouring matter (procyanidin trimer, procyanidin B2 and epicatechin) [77].

Case study 13

Objective: Cotton and silk has been dyed with tamarind fruit pods [78].

Umar et al. [76] dyed cotton and silk with aqueous extract, and purified methalonic and ethanolic extract of dried pod-skin of the tamarind fruit using alum, copper sulphate, ferrous (II) and (III) sulphate as mordants. All three methods of pre, post and simultaneous mordanting were used for comparison purpose.

Cotton and silk dyed with the aqueous extract of the tamarind pods, showed excellent to good light except when the sample was dyed with alum by the simultaneous mordanting process. All the dyed samples showed fair to good wash fastness with negligible staining. Rubbing fastness was also found to range from excellent to good, except when post mordanting process was used. Fastness to alkaline perspiration fastness was good (4–5) except in case of silk when iron (II) sulphate was applied by the simultaneous mordanting process or in case of cotton when copper (II) sulphate was applied by the post-mordanting process.

For cotton and silk dyed with methanolic extract of the purified dye, excellent to good light fastness was observed in all cases except when both cotton and silk were pre-mordanted with alum. The wash fastness of the dyed samples was found to be fairly good and negligible staining was observed. Rubbing fastness was found to range from excellent to good, while the perspiration fastness was fair.

For cotton and silk dyed with ethanolic extract of the purified dye, excellent to good light fastness was observed negligible colour staining. The wash fastness was fairly good (3–4) for cotton and silk. Perspiration fastness test was fairly good (3–4) in both acidic and alkaline medums, except in case of pre-mordanting process using iron (II) and iron (III) sulphates, where some loss in colour depth was visible specially when treated in an acidic medium. The extent of staining for both acidic and alkaline mediums was negligible.

Case study 14

Objective of the study: Sour and sweet tamarind seed husks or seed-coat has been used as a colourant to dye cotton and silk along [79].

Tepparin et al. [77] have dyed cotton, *Bombyx mori* (mulberry) silk and eri silk fabrics with aqueous extract of both sour and sweet tamarind seed husk/seed-coat (*Tamarindus indica* L.) varieties using zinc and iron sulphate and potassium dichromate as mordants. K/S values of silk dyed with tamarind seed husk in all cases was better than cotton with eri silk showing higher colour strength than mulberry silk under comparable conditions. Among the mordants, potassium dichromate gave highest colour strength on cotton and mulberry silk; while eri silk showed highest colour uptake by iron sulphate.

The colour fastness to washing of all the dyed fabrics ranged from good-excellent level and was stable even after five washing cycles. Colour extracted from sour tamarind seed-coat rendered a higher colour strength on cotton and silk than that obtained when sweet tamarind seed-coat extract was used under comparable conditions. Sour tamarind seed-coat yield redder, yellower and more intense shades than the sweet tamarind seed-coat.

2.7.3 Waste generated for tamarind

The reddish or purplish brown seeds rich in tannins [76] are generated as a by-product by the tamarind pulp industry. Also after removal of the pulp, the pods are mostly discarded. India alone produces about 0.3 million tons of tamarind annually.

2.8 Banana

Kingdom: Plantae; Family: Musaceae; Botanical Name: *Musa acuminate Colla*; Common Name: Banana; **Part used for dyeing**: Banana psuedostem, leaves and fruit peel.

2.8.1 Occurrence

Musa acuminata is an evergreen plant. The plant grows horizontally or obliquely from a false trunk also known as pseudo-stem and comprises of leaf sheaths packed in layers. The leaves are composed of a stalk called petiole and a blade or lamina. The individual flowers are white to yellowish-white in colour and grows upwards that is away from the ground. The fruits are slender berries whose size depends upon the number of seeds they contain. Each fruit bunch has about 160 fingers. As each plant produces only one bunch of bananas and the bare pseudo-stems cannot be used in the second harvest, the banana cultivation generates large amount of residues that are considered as organic wastes [49].

2.8.2 Chemical constituent of the dye

The holocellulose of banana pseudo-stem contain tetrahydroxyflavone (flavonoids) also known as luteolin and some tannins as the colouring components [49].

2.8.3 Waste generated for banana

Banana accounts for about 16% of the total world organic production [80]. After harvesting, almost 89% of banana plant (floral stem, outer part of pseudo-stem, peels, peduncle, underground parts, leaf stalk and leaves) remain as a waste.

2.9 Flower waste

Kingdom: Plantae; Family: Asteraceae; Botanical Name: *Tagetus erecta* L.; Common Name: Marigold; **Part used for dyeing**: Flower petals.

Case study 15

Objective of the study: Cotton has been dyed with banana floral stem sap without the use of any mordants [49].

Repon et al. [49] have retrieved sap from fresh banana floral stem using roller squeezer machine and used it to dye cotton by the exhaust method at varying temperatures. 100°C gave best result in term of the colour yield (K/S), colour levelness (ΔE), brightness index and colour fastness to wash, rubbing and perspiration. Assigning the role of temperature in boosting swelling of fibres to promote easier penetration of dye into interior of the fibre, the authors have explained the achievement of colour levelness at higher temperatures.

Except light fastness, all other colour fastness (washing, rubbing and perspiration) value ranged between 3 and 5, i.e., good to excellent. Wet rubbing properties were found to be lower than dry rubbing. Increase in temperature slightly improved the overall colour fastness properties. With increase in temperature, specimens showed a slight improvement in overall colour fastness properties. Authors have assigned poor light fastness of the dye to both lower photo stability of the natural pigments and also to the weaker dye-fibre interaction between the cotton fibre and the natural pigments in the banana floral stem sap.

Case study 16

Objective of the study: Alkaline and acetone medium have been used to extract colour from banana leaves to dye Egyptian cotton using metallic mordants [81].

Saleh [79] have extracted pigment from banana leaves using alkaline and acetone medium to dye Egyptian cotton fabrics in the presence of ferrous sulphate, copper sulphate and potassium dichromate as mordants. Dyeing of pre-mordanted cotton with the alkaline extracts of banana leaves was done at 80°C for 90 min at pH 9 using MLR 1:40; while dyeing with acetone extracted solution was carried out at 56°C for 5 min using MLR 1:40.

High performance thin layer chromatography (HPTLC) of the extracts identified pigment compounds, luteolin and apigenin in the alkaline extract and chlorophyll a, b and β-Carotene in the acetone extract. Though the K/S values for all dyeing were high, mordanted cotton displayed higher K/S value than unmordanted cotton irrespective of the mordant used. Mordanting with iron (ferrous sulphate) gave higher colour difference (ΔE) with yellowish brown colour probably due to a stable complex formed between iron and luteoline, a phenolic compound. Copper sulphate rendered a greener shade, while chromium produced pale yellow shades on cotton. Light fastness of the dyed samples was moderate, but good wash and rubbing fastness were obtained which improved when mordants were used.

Case study 17

Objective of the study: Alkaline fractions of banana fruit-peel have been used to dye cotton fabrics [47].

Salah et al. [47] extracted colour from banana peels using 0.1% NaOH. The extracted dye was applied to the ferric sulphate pre-mordated bleached and mercerized Egyptian cotton fabrics. The HPTLC analysis indicated the presence of pigment, Luteolin, in crude alkaline extract of banana peels.

Fe is known for its ability to readily chelate and form coordination complexes and several natural dyes. The authors have assigned the formation of a stabile complex between Fe of the mordant and luteoline of the colouring pigment in banana peel as the reason for maximum ΔE and corresponding lower L^* values of the pre-mordanted cotton fabric dyed with the alkaline extract of the peel. Interestingly they have denoted the cause of higher dye uptake by the pre-mordanted cotton fibre to the formation of a ternary complex between the mordant, dye and the fibre. The coordination numbers of iron causes some coordination sites to remain unoccupied when it interacts with cellulose of cotton and these unoccupied sites are taken over by the colour component in the dye thereby enhancing the interaction between the fibre and the dye.

Transformation of cotton from cellulose I to cellulose II during mercerization involves partial destruction of intermolecular bonds due to alkali-swelling. This allows the dye to penetrate the swollen amorphous region of the cellulose where they form hydrogen bonding with the fibre cellulose. This explains the better dyeability and corresponding higher dye uptake of the mercerized cotton compared to unmercerized cotton. Iron from the mordant also plays a role in increasing the dye uptake and shifting of the colour for mercerized cotton due to respective formation of a chelate with the sodium cellulose and sodium alcoholates and a complex formation with luteoline of the dye. Correspondingly the L^*, a^*, and b^* change for mercerized cotton.

The data obtained showed that the mercerized cotton have excellent antibacterial activity both in terms of zone of inhibition and quantitatively in terms of percentage reduction in bacteria. Swelling of cotton with its resultant shrinkage during mercerization causes reduction in the fabric interstices/porosity and this restricts the consequent UV transmittance giving high UV protection in terms of UPF to mercerized cotton. The UPF of the mercerized cotton fabrics was found to 64 compared to 18.0 for the unmercerized cotton.

The obtained results were also subjected to statistical analysis according to Sendcor and Cochran [82]. The experiments were in randomised complete design (RCD) with three replicates. Means were compared using the significant difference (LSD) at 5% level probability.

2.9.1 Occurrence

Marigold plant, a small perennial shrub that can grow up to 60 cm height belongs to Asteraceae family. It abundantly bears flowers varying in colour from yellow and gold to orange, red and mahogany depending on the species during the flowering season. Marigolds have been cultivated all over the world mostly South Asia [83].

2.9.2 Chemical constituent of the dye

Marigold flowers contain compounds called carotenoids containing lutein ($C_{40}H_{56}O_2$) and patulinin, a flavonoid. Lutein and patulinin are the two natural pigments responsible for colouring of textiles. Lutein is present along with its isomer, zeaxanthin also known as oxycarotenoids [84].

2.9.3 Waste generated for marigold

Marigolds, yellowish-orange coloured flowers are used in many religious ceremonies and offered in huge quantities to deities in temples throughout India.

Case study 18

Objective of the study: Cotton, silk and wool have been dyed with purified ethanolic extract of waste flowers collected from temples using metallic mordants [28].

Vankar et al. [28] have used temple waste comprising of tagetus flowers collected from the Kanpur city to dye cotton, silk and wool. NMR and mass spectra analysis of the crude extract from the flower identified the presence of patuletin as the flavanoid as the colouring matter present in *Tagetus erecta* L. Cotton was dyed, with colour extracted in an aqueous medium and also separately with colour extracted in ethanolic medium. Exhaust-dyeing of cotton fabric with ethanolic extract of the dye gave better colour strength (K/S of 141.73) than when dye extracted by the conventional method of aqueous extraction was used (K/S of 77.59).

Cotton (additionally pre-treated with tannic acid), silk and wool were pre-mordanted with different mordants (alum, copper sulphate, ferrous sulphate, stannic chloride and potassium dichromate) and then dyed with the purified ethanolic extract of the dye. Though the fastness properties improved with the use of metallic mordants, it rendered a slight change in the hue of the dyed fabric from yellow to yellowish-green to brown. Among the mordants used, ferrous sulphate gave the highest K/S on all fibres. Also the percentage of dye exhaustion due to metal mordanting varied with the different fabrics used; for cotton it varied between 45 and 52%; for silk it was 38–46% and for wool it ranged from 37 to 51%.

Case study 19

Objective of the study: New protein fibre derived from soyabean has been dyed with marigold flowers collected as temple waste [82].

Teli et al. [82] have dyed soyabean protein fibre (SPF) with marigold extract using natural mordants derived from tamarind seed-coat, *amla* (Indian gooseberry), *harda* (myrobolan) and compared the results with those obtained by using the more commonly used metallic mordant, alum, through pre-mordanting process. The K/S values of pre-mordanted SPF was higher than the unmordanted one. Alum rendered higher K/S value compared to the natural mordants and among the natural mordants, *amla* gave highest increase in the K/S values. Reportedly, increase in K/S values was observed with increase in concentration of mordants, and this was more so in case of the natural mordants. Results also indicated an increase in redness of the dyed fabric due to use of tannin-based natural mordants.

The concentration of the dye also impacted the K/S values, which improved with increase in the dye concentrations. Results also indicated that use of mordant improves fastness properties that were comparable for all natural and metallic mordants used in the study. Though the antibacterial properties of the samples dyed with marigold extract without mordanting was low, it improved after mordanting and all the three natural mordants rendered similar overall antibacterial activity. The antibacterial finish reduced on washing, but it was within the acceptable limit of 70% after 30 washes. Among the natural mordants, tamarind seed-coat and *amla* gave wash-durable antibacterial finish compared to *harda*.

Most of them are thrown in to the neighbouring ponds, lakes and rivers or simply dumped around the temple to decay naturally. This adds to pollution in and around most temples and has an undesirable effect on the river ecology [28, 81].

3. Conclusion

As ecological considerations become important factors in the selection of consumer goods all over the world consumers want quality product with no harm to the ecology, use of wastes from various sources for dyeing textiles can prove to be attractive for two reasons. One, they provide novel sources for dyeing textiles at lower costs. Two, they provide sustainable solution to the management of waste disposal.

Though a number of studies have been carried out to explore alternative source for natural dye, more specifically from wastes, more needs to be done in a systemic and scientific manner. Systematic collection of waste from the source and suitable alternatives to convert the waste into concentrates and appropriate recipes for dyeing to produce reproducible shades needs to be explored.

Authordetails

Deepali Singhee
Department of Textile Science, Clothing and Fashion Studies, J.D. Birla Institute, Kolkata, India

*Address all correspondence to: deepalisingheejdbi@gmail.com

References

[1] Alam MM, Rahman ML, Haque MZ. Extraction of henna leaf dye and its dyeing effect on textile fibre. Bangladesh Journal of Scientific and Industrial Research. 2007;**42**(2):217-222

[2] Samanta AK, Agarwal P, Dutta S. Dyeing of jute and cotton fabric using jackfruit wood extract. Part-1: Effects of mordanting in dyeing process variables on color yield and color fastness properties. Indian Journal of Fiber and Textile Research. 2007;**32**(4):466-476

[3] Saima N, Bhatit Ijaz A, Shahid A. Dyeing properties of cotton fabric using un-irradiated and gamma irradiate extracts of *Eucalyptus camaldulensis* bark powder. Indian Journal of Fibre & Textile Research. 2011;**36**(June):132-136

[4] Ajmal M, Adeel S, Azeem M, Zuber M, Akhtar N, Iqbal N. Modulation of pomegranate peel colourant characteristics for textile dyeing using high energy radiation. Industrial Crops and Products. 2014;**58**:188-193

[5] Bhargava D, Shahnaz J. Microbial dyes: A new dimension to natural dyes. Colourage. 2013;**60**(7):42-43

[6] Erdem IÖ, Leyla Y. Almond shell as a natural colorant. Indian Journal of Fibre & Textile Research. 2012;**37**(4):358-363

[7] Hana K. In: Křemenáková D, Militký J, Mishra R, editors. Natural Dyes: Their Past, Present, Future and Sustainability. Recent Developments in Fibrous Material Science. Czechoslovakia: Kanina POS; 2015. pp. 59-71

[8] Saxena S, Raja ASM. Roadmap to sustainable textiles and clothing. In: Natural Dyes: Sources, Chemistry, Application and Sustainability Issues. Singapore: Springer; 2014

[9] Dominique C. Natural Dyes Sources, Tradition, Technology and Science. Paris: Archetype Publications; 2007

[10] Prabhu KH, Bhute AS. Plant based natural dyes and mordants: A review. Journal of Natural Product and Plant Resources. 2012;**2**(6):649-664

[11] Bechtold T, Mussak R, Mahmud-Ali A, Ganglberger E, Geissler S. Extraction of natural dyes for textile dyeing from coloured plant wastes released from the food and beverage industry. Journal of the Science of Food and Agriculture. 2006;**86**:233-242

[12] Adeel S, Ali S, Bhatti I, Zsila F. Dyeing of cotton fabric using pomegranate (*Punica granatum*) aqueous extract. Asian Journal of Chemistry. 2009;**21**(5):3493-3499

[13] Kulkarni SS, Gokhale AV, Bodake UM, Pathade GR. Cotton dyeing with natural dye extracted from pomegranate (*Punica granatum*) peel. Universal Journal of Environmental Research and Technology. 2011;**1**(2):135-139

[14] Ali AM, Binder CF, Bechtold T. Aluminium based dye lakes from plant extracts for textile coloration. Dyes and Pigments. 2012;**94**:533-540

[15] Mongkholrattanasit R, Krystufek J, Wiener J. Dyeing and the fastness properties of natural dye extracted from eucalyptus leaves using padding techniques. Fibers and Polymers. 2010;**11**(3):346-350

[16] Mirjalili M, Karimi L. Extraction and characterization of natural dye from green walnut shells and its use in dyeing polyamide: Focus on antibacterial properties. Journal of Chemistry. 2013;**1**:1-9

[17] Erdem IÖ, Leyla Y, Esen O. Use of almond shell extracts plus biomordants as effective textile dye. Journal of Cleaner Production. 2014;**70**:61-67

[18] Lal C, Raja ASM, Pareek PK, Shakyawar DB, Sharma KK, Sharma MC. *Juglans nigra*: Chemical constitution and its application on pashmina (cashmere) fabric as a dye. Journal of Natural Product and Plant Resources. 2011;**1**(4):13-19

[19] Deepali S, Shraddha N. A study on dyeing of silk fabric with almond shells (*P. amygdalus* L.) waste. International Journal of Research. 2017;**4**(5):921-935

[20] Pandey R, Shweta P, Pandit P, Shanmugam N, Jose S. Colouration of textiles using roasted peanut skin—An agro processing residue. Journal of Cleaner Production. 2018;**172**:1319-1326

[21] Jitphusa A, Rattanankij T. Process of dyeing Chinese silk fabric using color of water betel nuts. Journal of Thai Interdisciplinary Research. 2016;**11**(6):20-26

[22] Mahanta D, Tiwari SC. Natural dye-yielding plants and indigenous knowledge on dye preparation in Arunachal Pradesh, Northeast India. Current Science. 2005;**88**(9):1474-1480

[23] Franca AS, Oliveira LS, Saldanha SA, Pedro IA, Santos, Salum SS. Malachite green adsorption by mango (*Mangifera indica* L.) seed husks: Kinetic, equilibrium and thermodynamic studies. Desalination and Water Treatment. 2010;**19**:241-248

[24] Prabhu KH, Teli MD. Eco-dyeing using *Tamarindus indica* L. seed coat tannin as a natural mordant for textiles with antibacterial activity. Journal of Saudi Chemical Society. 2014;**18**:864-872

[25] Shin Y, Son K, Yoo D. Dyeing properties and color of silk fabrics dyed with safflower yellow dye. Journal of the Korean Society of Clothing and Textiles. 2008;**32**(6):928-934

[26] Mohan R, Geetha N, Jennifer DH, Sivakumar V. Studies on natural dye (Pelargonidin) extraction from onion peel and application in dyeing of leather. International Journal of Recent Engineering Science. 2020;**7**(1):14-22

[27] Sivakumar V, Anna JL, Vijayeeswarri J, Swaminathan G. Ultrasound assisted enhancement in natural dye extraction from beetroot for industrial applications and natural dyeing of leather. Ultrasonics Sonochemistry. 2009;**16**:782-789

[28] Vankar P, Shanker R, Wijaypala S. Utilization of temple waste flower—*Tagetus erecta* for dyeing of cotton, wool and silk on industrial scale. Journal of Textile and Apparel, Technology and Management. 2009;**6**(1):1-15

[29] Doty K, Haar S, Kim J. Black walnut, Osage orange and eastern redcedar sawmill waste as natural dyes: Effect of aluminum mordant on color parameters. Fashion & Textiles. 2016;**3**:22-38

[30] Esfahlan AJ, Jamei R, Esfahlan RJ. The importance of almond (*Prunus amygdalus* L.) and is by-products. Food Chemistry. 2010;**120**(2):349-360

[31] Paul C, Cheng Y, Deng S, Lin X, Guangwei H, Roger R. Utilization of almond residues. International Journal of Agricultural and Biological Engineering. 2010;**3**(4):1-18

[32] Deniz F. Dye removal by almond shell residues: Studies on biosorption performance and process design. Materials Science & Engineering, C: Materials for Biological Applications. 2013;**33**(5):2821-2826

[33] Deniz F. Adsorption properties of low cost biomaterial derived from *Prunus amygdalus* L. for dye removal from water. The Scientific World Journal. 2013;**2013**:1-8. DOI: 10.1155/2013/961671

[34] Majid A, Imane K, Mohsen H, Shahrzad F. Removal of rhodamine B from aqueous solution by almond shell biosorbent. Journal of Biology & Environmental Sciences. 2012;**2**(9):39-44

[35] Laufenberg G, Kunz B, Nystroemm M. Transformation of vegetable waste in value added products: (A) the upgrading concept; (B) Practical implementations. Bioresource Technology. 2008;**87**(2):167-198

[36] Tiwari HC, Singh P, Mishra PK, Srivastava P. Evaluation of various techniques of extraction of natural colorants from pomegranate rind—Ultrasound and enzyme assisted extraction. Indian Textile Journal. 2010;**35**(13):272-276

[37] Davulcu A, Benli HS, Sen Y, Bahtiyari MI. Dyeing of cotton with thyme and pomegranate peel. Cellulose. 2014;**21**:4671-4680

[38] Al-Zoreky. Antimicrobial activity of pomegranate (*Punica granatum* L.) fruit peels. International Journal of Food Microbiology. 2009;**134**(3):244-248

[39] Negi PS, Jayaprakash GK. Antooxidant and antibacterial activities of *Punica granatum* peel extract. Journal of Food Science. 2003;**68**(4):1473-1477

[40] Sinnur HD, Samanta AK, Verma DK, Kaware R. Studies on coloration and UV protective action of anar peel (pomegranate rind) as an effective natural colorant for cotton khadi fabric. Journal of Institution of Engineers (India): Series E. 2018;**99**:9-26

[41] Sengupta S. Dyeing of wool fibre with onion skin as a natural dye. Man-Made Textiles in India. 2004;**47**(9):341-348

[42] Yildirim L, Erden IO. Banana peel in dyeing of polyamide/elastane blend fabric. Research Journal of Textile and Apparel. 2019;**23**(2):124-133

[43] Rabie SF, Shakra S, Adway H, Ali A. Physicochemical studies and application of some natural extracts and its metallic mordant complexes as dyes for textile fabrics. Al-Azhar Bulletin of Science. 2013;**24**(2):17-35

[44] Singhee D, Dhanania Y. Dyeing of silk fabric with onion peel waste using unconventional methods. Indian Journal of Natural Fibers. 2016;**3**(1):45-57

[45] Maryam A, Khan SI, Riaz A, Ali S, Noreen S. Effect of mordants with the application of natural dye extracted from *Allium cepa* on natural fabric. International Journal of Home Science. 2019;**5**(3):283-286

[46] Ozan A, Celik A, Gedik G, Yavas A. Natural dye extraction from waste barks of Turkish red pine (*Pinus brutia* Ten.) timber and eco-friendly natural dyeing of various textile fibers. Fibers and Polymers. 2013;**14**(5):866-873

[47] Salah SM. Antibacterial activity and UV protection property of some Egyptian cotton fabrics treated with aqueous extract from banana peel. International Journal of Clothing Science and Technology. 2012;**1**(1):1-6

[48] Tera Ferial M, Elnagar Khaled E, Mohammed SM. Dyeability and light fastness properties of onion scale dye on different fabric types for conservation applications. Journal of Textile and Apparel, Technology and Management. 2012;**7**(3):1-6

[49] Repon MR, Mamun MA, Islam MT. Eco-friendly cotton coloration using banana (*Musa sapientum*) waste: Optimization of dyeing temperature. Universal Journal of Engineering Science. 2016;**4**(1):14-20

[50] Shahid M, Mohammad F. Perspectives for natural products based

agents derived from industrial plants in textiles application—A review. Journal of Cleaner Production. 2013;**57**:2-18

[51] Rossi T, Silva PMS, De Moura LF, Araújo MC, Brito JO, Freeman HS. Waste from eucalyptus wood steaming as a natural dye source for textile fibers. Journal of Cleaner Production. 2017;**143**:303-310

[52] Win ZM, Swe MM. Purification of the natural dyestuff extracted from mango bark for the application on protein fibres. World Academy of Science, Engineering and Technology. 2008;**46**:536-540

[53] Ali S, Nisar N, Hussain T. Dyeing properties of natural dyes extracted from eucalyptus. The Journal of The Textile Institute. 2007;**98**(6):559-562

[54] Rossi T, Brito JS, Bittencourt EB, Almeida RSR, Faria PN, Dias CTS. Resíduo efluente de eucalipto como extrato corante para o tingimento têxtil de algodão. Redige. 2012;**3**:1-17

[55] Bechtold T, Mahmud-Ali A, Mussak R. Anthocyanin dyes extracted from grape pomace for the purpose of textile dyeing. Journal of the Science of Food and Agriculture. 2007;**87**:2589-2595

[56] Deng Q, Penner MH, Zhao Y. Chemical composition of dietary fiber and polyphenols of five different varieties of wine grape pomace skins. Food Research International. 2011;**44**:2712-2720

[57] Pinga L, Pizzib A, Guoa ZD, Brosse N. Condensed tannins from grape pomace: Characterization by FTIR and MALDI TOF and production of environment friendly wood adhesive. Industrial Crops and Products. 2012;**40**:13-20

[58] Baaka N, Ticha MB, Haddar W, Amorim MTP, Mhenni MF. Upgrading of UV protection properties of several textile fabrics by their dyeing with grape pomace colorants. Fibers and Polymers. 2018;**19**(2):307-312

[59] Rym M, Bechir E, Mhenni F. The use of response surface method to optimize the extraction of natural dye from winery waste in textile dyeing. The Journal of The Textile Institute. 2017;**108**(4):528-537

[60] Middha SK, Talambed U, Pande V. A review on anti-hyperglycemic and anti-hepatoprotective activity of eco-friendly *Punica granatum* peel waste. Evidence-based Complementary and Alternative Medicine. 2013;**2013**:1-10. DOI: 10.1155/2013/656172

[61] Jalal H, Pal MA, Ahmad SR, Rather M, Andrabi M, Hamdani S. Physico-chemical and functional properties of pomegranate peel and seed powder. The Pharma Innovation Journal. 2018;7(4):1127-1131

[62] FAO. Statistical Database. Food and Agriculture Organization of the United Nations. Tunis, Tunesia: Codex Alimentarius Commission; 2012

[63] Malviya S, Jha Arvind A, Hettiarachchy N. Antioxidant and antibacterial potential of pomegranate peel extracts. Journal of Food Science and Technology. 2014;**51**(12):4132-4137

[64] Gulrajani ML, Gupta D. Natural Dye and their Application to Textiles. New Delhi, India: Department of Textile Technology, Indian Institute of Technology; 1992

[65] Singh N, Singh V, Punia P. Antibacterial finish on silk fabric with pomegranate and onion peel extracts. Journal of Pharmacognosy and Phytochemistry. 2019;**SP5**:129-131

[66] Shahmoradi GF, Shams NA, Majid MS, Daryoush A, Javad M. The effect of mordant salts on antibacterial

activity of wool fabric dyed with pomegranate and walnut shell extracts. Coloration Technology. 2012;**128**:473-478

[67] Pirayesh H, Khazaeian A, Tabarsa T. The potential for using walnut (*Juglans regia* L.) shell as a raw material for wood-based particleboard manufacturing. Composites: Part B. 2012;**43**:3276-3280

[68] Li X, Liu Y, Hao J, Wang W. Study of almond shell characteristics. Materials. 2018;**11**:1782

[69] Samanta AK, Konar A, Chakraborti S, Dutta S. Dyeing of jute fabric with tesu extract: Part I—Effects of the different mordants and dyeing process variables. Indian Journal of Fibre & Textile Research. 2011;**36**(1):63-73

[70] Lou H, Yuan H, Yamazaki Y, Sasaki T, Oka S. Alkanoids and flavonoids from peanut skin. Planta Medica. 2001;**67**:345-349

[71] Helmy HM, Kamel MM, Hagag K, El-Hawary N, El-Shemy NS. Antimicrobial activity of dyed wool fabrics with peanut red skin extract using different heating techniques. Egyptian Journal of Chemistry. 2017;**Special Issue**:103-116

[72] Rustom IYS, Lopez-Leiva MH, Nair BM. Nutritional, sensory and physicochemical properties of peanut beverage sterilized under two different UHT conditions. Food Chemistry. 1996;**56**:45-53

[73] Wang D, Xu H. Biological activity components and functionality in peanut hull. Food and Nutrition in China. 2008;**12**:23-25

[74] Sobolev VS, Cole RJ. Note on utilization of peanut seed test. Journal of the Science of Food and Agriculture. 2003;**84**:105-111

[75] Kumari CM, Sandhya S, Neetee M. Study of dyeing of cotton fabric using peanut pod natural dyes using Al_2So_4 $CuSO_4$ and $FeSO_4$ mordanting agent. International Journal of Environmental & Agriculture Research. 2017;**3**(2):36-44

[76] Sydjaroen Y, Haubner R, Wurtele G, Hul WE, Erben G, Spiegelhalder B, et al. Isolation and structure elucidation of phenolic antioxidants from tamarind (*Tamarindus indica*) seeds and pericap. Food and Chemical Technology. 2005;**43**:1673-1682

[77] Kumar ANA, Joshi G, Ram HYM. Sandalwood: History, uses, present status and the future. Current Science. 2012;**103**(12):1408-1416

[78] Umar IA, bn nor Muh'd MN, Wong YC. Fastness properties of colorant extracted from tamarind fruits pods to dye cotton and silk fabrics. Journal of Natural Science Research. 2013;**3**(4):60-66

[79] Tepparin S, Sae-be P, Suesat J, Chumrum S, Hongmeng W. Dyeing of cotton, *Bombyx mori* and eri fabrics with the natural dye extracted from tamarind seed. International Journal of Bioscience, Biochemistry and Bioinformatics. 2012;**2**(3):159-163

[80] Mohapatra D, Mishra S, Sutar N. Banana and its by-product utilization: An overview. Journal of Scientific and Industrial Research. 2010;**69**:323-339

[81] Saleh SM, Abd-El-Hady YA, El-Badry K. Eco-friendly dyeing of cotton fabric with natural colorants extracted from banana leaves. International Journal of Textile Science. 2013;**2**(2):21-25

[82] Singh P, Borthakur A, Singh R, Awasthi S, Pal DB, Srivastava P, et al. Utilization of temple floral waste for extraction of valuable products: A close

loop approach towards environmental sustainability and waste management. Pollution. 2017;**3**(1):39-45

[83] Montazer M, Parvinzadeh M. Dyeing of wool with marigold and its properties, fibers and polymers. Fibers and Polymers. 2007;**8**(2):181-185

[84] Teli MD, Javed S, Maruti K. Ecofriendly dyeing and antibacterial finishing of soyabean protein fabric using waste flowers from temples. Textiles and Light Industrial Science and Technology (TLIST). 2013;**2**(2):78-84

Chapter 5

Application of Natural Dyes for Herbal Textiles

Anupama Mishra and Sapna Gautam

Abstract

Clothing, the second skin of our body, is of immense importance in human life and has been evolved continuously. Textile products having special properties are used for umpteen purposes, and one such product is herbal textiles. Changing trends in clothing, cosmetics, and aesthetics necessitates the development of newer technologies that can benefit end users by enhancing their health and aesthetics and provide a sense of relief and satisfaction. Herbal textiles are prepared by imparting essential oils and herbal extract finishes to textiles. These treatments provide medicinal value and aroma to the garment, which enhance its value. Herbal extracts and oils can be incorporated in textiles by microencapsulation, sensory perception technology, liposomes, dyes, and coated textile technology. The temperatures of the finish, the duration and concentration of plant extract, the blend of herbs, and the equipment used need to be carefully controlled. Herbal textiles are revitalizing and help in keeping the body fresh and healthy. These fabrics also shown to have therapeutic value in ailments like skin allergies, breathing problems, sleeping disorders, and blood pressure. In this chapter, emphasis is given on the significance of herbal clothing and the use of natural herbal extracts to produce them.

Keywords: natural dyeing, herbal finishes, herbal extracts, functional finishes, herbal wear

1. Introduction

Textiles are susceptible to micro-organisms and their products (e.g. toxins) known to cause infection and intoxication. Myriad textile products such as herbal textiles and clothing are available in the market to support healthy lifestyle and hygiene. Most textile products are meant to come in contact with the skin; hence, herbally treated fabrics are better choice as they do not harbor harmful chemicals and thus are ecofriendly. People of yore used to live in harmony with nature and passionately protected their environment; however, nowadays, excessive exploitation of natural resources has led to a phenomenal increase in environmental pollution. This is taking a serious toll on the lives of all living beings on the planet earth. Right kind of fabric is of paramount importance to a healthy lifestyle, and synthetic fabrics' teeming with chemicals and dyes poses severe health threat. One would not like to eat a bowl of pesticide-drenched food. Then, why wear clothing doused in chemicals? Textile dyes obtained from natural sources are usually friendly to human skin and are biodegradable and ecofriendly. Some natural dyes have medicinal properties and impart healing qualities to the fabric. They are ecofriendly, save energy, provide rural employment, and preserve traditional craftsmanship [1].

Clothing is our second skin and plays an important role in human life. Textiles finished or dyed from extracts of various medicinal herbs are called herbal textiles. Textiles and clothing infused with medicinal herbs are becoming popular, especially in urban India. The use of such products helps in providing protection against myriad biological toxins and pathogens. Customers demand for new ranges of technology, owing to upward trend in enhancing beauty, that provide better esthetics along with improved health and a sense of relief and satisfaction during use. This trend has produced Cosmetotextiles.

Among technical textiles, medical textiles are a promising product, which plays an important role in maintaining the human health. Medical textiles consist of textiles used in operative and postoperative tasks on and around the patient and medical practitioners. They are broadly classified as nonimplantable materials, implantable materials, extracorporeal devices, hygiene products, protective, and healthcare textiles. Healthcare, disposable, and nondisposable hygiene products mainly used in hospitals to reduce the chances of contamination by biological toxins and infectious organisms.

India has good potential to produce natural dyes commercially as it is considered as 1 of the 17 mega diversity countries globally [2]. Traditional knowledge facilitates the use of plants and their products for healing the wounds and burn injuries and treating the skin infections. Herbal extracts are being used successfully for imparting antimicrobial finish to textiles [3].

Herbal products derived from plants are gaining popularity worldwide for use in textiles on account of their easy availability, green approach, low toxicity, biocompatibility, and ecofriendly nature [2, 3]. Owing to market demand for textile products having added comfort and functional properties, herbal products may be explored for their potential benefits to produce such products. This mandates scientists to devise methods and techniques to use plants containing bioactive agent in producing value added and more attractive functional textile substrates. Insect repellent, deodorizing/aroma, UV protection, antimicrobial, and flame retardant are some of the new properties, which have been imparted recently to textiles. In recent years, the growing demand for herbal products has led to the idea of developing healthcare textile products using herbal extracts.

Currently, herbal colorants are the most attractive option because of high biodegradability, low toxicity, green chemistry, and potential application in the textile dyeing and finishing industry. Natural colorants from plant sources are considered as novel agents in imparting multifunctional properties such as antimicrobial, insect repellent, deodorizing, and UV protection to textiles. Many natural colorants possess some inherent functions in addition to their coloring attribute. These inherent properties are transferred to the textile substrate dyed with an herbal colorant. Alternatively, dyeing textiles with these herbal colorants can combine dyeing with functionality finishes, an ecofriendly process using less water and energy than currently used separate wet treatments. Recently, increased interest in the use of natural dyes in textile dyeing has been observed due to enforcement of the stringent environmental regulations by many countries because of toxic effects of synthetic dyes. This review gives an overview of the herbal methods available at present for dyeing textile materials with aim to provide other useful finishes [4].

2. The herbal textile concept

Textiles dyed exclusively with herbal extracts without using any chemicals are called herbal textiles. The herbs utilized in dyeing are different from vegetable dyes as they also have medicinal values. These herbs are applied to the fabric using natural ingredients in order to preserve medicinal properties. Further, bleaching of cloth is done by exposing it to sunlight without the use of any chemical bleach. The concept of herbal textiles has been derived from Ayurveda, the ancient Indian method of vedic healthcare. Ayurvastra is a branch of Ayurveda. In Sanskrit, *ayur* means health, and *veda* means wisdom, while *vastra* means clothing when it combines the meaning life cloth [5].

It is made by embedding essential oils and herbs into textiles. A wide range of herbs such as turmeric, tulsi, neem, rose, Indigo, lemon grass, castor, and ber can be used for this process providing medicinal value and aroma to the fabric. Ayurvastra (marketed as cosmetotextiles) has been used for curing several diseases such as diabetes, arthritis, skin infections, hypertension, and hay fever [5].

Herbal clothing is an ancient technique of dyeing textiles in medicinal herbs. This alternative method of treatment originated more than 5000 years ago in India. Its core principles are rooted in the Vedas, which are an ancient book of Hindus. The Vedas date back to around 3000 BC. The science of life or Ayurveda dates back to around 1000 BC. The origins of Ayurveda are also found in the Atharvaveda. It contains incantations and hymns that cure various diseases through mantra. Mythology says that Ayurveda was given to Dhanvantri by Lord Brahma. Charaka Samhita and Sushruta Samhita are the oldest known Ayurvedic texts [6]. About 100 years ago, people from different parts of India were using various forms of natural dyeing in which they repeatedly dipped clothes in herb-based preparations following each wash [7]. These herbs are used traditionally in Ayurvedic treatment and many medical applications. When this fabric is worn, the medicinal property of the fabric is then transferred to the skin. The process of herbal dyeing was developed through extensive research during the age-old dyeing methods practiced since the days of the Indus civilization. Herbs were used to create different colors and to provide various health benefits. The natural/herbal fabrics are 100% organic, completely free of synthetic chemicals, and toxic irritants and are biodegradable [8].

Herbal clothing is prepared from organic cotton fabric impregnated with special herbs and oils for health benefits. Since ancient times, herbal clothing (Ayurvastra) is believed to cure diabetes, skin infections, psoriasis, hypertension, asthma, arthritis, rheumatism, high blood pressure, eczema, and cancer. Antimicrobial property or natural dyes such as indigo, pomegranate, myrobalan, and Indian madder of against bacteria such as *Escherichia coli* and *Klebsiella pneumoniae* have been reported [9]. Herbs such as harad, rakta chandana, shweta chandana, and kumkuma are used in Ayurveda for protecting the skin from harmful rays of sun. The herbal preparations act on melanocytes present in the skin, modulate their function, and block the effect of UV rays on the skin. References in different texts of Ayurveda are replete with examples of wearing pure cotton clothes impregnated with herbs such as chandana, haridra usheera, manjista, and nimbi aragwadha to safeguard human body from inclement weather and to keep it healthy. They used herbal products such as soap nut and lime for washing cloths before soaps and detergents were discovered. Herbal textiles and dyes have great vista of applications in diverse walks of human life. These herbal dyes have been used for centuries to produce colors for fabrics, yarns, leather, food, and so on. Natural dyes exhibit better biodegradation and have a better compatibility with the environment. The application of natural dyes has the potential to earn carbon credit by consumption of fossil fuel-based synthetic dye [10].

3. Mode of action of herbal textiles on the body/herbal couture technology

Skin is the largest sense organ in the human body (**Figure 1**). It can act as a barrier as well as a medium for entry of certain substances in the body. Several

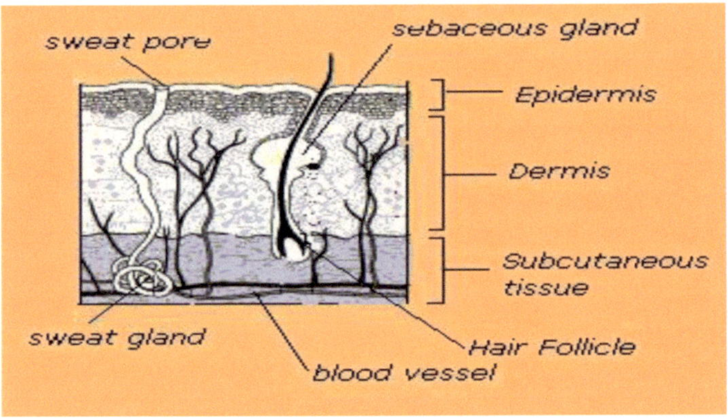

Figure 1.
Structure of the human skin [6].

toxins and chemicals present in the environment get accumulated in the conventional clothing and may enter into the body through the skin. Hence, enhancing skin's ability to resist entry of harmful chemicals and toxins into the body will improve the health. The radiated heat activates herbal molecules of the cloth and is reflected back to the body along with the herbal molecules. Through the blood, the herbal molecules enter deep into the body, show its effects at different levels of the body, and help in curing various diseases [5]. The skin can act not only as a protective barrier but also as a medium for outside substances to enter into the body. In the same way, the skin has the ability to absorb herbs found in our natural dyes. These herbs release their medicinal qualities in the body and strengthen the skin's ability to block and resist the harmful substances. The close contact of herbal clothing next to the skin enhances the body metabolism, which leads to effective elimination of body toxins. The herb-infused and herb-dyed organic fabrics act as healing agents after the herbs being absorbed through the skin [6].

3.1 Penetration of herbal molecules across skin

- Skin temperature helps the herbal molecules to be absorbed into the body through sweat pores to manifest the desired medicinal effect.

- The cloth should be in direct contact with the skin to manifest its curative effects.

3.2 Herbs and healing effects of herbal textile

Herbs are garden plants that are grown and harvested for culinary, aromatic, medicinal, and fibrous uses. Plant herbs are placed in the garden for their unmistakable fragrances, attractive textures, appealing colors, and variety of home uses. Due to current ecoconsciousness, there has been a revival of inherent in ecofriendly natural dyes throughout the world. The application of herbal products has given a new direction toward the treatments of various diseases through textile products. The herbs also do not pollute the environment through contamination of water resources in areas close to processing units. All kinds of shades of red, yellow, brown, orange, and green can be obtained with the help of the herbs.

Functional finishes using natural dyes have been in the vogue that promotes an ecofriendly lifestyle. Natural dyes are environmental friendly, nontoxic, noncarcinogenic, nonallergic, and renewable resources. Biosphere is being gifted with more than 500 plant species that yield the natural dyes [11]. The roots, flowers, leaves, seeds, and barks of more than 200 medicinal herbs, plants, flowers, roots, and barks are used to make the dyes. In herbal textiles, the color is gained from the medicinal preparation only, and no other colorants are used. As a result, its properties will last as long as the color is there. Since the herbs make beautiful shades, herbal wears are also becoming very popular for its primordial elegance.

Only herbal preparations devoid of any chemical are used in dyeing herbal textiles, whereas chemicals such as copper sulfate and ferrous sulfate are used as catalysts when dyeing is done with vegetable dyes. Thus, herbal dyes are different from vegetable dyes as they also impart some medicinal value. Herbal clothing is believed to help restore the balance within the body's systems and strengthen the immune system. Some of the medi-herbs used in herbal textiles give the magical healing quality to the dyed fabric or yarn (**Table 1**) [12–15]. **Table 1** shows the diseases cured by different herbs.

S. no.	Name of the herb	Medicinal properties	Diseases claimed to be cured
1.	**Neem** (*Azadirachta indica*)	Antibacterial and antifungal	Common skin diseases, skin allergy, controls blood sugar levels
2.	**Curry leaves** (*Murraya koenigii*)	Medicinal value	Beneficial in curing arthritis suppresses pain and reduces inflammation
3.	**Shoe flower** (*Hibiscus rosa-sinensis*)	Controls blood sugars	Diabetes
4.	**Fenugreek/ Methi** (*Trigonella foenum-graecum*)	Medicinal value	Controls cholesterol, hypertension
5.	**Amla** (*Emblica officinalis*)	Antifungal, antibacterial, antiviral	Boosts immunity
6.	**Acacia** (*Catechu*)	Antifungal, antibacterial	Used for treatment of parasitic infestation and itching
7.	**Manjistha** (*Rubia cordifolia*)	Medicinal value	Blood purifier and diuretic, leprosy
8.	**Cuscus/Khus khus grass** (*Chrysopogon zizanioides*)	Medicinal value	It helps in fighting asthma
9.	**Arjuna** (*Terminalia arjuna*)	Medicinal value	Cardiovascular support, hypertension, asthma
10.	Pomegranate (*Punica granatum*)	Antibacterial, antiviral, astringent	Antimicrobial treatment, hypertension
11.	**Harad/myrobalan** (*Terminalia chebula*)	Antiseptic, anti-inflammatory, controls cholesterol	Hypertension, anemia, asthma cures, wounds, ulcers
12.	**Aloe vera** (*Aloe barbadensis*)	Antibacterial, antifungal	Burns, wounds, common skin diseases, psoriasis, anemia
13.	**Gurhal** (*Hibiscus rosa-sinensis*)	Controls blood sugar levels	Diabetes
14.	**Jamun** (*Syzygium cumini*)	Controls blood sugar levels	Diabetes controls blood sugar levels

S. no.	Name of the herb	Medicinal properties	Diseases claimed to be cured
15.	**Mahuva** (*Madhuca longifolia*)	Controls blood sugar levels	Control diabetes
16.	**Champa flower** (*Magnolia champaca*)	Controls blood sugar levels	Diabetes, anemia
17.	**Touch-me-not** (*Mimosa pudica*),	Medicinal value	Controls blood sugar levels, diabetes
18.	**Eucalyptus** (*Eucalyptus globules*)	Antiseptic	Relieve congestion, prevent infection, ease muscle soreness
19.	**Tulsi/basil** (*Ocimum tenuiflorum*)	Antibacterial, antiseptic, antispasmodic, diaphoretic, febrifuge, nervine, controls cholesterol	Respiratory disorders, viral, bacterial infections. Controls blood sugar levels, boosts immunity, hypertension, HIV/AIDS
20.	Chitosan	Antimicrobial	Naturally occurring biopolymer having antimicrobial properties
21.	Balloon vine (*Cardiospermum halicacabum*)	Antiphlogistic, analgesic, blood refrigerant, anti-infectious	Cold, fever, renal edema, urinary tract infections,·furuncle, carbuncle, eczema, sprains, external wounds
22.	**Sandalwood** (*Santalum album*)	Alterative, antibacterial, anti-inflammatory, antiseptic, astringent, carminative, disinfectant, diuretic, expectorant, hemostatic, refrigerant, sedative, stimulant	Headache, acute dermatitis, bronchitis, cystitis, eye diseases, gonorrhea, herpes zoster, infection, palpitations, sunstroke, urethritis cools the skin, curing skin itching, burns, its mild fragrance has a soothing effect that helps in fighting stress
23.	**Onion** (*Allium cepa*)	Antioxidant, antimicrobial	Skin diseases, control diabetes
24.	**Indian Madder** (*Rubia cordifolia*)	Antibacterial, anti-inflammatory, antiseptic, astringent, sedative, stimulant	It removes blood impurities and cures various skin diseases, HIV/AIDS
25.	**Indian Mulberry** (*Morinda pubescens*)	Antibacterial, anti-inflammatory	It is blood purifier and used for wound healing, control diabetes
26.	**Saffron** (*Crocus sativus*)	Medicinal value	Controls heart disease
27.	**Indigo** (*Indigofera tinctoria*)	Antibacterial, antiallergy, antiseptic and aromatic, anti-inflammatory properties	Controls cholesterol, helps reducing the insulin resistance, thus controls diabetes, boosts immunity, anemia, skin diseases
28.	**Turmeric** (*Curcuma longa*)	Antihepatotoxic, antioxidant, antimicrobial antiseptic, antiallergic	Fighting skin diseases and helps fight skin disease, anemia, psoriasis, liver disorders
29.	**Henna** (*Lawsonia inermis*)	Blood purifier, anti-irritant, deodorant, antiseptic	Skin irritations such as heat rashes and skin allergies and to cool the body. Because of this cooling property, henna used as a prophylactic against skin diseases like burns, bruises, and skin inflammations, including sores from leprosy

S. no.	Name of the herb	Medicinal properties	Diseases claimed to be cured
30.	**Brazil wood** (*Caesalpinia echinata*)	Blood purifier, anti-irritant	Skin diseases such as burns, bruises, and skin inflammations
31.	**Castor** (*Ricinus communis*)	Blood purifier	Skin diseases
32.	**Baheda** (*Terminalia bellirica*)	Controls cholesterol	Hypertension, anemia
33.	**Guar** (*Cyamopsis tetragonoloba*)	Controls cholesterol	Hypertension
34.	**Safflower** (*Carthamus tinctorius*)	Controls cholesterol	Hypertension, HIV/AIDS
35.	**Reetha** (*Sapindus trifoliatus*)	Medicinal value	Psoriasis, skin disease
36.	**Amaltas** (*Cassia fistula*)	Therapeutic effect	Psoriasis
37.	**Cluster fig** (*Ficus racemosa*)	Therapeutic effect	Psoriasis
38.	**Winter cherry** (*Withania somnifera*)	Therapeutic effect	Psoriasis, rheumatism

Table 1.
Some of the medi-herbs used for herbal textiles.

Different parts of the plants namely root, bark, leaf, fruit, wood, seed, and flower are used to extract dyes. It is estimated that over 2000 pigments are found in various parts of plants; however, only about 150 have been commercially utilized [16]. Herbal cloth production by a particular method is accomplished by utilizing extracts of selected herbs for dyeing cotton/silk/linen, yarn, and fabric. Synthetic/chemical dyes are not used at any stage, and for washing, specific products are used [17].

The process of herbal dyeing was developed through extensive research during the age-old dyeing methods practiced since the days of Indus civilization. Manufacturing herbal textile begins with 100% hand-loomed organic cotton or silk wool, linen, jute, hemp, and so on and their natural blends that are dyed exclusively in herbal infusions for both color and health purposes. No chemical process is involved while dyeing and processing the fabric. The fabrics have to pass through various stages of treatment before they are turned into finished products. Herbal textiles are extra smooth and good for transpiration that helps in recovering from various diseases. Each fabric is infused with specific herbs, which are known to cure diseases. It may help to treat a broad range of diseases and skin infections such as eczema and psoriasis; the fabric also helps with conditions such as heart ailments, rheumatism, arthritis, blood pressure, diabetes, and respiratory conditions, such as asthma and sleeping disorders. It also helps in boosting immunity. The antibacterial and anti-inflammatory properties of herbal medicated clothing extend its use as dressings and bandages. It can also relieve the general body aches, stimulate the weight loss, and strengthen the immune system. It can also be used as energy booster; mood enhancer, for overall wellbeing, calming, and blood purification; and cooling [18]. The most effective time to wear herbal cured clothing is when the body is at rest, such as during sleep, relaxation, or meditation because this is when the body is naturally healing and re-establishing balance. Hence, most of herbal clothing products are sleepwear, bed sheets, towels, meditation clothes, and cotton mats [19].

4. Herbal dyeing process

The herbal textile production begins with pure organic yarn/fabric, and each step is carefully and meticulously controlled without the use of any chemical process in dyeing and processing. The process starts with the gray cloth going through various stages of treatment before it is dyed and becomes ready to wear. Only certified organic cotton, natural cotton, silk wool, linen, jute, hemp, and their natural blends should be used for herbal dyeing. The herbal dyeing process has been developed following the extensive research and is in practice since the Indus civilization. Machines are not used to prepare herbal clothing, and all the processes namely bleaching, dyeing, and other processes are accomplished in big vessels fabricated according to need. Further, no chemicals are added during spinning, weaving, or enhancing the appearance [20].

Most common fabric used is organic cotton, the cotton that is cultivated without chemical pesticides and fertilizers. Silk, wool, coir, linen, hemp banana, nettle, and bamboo are also used to prepare herbal clothing. For handloom, the process starts with the cotton yarn and goes through a process named desizing (to remove loose particles, debris, etc.). The common processes used in manufacturing of herbal clothing are detailed in the following paragraphs.

4.1 Desizing

It involves washing of the processed gray cloth with mineral-rich water and sea salt to remove sizing, gums, and oils used in the weaving.

4.2 Bleaching

This process is actualized with biodegradable, organic cleaning agents, and surfactants and sun dried. The bleaching of cloth is done naturally by exposing it to sunlight on a natural grass base and by using animal manure. The fabric or yarn is first bleached using cow urine, milk, honey, along with biodegradable, naturally derived, organic cleaning agents, and surfactants such as Saptala (*Acacia sinuata*) and Phenila (*Sapindus mukorossi*) [21]. Many Ayurvedic doctors consider cow's urine highly beneficial in balancing an individual's "doshas" or basic constituents of an individual's physiology and psychology, strengthening the immune systems and as an elixir in giving life [22]. The process ends with exposing the fabric or yarn directly to sunlight.

4.3 Mordanting

To make the colors look bright and retain color fastness, natural mordants such as myroballans, rubhabs leaves, oils, alum, fruit extracts of haritaki, bark of lodhra, minerals, and iron are used. Use of heavy metallic mordants such as copper, chrome, zinc, and tin is avoided as they are not ecofriendly. Mordanting is done before dyeing the fabric or yarn to make the colors look bright and retain fastness.

4.4 Dyeing/medication

The word medication is used instead of dyeing because the medicinal herbs are used to impart required color to fabric or yarn [20, 23]. The yarn or fabric is then medicated in a carefully controlled mixture of herbal medicinal preparations depending on the disease or ailment being treated. Required herbs are applied directly to the fabric with the help of natural ingredients so as to keep the medicinal value of herbs intact [7, 11]. Steps involved are as follows:

- The organic cotton yarn/fabric is dyed carefully with concoction of herbal dyes indicated for the treatment of desired disease(s).

- The herbal dyes are prepared by careful blending of medicinal herbs derived from plants, flowers, roots, and barks.

- The temperature, duration, number of the soaks, blend of herbs, and equipment are carefully controlled.

- Desired herbs with the help of natural ingredients are applied directly on fabric in order to preserve the medicinal value of herbs.

- Shades of red, yellow, brown, orange, and green can be obtained with careful use of herbs.

- Finally, the medicated cloth is cooled and washed several times to remove the nonabsorbed particles and always dried in shade.

4.5 Finishing

The finishing process used after herbal dyeing is also organic and is given by sprinkling pure water on the cloth and then stretching under pressure, using hand rolls, *Aloe vera*, castor oil, and so on [7, 11].

4.6 Recycling residue

After dyeing, residue can be recycled. Solid and liquid wastes are separated by the filtration and used for farming purposes—as biomanure and for watering the fields and also to generate the biogas [11].

5. Technologies used for herbal finishing

Herbal textile is made by embedding essential oils and herbal extract finishes into textiles. Such textiles have their applications in providing medicinal value and aroma to the garment. Microencapsulation, sensory perception technology, liposomes, dyes, and coated textile technology are used for incorporating the herbal drugs and oils in textiles. The temperatures of the finish, the duration and concentration of plant extract, the blend of herbs, and even the equipment used are carefully controlled [24].

Spray drying, air suspension coating, solvent evaporation, sol-gel microencapsulation, and in-situ polymerization are some commonly used techniques used for textile finishing by microencapsualtion. For encapsulation of flavor, fragrance, and oils, spray drying is used in which core particles dispersed in a polymer solution and sprayed into a hot chamber leading to solidification of shell particles on core material and polynuclear or matrix-type microcapsules are obtained. Air suspension coating is done by suspending solid particles of core material in a hot or cold upward moving steam of air. Coating material may be derived from cellulose derivatives, dextrins, emulsifiers, lipids, protein derivatives, and starch derivatives, which is atomized through nozzles into the chamber and deposits as a thin layer on the surface of suspended particles. Solvent evaporation is a process in which the drug is dissolved, dispersed, or emulsified in an external aqueous or oil phase; it has been used to impart antifungal finish to textiles. Sol-gel microencapsulation is used to produce flame retardant, water/oil repellent, UV protecting, antiwrinkle, self-cleaning, flavor and odor containing, and antibacterial textiles. Sol-gel synthesis involves the hydrolysis of silicone monomer and the condensation of silica into a porous structure with a three-dimensional networked structure, which can be tailored for a wide range of useful properties.

Microencapsulation, widely used in textile finishing, is a process in which small capsules having desired properties are prepared using small particles/droplets surrounded by a coating material. The material inside the microcapsule is known as core material, whereas outer coating is called a shell. Microcapsules may have diameter ranging a few micrometers to a few millimeters. Many special and functional properties can be imparted to the fabrics by microencapsulating the core material. This core material can be any substance having a special function to perform for the fabric. Encapsulation has allowed moisturizers, therapeutic oils, and insecticides to be incorporated into fabrics. Microencapsulation of antimicrobial agents is also gaining popularity in sportswear and medical textiles [25].

For preparation of herbal textile by microencapsulation, the "core material" along with the wall material should give a durable finish. This wall material can be certain substances like gum acacia, and the core material is the desired herbal extract. The microencapsulated herbal extracts have shown antimicrobial activity and wash durability up to 20 washes [26]. Microencapsulation is a very popular technique for incorporating fragrance into fabrics. These capsules applied are unbreakable under the normal conditions, and the application of pressure on fabric releases fragrance, which causes healing in aromatherapy [27]. Microencapsulated herbal extracts such as *Aloe barbadensis* Mill, Bitter Gourd, Ginger, and *Cuminum cyminum* Linn have successful application in which the herbal extracts are used as core material and gum acacia as wall material [28]. According to the study, microencapsulated extracts of *Andrographis paniculata*, when applied on the fabric to impart mosquito repellency finish, gave good mosquito repellent activity up to 30 washes [29].

5.1 Microencapsulation in cosmetic textiles

New textile technologies have enabled the application of cosmetic ingredients on fabric to provide its functional benefit to the end-use product, and therefore, cosmetic textiles are moving from laboratory to showrooms. Fragrance finish is one such finish that falls under this category. A fragrance is made from a pleasant smelling aroma compound. Aromachology is a science that studies the effects of fragrances on the human body and mind. It researches how scents can be used to induce relaxation and make life more pleasant. Fragrances can be used in various forms such as essential oils, perfumes, colognes, household products, and potpourri. Some areas in textiles where fragrance finish can find application include Cosmetotextiles.

The major use of microencapsulation in cosmetic textiles is in the application of vitamins, essential oils, skin moisturizing agents, skin cooling, and antiaging agents [30]. Yamato et al. studied microcapsules having active substances that can improve the physiological conditions of human skin. The microcapsule gradually released its content when the textile structure was subjected to light pressure created due to the movement of human body [31].

Aroma finish is a process by which textile material is treated with the pleasant odor producing essential oils and aromatic compounds so that the wearer gets beneficial effects. Various essential oils such as lavender, rosemary, and jasmine are used.

The fragrance compounds and the essential oils are volatile substances, and prolonging life span of their odor is most difficult task in preparation of the textiles for aromatherapy. Microencapsulation is an effective technique to achieve this task. Microcapsules are tiny spherical enclosures containing a liquid or gas and assume the shape of the enclosed particle if containing a solid [32]. With microencapsulation method, fragrance lasts for more number of washes since microencapsulation makes the finished particles more securely attached to the internal structure of the fiber and thus releasing them slowly [33].

5.2 Microencapsules for antimicrobial finishes

Bacteria often cause decay of fabrics leading to loss of various useful properties of fabrics. This problem can be prevented by the use of antimicrobial finishes that can be applied with the help of microencapsulation. This finish is especially beneficial for textiles for medical and technical uses [34].

Herbal extracts prepared from Chamomile, sage, and green tea were applied to cotton fabric using pad-dry-cure method. Treated fabrics exhibited antimicrobial activity against Gram-negative, Gram-positive bacteria and yeasts. This antimicrobial activity was retained even after 10 cyclic washes [35].

Textile materials having antimicrobial activity are used in hygienic, health, personal care, and surgical products. Some products include baby and adult diapers and sanitary pads. Most of the diapers are made by synthetic materials, which are nonbiodegradable and poor antimicrobial properties, and may cause rashes and dermatitis. Natural antimicrobial coatings on diapers obtained by using curcumin, *Aloe vera*, tulsi, and neem have same antimicrobial effect when compared with synthetic antimicrobial coatings [36]. Curcumin or diferuloylmethane (1,7-bis(4-hydroxy-3-methoxyphenyl)-1,6-heptadiene-3,5-dione) is the main phytochemical present in turmeric. This polyphenolic compound has several biological activities including antimicrobial activity. Curcumin is suspected to affect the bacterial cytoskeleton, thus producing the antimicrobial effect [37]. The *Aloe vera* leaves secrete a colorless gel-like substance, which contains about 98–99% water and 1–2% active compounds such as aloesin, aloin, aloe emodin, aloe mannan, flavonoids, saponin, sterols, amino acids, and vitamins. Anthraquinones and their derivatives such as barbaloin-IO-aloe emodin-9 anthrone, isobarbaloin, and chromones, which inhibit the bacterial protein synthesis, are potent antimicrobial agents [38]. Antimicrobial property of tulsi (holy basil) may be attributed to essential oils found in leaves. These essential oils contain germacrene-A, clemene, caryophyllene, eugenol, and caryophylline oxide. Additionally, tulsi leaves also have other biologically active phytochemicals such as rosmarinic acid, oleanolic acid, and ursolic acid. These essential oils and phytochemicals have antibacterial properties owing to their damaging effect of bacterial cell membrane and thus causing leakage of cellular potassium [39]. Neem leaf, bark, and seeds have antimicrobial properties, and it is high, moderate, and low, respectively. Neem contains several bioactive constituents like azadirachtin and other compounds such as nimbolinin, nimbin, nimbidin, nimbidol, sodium nimbinate, gedunin, salannin, quercetin, nimbanene, 6-desacetylnimbinene, nimbandiol, nimbolide, ascorbic acid, n-hexacosanol and amino acid, 7-desacetyl-7-benzoylazadiradione, 7-desacetyl-7-benzoylgedunin, 17-hydroxyazadiradione gedunin, and nimbiol. Quercetin and β-sitosterol, polyphenolic flavonoids purified from leaves of neem have antifungal and antibacterial activities. Neem (*Azadirachta indica*) in plant parts shows antimicrobial property through bacterial cell wall breakdown [40].

6. Herbal finishes by natural dyeing

Wet processing of textile substrates from its preparatory stage to dyeing and then finishing is important for enhancing its esthetic value, removal of impurities, color shade, color pattern, and requisite functionality. Some of the traditional processes used in herbal finishing involve an excessive use of water, energy, and chemicals. Recently, due to global concerns on environmental pollution, sustainability mandates the development of sustainable dyeing and finishing processes using plant waste and nonfood plant extracts [41]. Based on environmental friendly, plant-based products having biocompatibility, biodegradability, and nontoxicity along with insect repellent, deodorizing, flame retardant, UV protection, and antimicrobial activity are in great demand all around the world for producing more appealing and highly functional value-added textiles [42–45]. A wide variety of finishing chemicals from plants are now available in the market that matches the expectations of consumers [46]. Various plants reported as source of natural dyes are teak, mahogany, ketapang, tamarind, mangosteen, mango, suji, pandan, indigofera, guava, banana, and onion [47]. Plant parts including roots, leaves, twigs, stems, heartwood, bark, wood shavings, flowers, fruits, rinds, hulls, husks, and the like used to produce herbal dye. Additionally, most of the herbal dyes have inherent medicinal properties [48].

6.1 Herbal textiles having antimicrobial finish

Natural fibers such as cotton and wool are susceptible to microbial growth and even dust mites because they retain oxygen, water, and nutrients. Hence, antimicrobial finishes should be applied to textiles to destroy or suppress the growth of microorganisms and also to protect the textiles from strength and color loss, unpleasant odor, and quality deterioration [49]. Micro-organisms may deteriorate the clothes in a closet, curtains, carpets, bed, bath and kitchen linens, pillows, and mattresses. Several microorganisms also thrive on the skin, while dust mites live on skin cells sheded on sheets, towels, and clothing. A hospital houses an immense amount of textiles with high volumes of traffic. Because of the constant flow of people, especially those with infectious diseases, specific finish hospital uses are required. Patients, caretakers, and hospital staff are at risk of acquiring infection as inherent properties of the textile fibers susceptible to the growth of micro-organisms.

In past, natural dyes were applied to textiles for simultaneous coloration and antimicrobial finishing successfully. Finish imparted by catechu on wool was found to be effective against *Escherichia coli*, *Staphylococcus aureus*, *Candida albicans*, and *Candida tropicalis*. Observed antimicrobial characteristics and negligible cytotoxicity of catechu indicated the dye as a promising antimicrobial agent for developing bioactive textile materials and herbal clothing [50, 51]. Several natural, nonmetallic, and antimicrobial finishes exist, and one of such antimicrobial finish is chitosan, a deacetylated form of chitin, which is a main component in crustacean shells and reported to be effective against both Gram-positive and Gram-negative bacteria [52]. Many antimicrobial textiles are produced with the combinations of bioactive substances to enhance the antimicrobial efficacy of the finishes and counteract the negative aspects of the treatments. By combining finishes, the occurrence of drug-resistant strains arising from the finish is reduced.

Pure cotton woven/knitted fabrics treated with the selected medicinal 16 medicinal herb extracts such as neem, turmeric, holy basil, and sandal wood have been evaluated for curing seven different diseases such as allergic dermatitis, psoriasis, asthma, liver disorders, headache, joint pain, and sinus trouble/cold. Seven different types of curative garments have been developed. The antibacterial assessments of the medicinal herb extracts treated fabrics and clinical trials have confirmed the correlation between the curative performance and its antibacterial activity. The curative property of the garments in all the cases was found to be significant and lasted for 10–15 washes [53].

Although known since long for dyeing and medicinal value, the protective properties of herbal dyes have been noticed only recently. Several plants used for dye production are classified as medicinal, and some have recently been shown to possess significant antimicrobial activity. Several natural dyes have demonstrated to possess antimicrobial activity like curcumin from turmeric; naphthoquinones such as lawsone from *Lawsonia inermis*, juglone from walnut, lapachol from taigu, and catechin from *Acacia catechu*; and several anthraquinones from *Rubia tinctorum*, *Rubia cordifolia*, *Rheum emodi*, *Punica granatum*, and *Quercus infectoria* [54, 55]. These natural antimicrobial agents may significantly reduce the risk of infections especially when they are used in close contact. Natural bioactive compounds present in natural dyes/pigments have also been reported as significant antimicrobial agents for finishing in green dyeing procedures.

6.2 Herbal textiles with UV protection

Ultraviolet (UV) rays, electromagnetic spectrum between visible light and X-rays, influence the physiology of living organisms. Exposure may cause tanning to skin cancers. Sunscreen lotions and clothing provide protection from dangers of UV rays. Change in construction parameters of fabrics with appropriate UV absorbers and adequate finishing methods may be used to prepare the UV protective fabrics.

Textiles, as a protective shield against solar radiation, have been since ancient times. Textile structures provide the desired characteristics such as pliability, good mechanical strength, softness, esthetics, and other engineered properties, which are required for preparing good sunscreening apparels. Textiles themselves are not able to provide effective protection against UV rays; hence, UV blocking agents should be applied over the fabric to make them UV protective. Quantum of protection required for different skin types is determined by UV radiation intensity and distribution subject to geographical location, time of day, and season. Several UV protection agents have been developed to enhance the UV protection offered by different textiles. Both organic and inorganic UV blocking agents are available. The organic blockers are also termed as UV absorbers because they absorb UV rays, whereas inorganic blockers scatter them [47].

Three yellow dyes obtained from *Rheum emodi*, *Gardenia yellow,* and *curcumin* have been successfully used for simultaneous dyeing and functionalization of silk to render its UV protective [56]. Wool fabric dyed with dye obtained from the eucalyptus leaves using pad-batch and pad-dry techniques showed ultraviolet protection factor (UPF) values ranging between very good and excellent [57]. UV protection properties of chlorogenic acid, the main ingredient of water extract from honeysuckle, on wool have been studied. The wool treated with honeysuckle extract showed good UV absorbance; thus, extract of honeysuckle may be developed as a natural UV-absorbing agent [58]. Natural plant colorants such as madder (*Rubia tinctorum*) and indigo (*Indigofera tinctoria*) and of insect origin cochineal (*Dactylopius coccus*) were applied on cotton fabrics, and indigo was found to have higher UPF values [59]. UV rays induce DNA damage, oxidative stress, and inflammatory processes in skin. *Rheum emodi* contains anthraquinones (rhein, emodin, aloe-emodin, chrysophanol, physcion, and rhein), stilbenoids (resveratrol and piceatannol), and flavonoids. UV protective effect of *Rheum emodi* is supposed to be due to flavonoids, which absorb the UV radiation [60]. Main coloring ingredient of the dye extracted from *Gardenia yellow* is crocin, which is a water soluble carotene, which absorbs the UV rays [61]. Curcumin alleviates the effect of UV rays by reducing the transmittance of UV light across fabric [62]. Madder roots contain anthraquinone derivatives mainly purpurin and munjistin and also small amounts of rubiadin, nordamncanthal, pseudopurpurin, and xanthopurpurin. Anthraquinones are considered to be strong UV absorbers [22].

6.3 Deodorizing herbal textiles

Bacterial growth and/or waste released from human body are the main causes for odor in garments. Recent advances in textile manufacturing have improved the performance of textiles with respect to odor with antimicrobial and UV protection properties. To meet the consumer's growing demand for hygienic clothing, extensive research has been done to deodorizing property to textiles with the application of natural colorants. The deodorizing performance of fabrics dyed with herbal colorants has been compared, and pomegranate was found to be best among gardenia, *Cassia tora* L., coffee sludge, and pomegranate rind [23]. Fabric dyed with gallnut recorded a better deodorizing function against ammonia, trimethyl amine, and acetaldehyde than control fabrics. Dyed fabrics also found to possess an excellent antimicrobial activity against *Staphylococcus aureus* and *Klebsiella pneumoniae* [20]. Young-Hee Lee and co-workers dyed cotton, silk, and wool fabrics with pomegranate (*Punica granatum*) and showed excellent deodorizing property in the dyed fabrics [21]. Cotton fabrics dyed with C.I. Direct Blue 200, a copper complex direct dye, and pre- and post-mordanted with Cu(II) sulfate for deodorization of ethyl mercaptan showed that deodorization effects increased quadratically with the copper ion uptake [63]. These studies indicate that natural and synthetic dyes can be utilized for deodorizing functionalization of textiles.

6.4 Moth resistant and insect repellent textiles

Carpets, blankets, and shawls are prepared from wool, and other similar fibers have properties such as warmth, softness, and flame retardancy protein content of the fiber, which are susceptible to attack by moth as its larvae thrive on the protein present in the wool. Cloth moth (*Tineola bisselliella*) and carpet beetle (*Anthrenus verbasci*) are common moths, which attack the wool items. Dichlorodiphenyltrichloroethane (DDT), permethrin, permethrin/hexahydro pyrimidine derivative, and cyhalothrin are some of the insecticides used to provide antimoth finish. Nano titanium oxide particles have also been utilized as an antifeeding compound on wool fabric to kill larvae of the carpet beetle, *Anthrenus verbasci*, which feeds on wool fibers [64]. Chemicals used for antimoth finishing are detrimental to the environment; hence, natural colorants may be good alternatives. Shakyawar et al. used saffron flower waste, onion skin, henna, myrobolan, silver oak leaf, madder, walnut, dholkanali, and yellow root natural dyes for antimoth finishing and recorded the best results for silver oak leaves, walnut husk, and pomegranate rind [65]. Natural dyes from cochineal, madder, walnut (quinines), chestnut, fustic, indigo, and logwood (flavonoids) were also applied on wool, and their antimoth properties against black carpet beetles were tested, all except indigo, which increased the insect resistance of the wool fabric. Metallic mordants were found to have no significant effect on insect resistance with all natural dyes used. The anthraquinone dyes including cochineal, madder, and walnut were found to be quite effective in protecting wool fabric against black carpet beetles [66].

7. Production and consumption of herbal textiles

Synthetic/chemical dyes are increasingly being banned for use in the textile industry, which have led to severe blow to the handloom industry. At the same time, textiles dyed with natural herbal dyes having medicinal attributes are commanding a huge market due to their inherent advantages. Nowadays, the use of herbal dyes in the textile industry is preferred owing to the advantage of dyes extracted from the medicinal plants, origin from the renewable resources, limited chemical reactions involved in their preparation, biodegradable properties, disease-curing properties, and ecofriendly in nature [18, 67].

Herbal textiles are of two kinds: one category is for curing diseases and the other is for wellbeing. The first category of cloths, which are used for curing diseases, includes sleep wear, bed sheets, towels, meditation clothes, and coir mats. Herbal textiles are mostly used in making such garments that stay close to human skin, so that all the medicinal components could be absorbed through the skin. Herbal textiles can fetch more competitive than conventional fabrics owing to low production cost. For example, a common chemical dye costs around $75 per kg, and a herbal dye costs approximately $5 per kg. Most of the herbs used in dyeing are cultivated in South Asian countries, namely India, Pakistan, and Bangladesh, India being the largest producer. The Handloom Weavers Development Society (HLWDS) of Kerala, India has exported herbal and organic textile worth Rs. 50 Lakh to the United States, Europe, and Japan. The herbal textile is gaining such popularity that the Japanese Government has accorded HLWDS a $40,218 grant. Large consignments of herbal textiles have been shipped to international markets of the United States, Canada, France, Denmark, Italy, Poland, Maldives, Mauritius, Japan, and Sri Lanka [17, 68–70].

Most effective time to wear the herbal-infused clothing is while resting, sleeping, or meditating, when the body is naturally healing and re-establishing balance, so many of the products are created with these factors in mind. The technique for producing the herbal textiles is also used for floor coverings and coir mats. For coir mats, the fibers are first soaked in herbal dyes and then woven into coir mats. Apart from medicated handloom clothes such as sarees, T-shirts, shirts, trousers, kurtas, dhods, chudidhars, nightwear, fashion wear, sportswear, and inner wears, caps for healing headaches, goosery caps for head balance, cooling caps, hair shining caps, hair strengthening caps, bandages, and mask are also prepared [68, 71].

Plant dyes are increasingly being incorporated by designers into their designs contributing to revival of Ayurvastra or herbal clothing. Ayurganic is a line of loungewear inspired from the concept of Ayurvastra. Designer duo Lecoanet and Hemant have revived the tradition of Ayurvastra in Ayurganic, their line of medicinal clothing. The multi-award winning designers who began their journey as couturiers in Paris are now refining the concept of Ayurvastra. Anjelika dreams organic produces fair trade, organic handmade clothing that follows this tradition. Gibie towels specialize in Ayurvastra towels, Ayurvastram in Ayurvastra textiles and yoga mats, and niraamaya in Ayurvastra yoga mats. The handloom weavers' development society in India is producing a wide range of herbal fabrics including sarees, bed sheets, and dress materials using medicinal dyes and exporting them to the United States, Germany, the United Kingdom, Singapore, and Taiwan [6, 11].

"Herbalfab" ecofriendly fabric brand has developed a unique technique to dye with flowers, fruits, roots, and so on. Turmeric, myrabolams, onion, kesu and dhavadi flowers, and natural indigo are used for preparing the natural dyes. This prevents the water pollution by replacing the petrochemical dyes with herbal dyes and imparts the medicinal value to cloth. The colors obtained are unique, which can kick start a new fashion trend. Herbalfab also encompasses a range of other ecofriendly fabrics like khadi, which is handspun and handwoven, nonviolent silk, organic denim, and so on [72].

The Handloom Weavers' Development Society, India produces a variety of home textiles using herbal fabrics in India. They also export herbal clothing such as sarees, bed sheets, and dress materials to other countries. With increasing demand for herbal clothing, companies such as Ayurvastra online have started offering the online products. Many textile industries are getting inclined toward Cosmetotextiles. The brands making herbal clothes are Aura, Cognis, Pantaloons, Quiospheres, and so on. These brands have a range of clothing namely bandee, vest, camisole, men and women's tee, coirs, and so on [73].

8. Benefits and drawbacks of herbal textiles

8.1 Health benefits

- It acts as a revitalizing tonic and helps in keeping the body fresh and healthy. The uses of proper herbs in the textiles have proven to cure diseases such as arthritis and fever. We breathe through our body more than we do through nose. It can improve the skin's natural ability to block and resist the harmful chemicals and toxins from entering the body, which will be beneficial to health.

- Herbal cloths have the ability to protect us from various skin diseases and provide relief from infectious diseases and mental ailments.

- No synthetic/chemical dye is used at any stage of herbal cloth production. For washing and cleaning purpose, the nut or nut powder of *Sapindus laurifolia* is recommended [74].

- Medicinal herbs treated fabrics also have a lot of therapeutic value; thus, the fabric has been found very helpful for people suffering from ailments such as skin allergies, breathing problems, sleeping disorders, and blood pressure.

- The health benefits of herbal clothing and its usage depend on the theory of touch. The body loses toxins when it comes in contact with herbal clothing, and this improves the metabolism. Herbal clothing is also known to help fight against many common diseases such as diabetes, hypertension, skin allergies, asthma, and heart ailments.

- The human body naturally heals itself during sleep or meditation. Thus, when the body is at rest, herbal clothing will work most effectively. The skin is known as the largest organ of the body. Not only it acts as a fence and protects the body, but it can also be the channel for outside germs and toxins to enter into the body. Herbal clothing guards against the harmful toxins trying to enter the body through the skin [75, 76].

- Herbal textiles are mainly used in making sleepwear, undergarments, bed coverings, towels, meditation clothing, and so on, which remain close to the skin absorbing all the benefits it gives out. Herbal textiles are also used in home textile products such as mattresses, coir mats, door mats, bath towels, bed spreads, and carpets.

- Some of herbal constituents are antiallergens and hence are safe for skin contact and are mostly nonhazardous to human beings.

- Natural dyes are usually moth proof and can replace the synthetic dyes in kids' garments.

- We breathe through our skin more than we do through nose, and chemically dyed textiles having carcinogenic amines and chemicals may be allergic and dangerous to human skin. Organic clothing can help reduce the exposure to allergens and other irritants and give a comfortable feeling [77–79].

8.2 Environment benefits

- Preparation of herbal cloths is a labor-intensive industry and hence will provide the job opportunities for manpower engaged in cultivation, extraction, and application on textile, food, leather, and so on.

- Use of natural dyes may earn the carbon credit as it reduces the consumption of petroleum-based synthetic dyes.

- It promotes the agriculture and balances the ecological cycle.

- Waste can be used as manure.

- No heavy metals such as chrome and copper are used in natural/herbal Ayurvedic dyeing.

- Manual farming and organic practices have a lower carbon footprint as the entire process consumes less fuel and energy and emits fewer greenhouse gases than chemical textiles.

- Herbal dyes are extracted from the plants cultivated without the use of synthetic pesticides and fertilizers, and also no chemical defoliants are used.

- Processing is in green environment; hence, workers' health is not compromised, and also water and electric use and toxic runoff are reduced.

- Stringent testing ensures that product is free from contaminants such as nickel, lead, formaldehyde, amines, and pesticides.

- It improves the soil fertility and soil structure by increasing the soil activity, thus reducing the risk of erosion.

- It is the massive saving of precious water.

- It promotes the development of earthworms and other arthropods leading to improvement in the growth conditions of the crop. Furthermore, organic crops thrive on root symbioses and are better equipped to exploit the soil; hence, fields can accommodate a more number of plants, animals, and microorganisms [80–82].

Natural fibers and natural dyes used in the preparation of herbal clothing allow its natural breakdown without damaging the environment compared to the synthetic material, which takes hundreds of years to breakdown and leaves the pollutant material in soil [66, 67, 83].

8.3 Other benefits

- Herbal textiles are ecofriendly, and also residues they produce can be further used for making the other green products.
- Solid and liquid wastes from herbal dyeing are amenable to recycling producing the organic manure.
- Some of the ecofriendly industries producing the herbal textiles produce organically recycled paper bags by utilizing the residual organic and herbal dyed fabrics that can reduce the pollution due to plastic bags [84].
- It also promotes the cultivation of herbs required for the production of herbal textile products.
- Herbal dyes add value to the cotton fabrics in ecofriendly manner.
- Herbal fabrics are lightweight and so used to construct the perfect breezy clothes. The biggest advantage of cotton herbal cloth is that it is cool in summer and warm in winter [85, 86].
- The color shades produced by herbal dyes are unique, hence commanding high demand in market.

8.4 Drawbacks

- Herbal fabrics and dyes are not good substitutes for synthetic dyes offering broader range and variations.
- Retaining color and its herbal benefits demands utmost care in washing; herbal cloths should be washed separately, preferably hand wash or gentle machine wash with bleach-free detergents and should be shed dried. This causes inconvenience to the user.
- The dyeing process is cumbersome and time taking, and each stage should be carefully controlled.
- Medicinal herbs impart colors; hence, color choice for consumer is very limited as only few herbs can be used for dyeing a fabric that meant to be used in a particular disease condition.
- Herbal cloths show the curative effects, but progress is slow.
- Though there are no complaints on allergic reaction by the consumers, some people think that the herbal clothing shows side effects on children [87–89].

9. Conclusion

Textile industry has realized its negative contribution to the environmental pollution through the harmful chemicals used in manufacturing processes and is continuously trying to find solution for this problem. Herbal dyeing is one important step toward achieving the organic lifestyle, thus reducing the environmental pollution. Herbal textile products are devoid of pesticides, fertilizers, antibiotics, growth hormones, genetically modified organisms, additives, irradiation, or sewage sludge, hence avoiding many ailments that are common with conventional clothing.

Modern consumer's demand for novel ecomaterials is expected to increase in future. Recently, efforts have been made toward the development of commercially viable natural colorants by making advances in aspects such as identification of new sources, formulations, extraction, purification, and stability techniques. In spite of enthusiastic studies discussing the socioeconomic viability of natural dye production and applications at commercial scale for sustainable utilization of bioresources, there is a need for investigations related to hygiene and ecosafety, which have great future for the discovery of relatively better and more stable natural pigments that may have wider industrial applications.

Herbal textiles have tremendous scope in world textile market and may become a major textile product in future. The blend of herbs and textiles to achieve the health in an ecofriendly manner is the great way of adopting the healthy life. Herbal clothing is receiving the increased awareness all over the world. Herbally treated textiles are one of the great solutions to revive and increase the share of Indian handloom industry in world textile market. Furthermore, herbal clothing is nontoxic, noncarcinogenic, biodegradable, and ecofriendly; these characteristics make them an obvious choice for modern informed costumer looking for a healthy product. Additionally, the production of biocolorants to meet the rising demand shall force entrepreneurs to take up this venture for greater profits leading to more employment generation.

Author details

Anupama Mishra[1*] and Sapna Gautam[2]

1 Department of Textile and Apparel Designing, College of Community Science, Central Agricultural Universities, Tura, Meghalaya, India

2 Department of Clothing and Textiles, College of Community Science, PSKHPKV, Palampur, Himachal Pradesh, India

*Address all correspondence to: anupamamishra8@gmail.com

References

[1] Hana K. Natural dyes: Their past, present, future and sustainability. In: Recent Developments in Fibrous Material Science. Czech Republic: O.p.s. Kanina; 2015. pp. 59-71

[2] Mukherjee S, Kanakarajan S. Extraction, optimisation and dyeing of silk yarn using natural dye from *Cosmos* sp. International Journal of Development Research. 2017;**7**(7):13865-13871

[3] Shafei AEI, Shaarawy S, Motawe FH, Refaei R. Herbal extract as an ecofriendly antimicrobial finishing of cotton fabric. Egyptian Journal of Chemistry. 2018;**61**(2):317-327

[4] Hasan KMF, Deb H, Rahman MDM, Morshed MN, Azad SAI, Genyang C. A review on antibacterial coloration agent's activity, implementation & efficiency to ensure the ecofriendly & green textiles. American Journal of Polymer Science & Engineering. 2016;**4**(1):1-20

[5] Jyothirmai S, Panda S. Ayurvastra—Herbal clothing (A new technology to heal naturally). IJARIIE; 2016;**2**(4): 2395-4396

[6] Rockefeller JD. An Introduction to Ayurveda and Its Benefits. Kindle ed. California, USA: J.D. Rockefeller Book Club, Scotts Valley; 2018. p. 24

[7] Gupta C, Prakash D, Gupta S. Functional clothing: A novel concept for disease therapeutics. African Journal of Basic and Applied Sciences. 2013;**5**(4):160-166

[8] Geissler S. Extraction of natural dyes for textile dyeing from coloured plant wastes released from the food and beverage industry. Journal of the Science of Food and Agriculture. 2006;**86**:233-242

[9] Machado TB, Pinto AV. Antimicrobial activity of selected plants for medical textiles. International Journal of Antimicrobial Agents. 2003;**21**(3):279-284

[10] Gupta Gulrajani ML, Agarwal Jain M. Application of natural dyes on bleached coir yarn. Indian Textile Journal. 1992;**102**:78

[11] Mahanta D, Tiwari SC. Natural dye yielding plants and indigenous knowledge on dye preparation in Arunachal Pradesh, North East India. Current Sciences. 2005;**88**:1474-1480

[12] Deepak K, Abhishek S, Raghav V, Dharmendra G, Abhishek K. Herbal textiles: Green business, green earth!!! Colourage. 2011;**58**(4):54-60

[13] Kolte PP, Shivankar VS, Ramachandran M. Herbal clothing—An Ayurveda doctor rest. Journal on Emerging Trends in Modelling and Manufacturing. 2015;**1**(1):26-31

[14] Elayarajah B, Rajendran R, Balakumar C, Venkatrajah B, Sudhakar A, Janiga PK. Antibacterial synergistic activity of ofloxacin and ornidazole treated biomedical fabrics against nosocomial pathogens. Asian Textile Journal. 2011;**1**:87-97

[15] Mader LS. Ayurvastra: Dyeing fabric with medicinal ayurvedic plants. American Botanical Council. 2011;**92**:16-18

[16] Kumar D, Srivastava A, Vidyarthi R, Gupta D, Kumar A. Herbal textiles: Green business, green earth. Colourage. 2011;**58**(4):54

[17] Jain M. Ayurvedic textiles: A wonderful approach to handle health disorders. Colourage. 2010;**57**(10):45-52

[18] Rangari NT, Kalyankar TM, Mahajan AA, Lendhe PR, Puranik PK.

Ayurvastra: Herbal couture technology in textile. International Journal of Research in Ayurveda and Pharmacy. 2012;**3**(5):733-736

[19] Venkatesh CS. Our Tree Neighbours. New Delhi: National Council of Educational Research and Training; 1976. pp. 291-292

[20] Lee YH, Hwang EK, Baek YM, Lee MS, Lee DJ, Jung YJ, et al. Deodorizing and antibacterial performance of cotton, silk and wool fabrics dyed with *Punica granatum* L. extracts. Fibers and Polymers. 2013;**14**(9):1445-1453

[21] Kobayashi Y, Kamimaru M, Tsuboyama K, Nakanishi T, Komiyama J. Deodorization of ethyl mercaptan by cotton fabrics mordant dyed with a direct dye and copper sulfate. Textile Research Journal. 2006;**76**(9):695-701

[22] Gawish SM, Helmy HM, Ramadan AN, Farouk R, Mashaly HM. UV protection properties of cotton, wool, silk and nylon fabrics dyed with red onion peel, madder and chamomile extracts. Journal of Textile Science and Engineering. 2016;**6**:266. DOI: 10.4172/2165-8064.1000266

[23] Lee YH, Hwang EK, Baek YM, Kim HD. Deodorizing function and antibacterial activity of fabrics dyed with gallnut (*Galla chinensis*) extract. Textile Research Journal. 2015;**85**(10):1045-1054

[24] Shrimali K, Dedhia EM. Microencapsulation for textile finishing. Journal of Polymer and Textile Engineering. 2015;**2**(2):1-4

[25] Hammad H. Microencapsulation: Process, techniques and applications. International Journal of Research in Pharmaceutical and Biomedical Sciences. 2011;**2**(2):474-481

[26] Barari M, Majidi RF, Madani M. Nanoscience Nanotechnolology. 2009;**3**:43-48

[27] Sudha S, Devagiri S, Neelkandan R. Microencapsulation: An overview. Indian Textile Journal. 2005;**115**(12):25-29

[28] Ganesan P. Microencapsulation of copper enriched herbals for curative garments. Indian Journal of Traditional Knowledge. 2012;**11**(3):532-536

[29] Ramya K, Maheshwari V. Development of eco friendly mosquito repellent fabric finished with *Andrographis paniculata* plant extracts. International Journal of Pharmacy and Pharmaceutical Sciences. 2014;**6**(5):115-117

[30] Cheng SY, Yuen CWM, Kan CW, Cheuk KKL. Development of cosmetic textiles using microencapsulation technology. Research Journal of Textile and Apparel. 2008;**12**(4):41-51

[31] Microcapsule, treating liquids containing the same, and textile structure having microcapsules adhering thereto. US Patent 005232769. 1993. Available from: http://en.wikipedia.org/wiki/Micro-encapsulation.

[32] Sen K, Thomas KA. Microencapsulation in Textile Finishing: Scope and Challenges. IIT Delhi: Department of Textile Technology; 2018

[33] Srivastava A, Rohini, Parmar MS. Innovative fragrance finishes on bamboo fabric. The Indian Textile Journal. 2013;**123**(9):43

[34] Ramachandran T, Kumar R, Radhai R. Antimicrobial textiles - An overview. Journal of the Institution of Engineers. India: Part TX: Textile Engineering Division; 2004;**84**:42-47

[35] Sathianarayanan MP. Antibacterial finish for cotton fabric from herbal products. Indian Journal of Fibre & Textile Research. 2010;**50**:35

[36] Rajput A, Ramachandran M, Gotmare VD, Raichurkar KK. Recent

bioactive materials for development of eco-friendly dippers: An overview. Journal of Pharmaceutical Sciences and Research. 2017;**9**(10):1844-1848

[37] Moghadamtousi ZS, Kadir HA, Hassandarvish P, Tajik H, Abubakar S, Zandi K. A review on antibacterial, antiviral, and antifungal activity of curcumin. Biomed Research International. 2014;**2014**:1-12. DOI: 10.1155/2014/186864

[38] Goudarzi M, Fazeli M, Azad M, Seyedjavadi SS, Mousavi R. Aloe vera gel: Effective therapeutic agent against multidrug-resistant *Pseudomonas aeruginosa* isolates recovered from burn wound infections. Chemotherapy Research and Practice. 2015;**2015**:1-5. DOI: 10.1155/2015/639806

[39] Mallikarjun S, Rao A, Rajesh G, Shenoy R, Pai M. Antimicrobial efficacy of Tulsi leaf (*Ocimum sanctum*) extract on periodontal pathogens: An in vitro study. Journal of Indian Society of Periodontology. 2016;**20**(2):145-150. DOI: 10.4103/0972-124X.175177

[40] Alzohairy MA. Therapeutics role of *Azadirachta indica* (Neem) and their active constituents in diseases prevention and treatment. Evidence Based-Complementary and Alternative Medicine. 2016:1-11. DOI: 10.1155/2016/7382506

[41] Samanta KK, Basak S, Chattopadhyay SK. Sustainable dyeing and finishing of textiles using natural ingredients and water-free technologies. In: Muthu S, editor. Textiles and Clothing Sustainability. 2017

[42] Wai CT, Pui FN, Qun CBF. Dyeing performance of kapok flower colourant on wool fabrics. World Journal of Textile Engineering and Technology. 2015;**1**:23-27

[43] Zhou Y, Tang R-C. Natural flavonoid-functionalized silk fiber presenting antibacterial, antioxidant, and UV protection performance. ACS Sustainable Chemistry & Engineering. 2017;**5**(11):10518-10526

[44] Shahid-ul-Islam M. Perspectives for natural product based agents derived from industrial plants in textile applications—A review. Journal of Cleaner Production. 2013;**57**:2-18

[45] Kartick, Samanta K, Basak S, Chattopadhyay SK. Roadmap to sustainable textiles and clothing. In: Muthu SS, editor. Textile Science and Clothing Technology: Environmental and Social Aspects of Textiles and Clothing Supply Chain. Singapore: Springer; 2006. pp. 263-287

[46] Baliarsingh S, Jena JR, Das T, Das NB. Role of cationic and anionic surfactants in textile dyeing with natural dyes extracted from waste plant materials and their potential antimicrobial properties. Industrial Crops and Products. 2013;**50**:618-624

[47] Hou X, Chen X, Cheng Y, Xu H, Chen L, Yang Y. Dyeing and uv-protection properties of water extracts from orange peel. Journal of Cleaner Production. 2013;**52**:410-419

[48] Yi E, Yoo ES. A novel bioactive fabric dyed with unripe *Citrus grandis* Osbeck extract Part 1: Dyeing properties and antimicrobial activity on cotton knit fabrics. Textile Research Journal. 2000;**80**:2117-2123

[49] Fauziyah N, Hakim L. Plants as natural dyes for Jonegoroan batik processing in Jono cultural Tourism Village, Bojonegoro, East Java. Journal of Indonesian Tourism and Development Studies. 2015;**3**(2) 2338-1647

[50] Paul R. Functional Finishes for Textiles, Improving Comfort Performance and Protection. UK: Wood head Publishing; 2015. pp. 1-14

[51] Chandrasekaran K, Ramachandran T, Vigneswaran C. Effect of medicinal herb extracts treated garments on selected diseases. Indian Journal of Traditional Knowledge. 2012;**11**(3):493-498

[52] Çalis A, Çelik GY, Katircioglu H. Antimicrobial effect of natural dyes on some pathogenic bacteria. African Journal of Biotechnology. 2009;**8**(2):291-293

[53] Karekar N, Lakhani P, Patel U. Ayurvastra: A Novel Use of Herbal Drugs. Mumbai Educational Trust Institute; 2013. p. 36. Available from: articles@pharmatutor.org

[54] Dev VG, Venugopal J, Sudha S, Deepika G, Ramakrishna S. Dyeing and antimicrobial characteristics of chitosan treated wool fabrics with henna. Carbohydrate Polymers. 2009;**75**(4):646-650

[55] Prusty AK, Das T, Nayak A, Das NB. Colourimetric analysis and antimicrobial study of natural dyes and dyed silk. Journal of Cleaner Production. 2010;**18**(16):1750-1756

[56] Rungruangkitkrai N, Tubtimthai N, Cholachatpinyo A, Mongkholrattanasit R. UV protection properties of wool fabric dyed with eucalyptus leaf extract by the padding techniques. In: RMUTP International Conference: Textiles and Fashion (Bangkok). 2012

[57] Chattopadhyay SN, Pan NC, Roy AK, Saxena S, Khan A. Development of natural dyed jute fabric with improved colour yield and UV protection characteristics. The Journal of the Textile Institute. 2013;**104**(8):808-818

[58] Zhou Y, Zhang J, Tang RC, Zhang J. Simultaneous dyeing and functionalization of silk with three natural yellow dyes. Industrial Crops and Products. 2015;**64**:224-232

[59] Hwang EK, Lee YH, Kim HD. Dyeing, fastness, and deodorizing properties of cotton, silk, and wool fabrics dyed with Gardenia, coffee sludge, *Cassia tora* L., and pomegranate extracts. Fibers and Polymers. 2008;**9**(3):334-340

[60] Pandith SA, Dar RA, Lattoo SK, et al. *Rheum australe*, an endangered high-value medicinal herb of North Western Himalayas: A review of its botany, ethnomedical uses, phytochemistry and pharmacology. Phytochemistry Reviews. 2018;**17**:573-609. DOI: 10.1007/s11101-018-9551-7

[61] Park J, Seok KJ, Suh HJ, Boo YC. Gardenia jasminoides extract attenuates the UVB-induced expressions of cytokines in keratinocytes and indirectly inhibits matrix metalloproteinase-1 expression in human dermal fibroblasts. Evidence-Based Complementary and Alternative Medicine. 2014;**2014**:1-10. DOI: 10.1155/2014/429246

[62] Liu X, Zhang R, Shi H, et al. Protective effect of curcumin against ultraviolet A irradiation-induced photoaging in human dermal fibroblasts. Molecular Medicine Reports. 2018;**17**(5):7227-7237. DOI: 10.3892/mmr.2018.8791

[63] Nazari A, Montazer M, Dehghani-Zahedani M. Chemistry of natural colorants on textiles also discussed with relevance to adsorption in bioresourced colorants, have created a revolution in textile research and development. Industrial & Engineering Chemistry Research. 2013;**52**(3):1365-1371

[64] Shakyawar DB, Raja AS, Kumar A, Pareek PK. Antimoth finishing treatment for woollens using tannin containing natural dyes. Indian Journal of Fibre & Textile Research. 2015;**40**(2):200-202

[65] Park JH, Gatewood BM, Ramaswamy GN. Naturally occurring quinines and flavonoid dyes for wool: Insect feeding deterrents. Journal of Applied Polymer Science. 2005;**98**(1):322-328

[66] Mapari SA, Nielsen KF, Larsen TO, Frisvad JC, Meyer AS, Thrane U. Exploring fungal biodiversity for the production of water-soluble pigments as potential natural food colorants. Current Opinion of Biotechnology. 2005;**16**(2):231-238

[67] Gupta D, Laha A. Antimicrobial activity of cotton fabric treated with *Quercus infectoria* extract. Indian Journal of Fibre and Textile Research. 2007;**32**(1):88-92

[68] Baid AM. Method of dyeing textile article from medicinally rich herbs. 2014. Available from: http://www.patentgenius.com. [Accessed: 10 February 2014]

[69] Kumar D, Srivastava A, Vidyarthi R, Shivam A, Kumar A. Herbal Textiles: Green Business, Green Earth!!! Available from: http://www.fibre2fashion.com/industry-article/34/3357/herbal-textiles-green-business-green-earth1.asp [Accessed: 05 March 2014]

[70] Bagyiilakshmi G. Developing Herbal Antimicrobial Finished Cotton Fabric for Wound Dressing. Coimbatore: Department of Textiles and Clothing, Avinashilingam University for Women; 2019

[71] Singh K, Jain A, Panwar S, Gupta D, Khare SK. Antimicrobial activity of some natural dyes. Dyes and Pigments. 2005;**66**(2):99-102

[72] Mussak RAM, Bechtold T. Natural colorants in textile dyeing. In: Bechtold T, Mussak R, editors. Handbook of Natural Colorants. Chichester: Wiley; 2009. pp. 315-338

[73] Khan SA, Ahmad A, Khan MI, Yusuf M, Shahid M, Manzoor N. Dyeing, fastness and antimicrobial properties of woollen yarns dyed with gallnut (*Quercus infectoria* Olive) extract. Dyes and Pigments. 2012;**95**(2):206-214

[74] Khan MI, Ahmad A, Khan SA, Yusuf M, Shahid M, Manzoor N. Assessment of antimicrobial activity of catechu and its dyed substrate. Journal of Cleaner Production. 2011;**19**(12):1385-1394

[75] Hill DJ. Is there a future of natural dyes? Review of Progress and Coloration. 1997;**27**(1):18-25

[76] Shabbir M, Rather LJ, Bukhari MN, Shahid M, Khan MA, Mohammad F. An eco-friendly dyeing of woolen yarn by *Terminalia chebula* extract with evaluations of kinetic and adsorption characteristics. Journal of Advanced Research. 2016;**7**(3):473-482

[77] Hou X, Yang R, Xu H, Yang Y. Adsorption kinetic and thermodynamic studies of silk dyed with sodium copper chlorophyllin. Industrial and Engineering Chemical Research. 2012;**51**(25):8341-8347

[78] Ramachandran T, Rajendrakumar K, Rajendran R. Antimicrobial textiles—An overview. Journal of the Institution of Engineers (India). Part TX: Textile Engineering Division. 2004;**84**(2):42-47

[79] Joshi M, Wazed AS, Purwar R, Rajendran S. Ecofriendly antimicrobial finishing of textiles using bioactive agents based on natural products. Indian Journal of Fibre and Textile Research. 2009;**34**(3):295-304

[80] Khushwaha A, Singh RP, Gupta V, Singh M. Antimicrobial properties of peels of citrus fruits. International Journal of Universal Pharmacy and Life Sciences. 2012;**2**(2):24-38

[81] Kumar Sampath KP, Debjit B, Duraivel S, Umadevi M. Traditional and medicinal uses of banana. Journal of Pharmacognosy and Phytochemistry. 2012;**1**(3):51-63

[82] Arora M, Kaur P. Antimicrobial & antioxidant activity of orange pulp and peel. International Journal of Science and Research. 2013;**2**(11):412-415

[83] Ursache M. Functional finishes of cotton knitted fabrics. Buletinul Institutului Politehnic Din Iasi. 2010;**4**:49-56

[84] Paul R. Functional finishes for textiles: An overview. In: Improving Comfort, Performance and Protection. Woodhead Publishing; 2015. p. 678

[85] Salah M, Abd-El-Hady A. Eco-friendly dyeing of cotton fabric with natural colorants extracted. In: Textile Science and Clothing Technology. Singapore: Springer; 2013

[86] San YO, Khin O. Investigation on flame retardant effect of Myanmar Banana (*Musa sapientum* Linn) pseudostem sap on cotton fabric. International Journal for Innovative Research in Multidisciplinary Field. 2017;**3**(8):139-144

[87] Saravanan M, Dhurai B. Comparison of natural agents to develop antibacterial kitchen. Textiles International Research Journal of Pharmacy. 2017;**8**(7):93

[88] Teli MD, Sheikh J, Shastrakar P. Eco-friendly antibacterial printing of wool using natural dyes. Journal of Textile Science. 2014;**4**:151

[89] Ortega Z, Morón M, Monzón MD, Pere B, Paz R. Production of banana fiber yarns for technical textile reinforced composites. Materials. 2016;**9**:370

Section 3

Pantone-Like Shade Book with Natural Dyes and Pigments

Chapter 6

A Practical Guideline of Few Standardized Ready Made Shades of Natural Dyed Textiles

Anowar Hossain

Abstract

Marigold flower *Tagetes erecta* L., Arjuna Bark *Terminalia Arjuna*, Eucalyptus leaves *Eucalyptus Radiata*, Peach/Jam Leaves *Acacia acuminata*, Pecker leaves *Cinnamomum tamala*, Guava leaves *Psidium guajava*, Basil leaves *Ocimum basilicum*, Jackfruit wood *Artocarpus heterophyllus*, Catachu fruit *Senegalia Catechu*, Bohera fruit *Terminalia bellirica*, Betel nut fruit *Areca catechu*, Haritaki fruit *Terminalia chebula*, Mahogany fruit peel, Mahogany seed peel, and Mahogany seed *Swietenia macrophylla* are the common natural sources in Bangladesh, an Asian country which were experimented in terms of mordanting free natural coloration on cotton fabric under conceptual confirmation of referred journal where author has been picked the idea from the generation of available shade in his research laboratory and tested from different laboratories and it may be establish as mordant free natural dyeing for specific colorant on the basis of color fastness and shading behaviors. Fifteen standardized Ready Made Shade (RMS) has been presented with CIE color parameters, color fastness, wash fastness, and light fastness grading. A reproducing guideline for every Ready Made Shade (RMS) has been mentioned in this chapter.

Keywords: readymade shade (RMS), natural dyeing, eco-friendly dyes, natural extraction, mordant free dyeing

1. Introduction

A concept of readymade shade (RMS) has been developed by using numerous number of natural dyes which were collected from agri-production unit and different village sources in Bangladesh prepared and powdered in color processing unit, extracted with water solvent process, dyed and experimented the feasibility of coloration in textile chemistry laboratory to generate different hues of natural dyes after effective implementation on natural fiber based textiles which outcome of color may be beneficial for primary selection of color tone by the textile technologist of natural coloration cum customers for the decision-making of specific color tone both for product development and fashion concerns who are genuinely searching an eco-friendly dyes under the consideration of repeated hue without which the real output of natural coloration, sustainability of dyes and natural dyeing process as well as its actual production of different hue is being a challenged now as the textile technologists are habituated the essence of readymade shading behavior of synthetic dyes whereas toxicity is the main endangered for the human being.

2. Background

Environment pollution is the great challenge of color scientists in dyeing and finishing plant, so eco-friendly coloration is the key target of recent researchers and manufacturers. Natural coloration is being accepted by the environment scientist and related committee [1–3]. Researchers in the area of natural coloration have an immense intension to minimize pollution [4]. Focus on science and engineering on natural dyes based research, extractions, purifications, and implementations are rapidly climbing [5–9]. In this chapter, mordant free natural coloration and its feasibility was experimented to minimize environment pollution load which is not limited to synthetic dyeing but also natural dyeing in terms of mordant free coloration concepts for specific standardized shade. Author also experimented to establish an approach of green mordanting [10, 11] in natural coloration [12–16] although mordant [17] has an effect for the augmentation of fiber surface color and surface chemistry [18] and its appropriate affinity in terms of fastness properties [1, 19–24]. Light fastness [21–24] of natural dyed fabric is a critical issue for natural colorant if dyes sources and collection processes are not maintained accurately. There are many sources of natural colorant, using by the researcher as per availability of origin to origin in the world. Researchers are already invented and proposed like Jatropha flower [25], red sandal wood [26], wood of *Artocarpus heterophyllus* [27], Acalypha [28], areca nut [29], *Butea monosperma* [30], neem leaves [31, 32], Gomphrena globosa [33], Onosmaechiodes [34], *Nerium oleander* flower [35], Kesula flowers [36], Parijataka (Nyctanthes Arbor) flowers pigment [37], Hibiscus ovalifoliusand *Sesbania aegyptiaca* [38], Bark of Macaranga Peltata [39], Cutch, Ratanjot and Madder [40], Mesta Calyx [41], Kapila, Onion, Tesu [42], myrobalan, gallnut, pomegranate [43], Marigold flower [44, 45], *Areca Catechu* [46], Eucalyptus leaves [47], Eucalyptus bark [48, 49], Jackfruit wood [50], Peach [51], Arjuna [52, 53], Catechu [54] and others. Developments of natural indigo shade in comparison with synthetic dyes [55] are not only prime objects for coloration but also maximum natural dyes have medicinal value [56, 57], noncarcinogenic, production-friendly, and environment-friendly. Research and development of natural dyeing on cotton, jute, and silk fabric was done by Samanta and his researchers team at DJFT, University of Calcutta, India have an technical output, research impacts and motivation for future researchers, and manufacturing unit for the contributions of dyes selection, extraction, analysis, standardization, and commercialization [58–68]. Color matching [69] and reproducing of shade is another challenging issue for the manufacturing company and colorists when natural dye is compared with synthetic dyes but there are so many colorants of natural sources that can be possible to reproduce almost nearest shade. Tolerating percentage/acceptable range of color differences, ΔE value, and other properties can be standardized in a convincing way with local and foreign customer of any country as the real customer of any textile product is not going to sell the product after shade matching with spectrophotometer without enjoying its esthetic value, so a little bit lightness and darkness matter is not a trading challenge of dyer and manufacturing plant. Comparing to synthetic dyes, natural dyed fabric has a great market demand as the people who are concern about the intimidation of environment. Thus natural dyes have diverse applications and multi fiber based production options that are already available in literature for cotton [68], jute [66], wool [70], silk [34] based dyeing, processing in both laboratory stage and bulk production (**Figures 1–15**).

RMS of Marigold (Figure 1): Waste Marigold flowers were collected from flower garden, washed and 5 days dried with summer sun light, and made it crispy and fine powdered with grinding machine. Two gram powder was mixed with 100 ml water, heated at boiling temperature for 40 min and filtered properly. The dyes solution was used for cotton fabric dyeing in open bath medium for 30 min at 80°C.

3. Standardized Ready Made Shade (RMS) of natural dyed textiles

3.1 Natural coloration of cotton fabric with Marigold flower, types of dyes: natural, source of dyes: Marigold tree. scientific name: *Tagetes erecta*

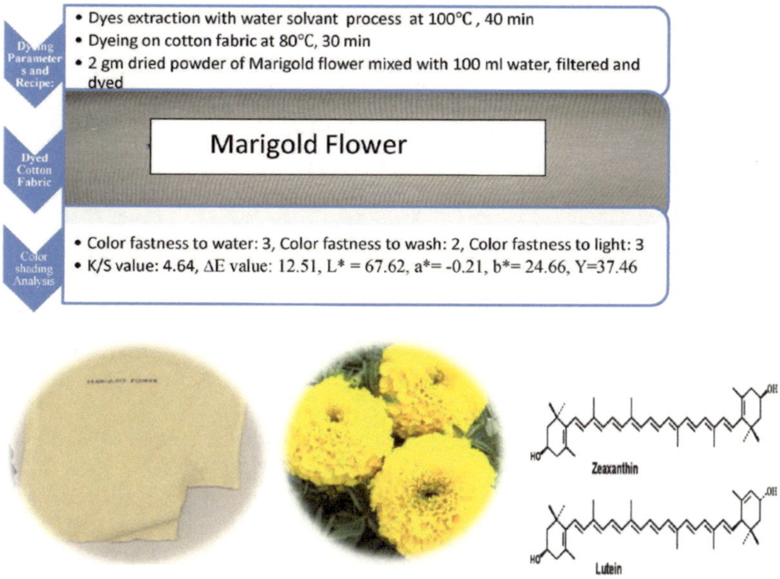

Figure 1.
RMS of Marigold.

3.2 Natural coloration of cotton fabric with Arjuna Bark, types of dyes: natural, source of dyes: Arjuna tree, Scientific name: *Terminalia arjuna*

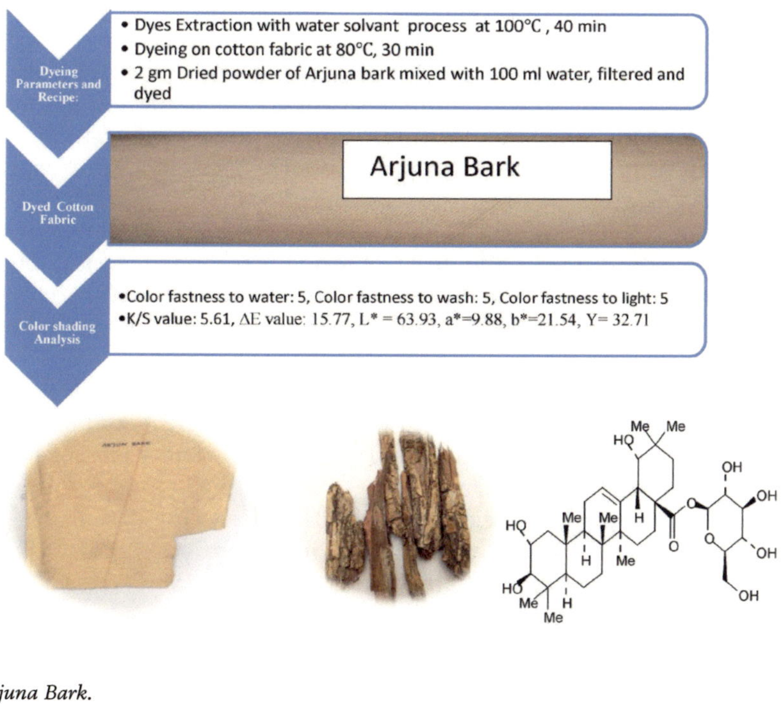

Figure 2.
RMS of Arjuna Bark.

3.3 Natural coloration of cotton fabric with Eucalyptus leaves, types of dyes: natural, source of dyes: scientific name: *Eucalyptus radiata*

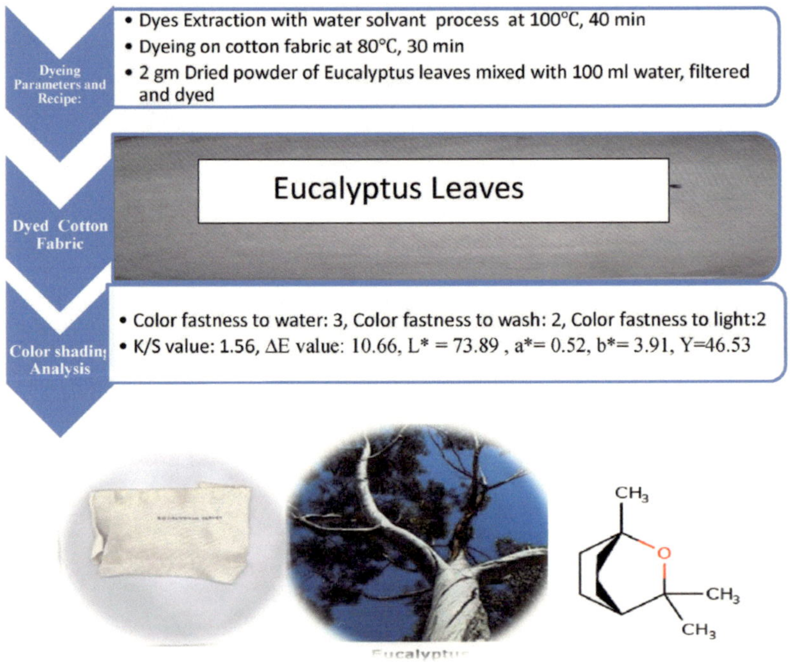

Figure 3.
RMS of Eucalyptus leaves.

Coloring components Zeaxanthin and Lutein structural group of marigold flower may be responsible for good color combination has OH group may be showed good color fastness properties with cellulosic fiber. Lutein ($C_{40}H_{56}O_2$) and Zeaxanthin ($C_{40}H_{56}O_2$) molecules may prevent UV damage for its strong antioxidant properties.

RMS of Arjuna Bark (Figure 2): Dried Arjuna bark were plucked from Arjuna tree, washed and 5 days dried with summer sun light, and made it crispy and fine powdered with grinding machine. Two gram powder was mixed with 100 ml water, heated at boiling temperature for 40 min, and filtered properly. The dyes solution was used for cotton fabric dyeing in open bath medium for 30 min at 80°C.

Ferrous ion of Arjuna Bark may be influencing the L*, a*, b*, Y value for providing outcome of deeper color and OH group is responsible for good fastness properties. Arjuna bark has phytosterol, lactones, flavonoids, phenolic compounds, and tannins, glycosides where tannin may be responsible for coloring agent.

RMS of Eucalyptus Leaves (Figure 3): Semi-dried leaves were plucked from Eucalyptus tree, washed and 5 days dried with summer sun light, and made it crispy and fine powdered with grinding machine. Two gram powder was mixed with 100 ml water, heated at boiling temperature for 40 min, and filtered properly. The dyes solution was used for cotton fabric dyeing in open bath medium for 30 min at 80°C.

OH group of cellulose can easily make a bonding with dyes group of Eucalyptus. Eucalyptol is a colorless compound which is responsible for antibacterial properties but may be remaining pigment or other phytochemical compound is making color on the fabric surface.

DOI: http://dx.doi.org/10.5772/intechopen.92360

3.4 Natural coloration of cotton fabric with Pecker leaves, types of dyes: natural, source of dyes: Pecker tree, scientific name: *Cinnamomum tamala*

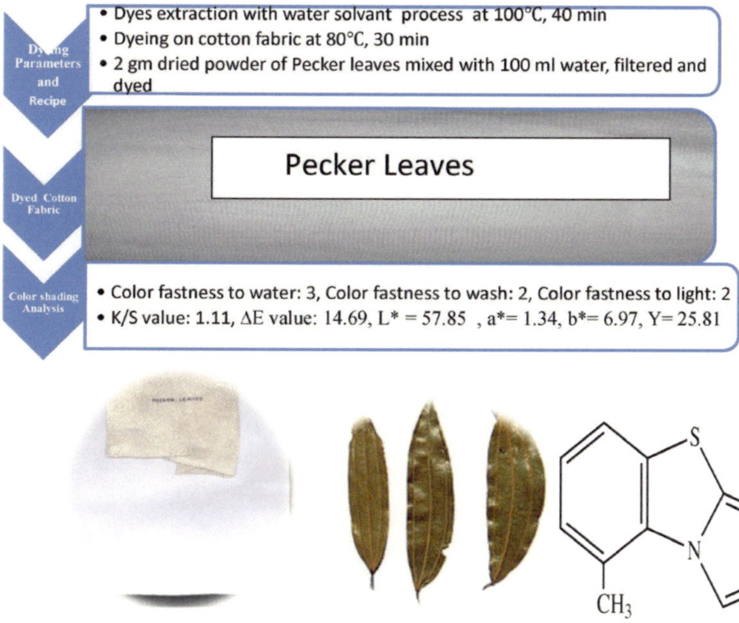

Figure 4.
RMS of Pecker leaves.

RMS of Pecker Leaves (Figure 4): Semi-dried leaves were plucked from pecker tree, washed and 5 days dried with summer sun light, and made it crispy and fine powdered with grinding machine. Two gram powder was mixed with 100 ml water, heated at boiling temperature for 40 min, and filtered properly. The dyes solution was used for cotton fabric dyeing in open bath medium for 30 min at 80°C.

OH group of cellulose can easily make a bonding with dyes group of Pecker leaves. Eugenol is the main constituent of antimicrobial properties and other components like pinene, camphene, and limousine may be responsible for color formation.

RMS of Guava Leaves (Figure 5): Semi-dried leaves were plucked from tree, washed and 5 days dried with summer sun light, and made it crispy and fine powdered with grinding machine. Two gram powder was mixed with 100 ml water, heated at boiling temperature for 40 min, and filtered properly. The dyes solution was used for cotton fabric dyeing in open bath medium for 30 min at 80°C.

OH group of dyes is responsible for good color fastness properties. Mixed components of flavonoid and tannin are highly responsible for bluish color of cotton fabric. Acidic pH and phenolic compound of guava leaves may generate flammable characteristics of guava leaves colored fabric.

RMS of Basil Leaves (Figure 6): Fresh and green Basil leaves were plucked from tree, washed and 5 days dried with summer sun light, and made it crispy and fine powdered with grinding machine. Two gram powder was mixed with 100 ml water, heated at boiling temperature for 40 min, and filtered properly. The dyes solution was used for cotton fabric dyeing in open bath medium for 30 min at 80°C.

α terpineol remaining in dyes of Basil leaves may be responsible for dyeing of cotton fiber and OH group is making strong affinity with cellulose. Phytochemical constituents and medicinal properties of Tulsi leaves may create pathogen protective finish of cotton fabric like antimicrobial, COVID 19, and other viruses.

3.5 Natural coloration of cotton fabric with Guava leaves, types of dyes: natural, source of dyes: Guava tree, scientific name: *Psidium guajava*

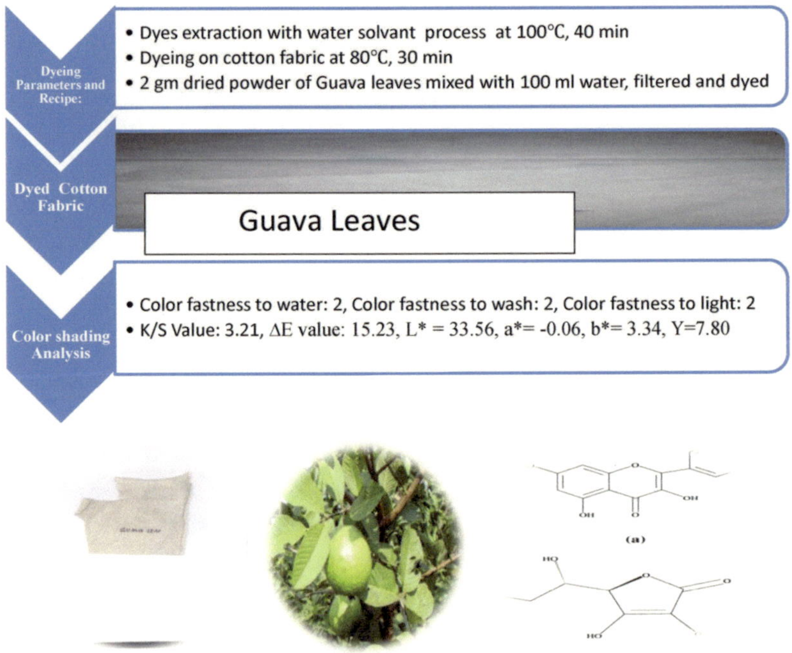

Figure 5.
RMS of Guava leaves.

3.6 Natural coloration of cotton fabric with Basil leaves, types of dyes: natural, source of dyes: Basil tree, scientific name: *Ocimum basificum*

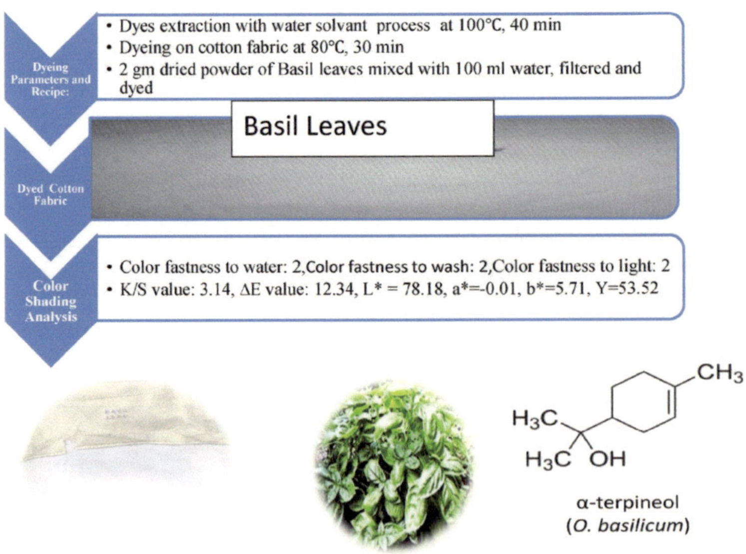

Figure 6.
RMS of Basil leaves.

3.7 Natural coloration of cotton fabric with Jackfruit wood, types of dyes: natural, source of dyes: Jackfruit tree, scientific name: *Artocarpus heterophyllus*

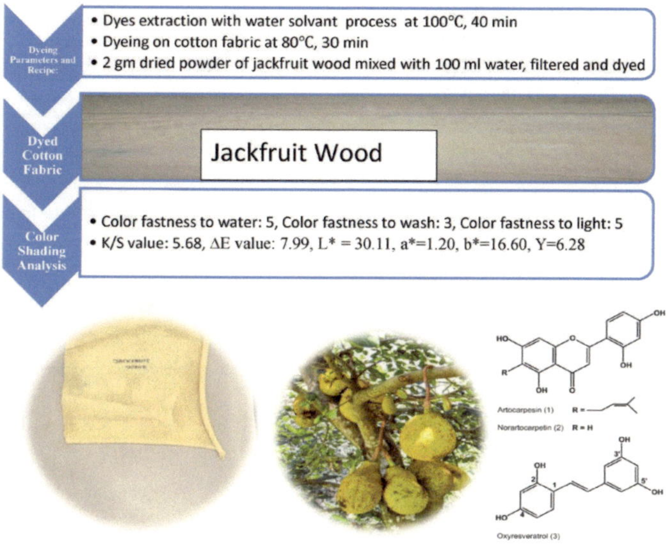

Figure 7.
RMS of Jackfruit wood.

3.8 Natural coloration of cotton fabric with Catechu fruit peel, types of dyes: natural, source of dyes: Catechu tree, scientific name: *Senegalia catechu*

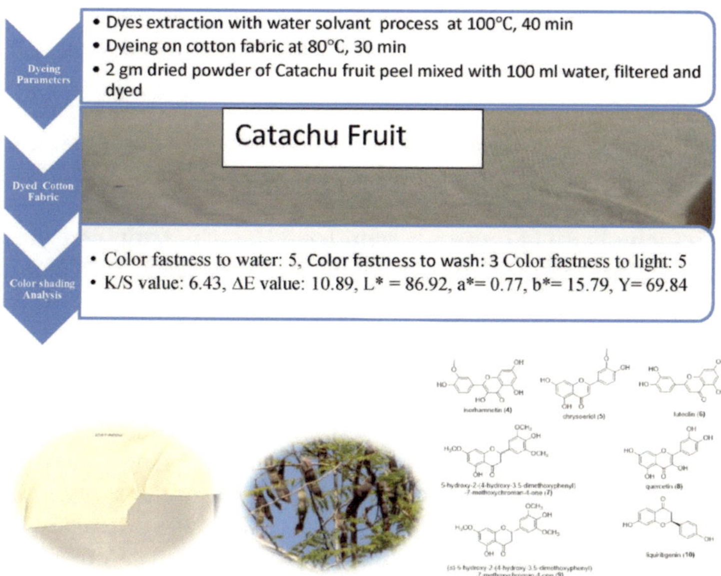

Figure 8.
RMS of Catachu fruit.

3.9 Natural coloration of cotton fabric with Bohera fruit peel, types of dyes: natural, source of dyes: Bohera tree, scientific name: *Terminalia bellirica*

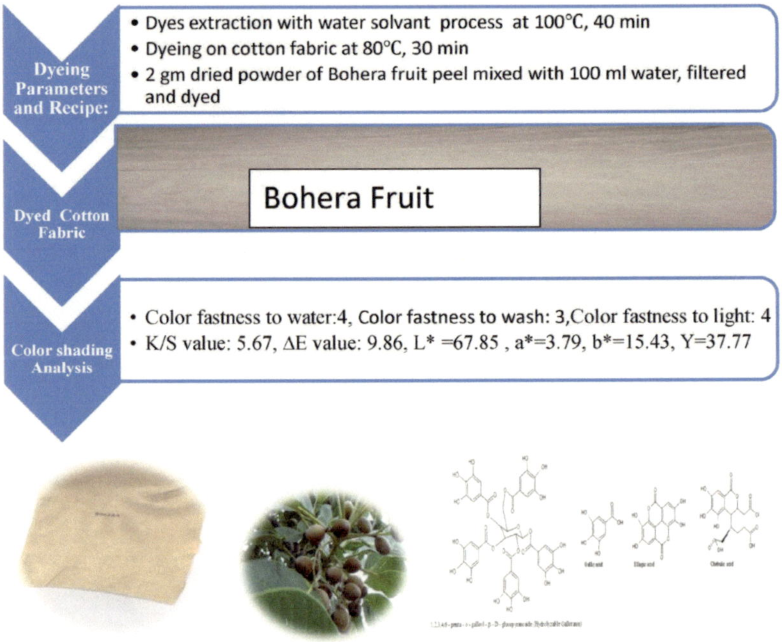

Figure 9.
RMS of Bohera fruit.

RMS of Jackfruit Wood (Figure 7): Powder of Jackfruit wood were collected from saw mill, filtered and 5 days dried with summer sun light. Two gram powder was mixed with 100 ml water, heated at boiling temperature for 40 min, and filtered properly. The dyes solution was used for cotton fabric dyeing in open bath medium for 30 min at 80°C.

Artocarpesin and Norartocarpetin group of golden yellow color jackfruit tree wood has phenolic compounds, apigenin, curcumin may be responsible for making a yellow color having an optimum value of L*, a*, b*, Y as well as OH group is improving color fastness properties.

RMS of Catachu Fruit (Figure 8): Semidried Catachu fruit were plucked from tree, washed, fruit peel were cut into small piece, 5 days dried with summer sun light, and made it crispy and fine powdered with grinding machine. Two gram powder was mixed with 100 ml water, heated at boiling temperature for 40 min, and filtered properly. The dyes solution was used for cotton fabric dyeing in open bath medium for 30 min at 80°C.

Catachu is making off white color with soft hand feel and may be remaining natural pigment in the chemical components is responsible for dyeing without mordanting. Coloring component in the catechu is catechin having molecular formula $C_{15}H_{14}O_6$.

RMS of Bohera Fruit (Figure 9): Bohera fruit powder was purchased from herbal medicine shop, Tongi, Dhaka, Bangladesh. Two gram powder was mixed with 100 ml water, heated at boiling temperature for 40 min, and filtered properly. The dyes solution was used for cotton fabric dyeing in open bath medium for 30 min at 80°C.

3.10 Natural coloration of cotton fabric with Mahogany seed, types of dyes: natural, source of dyes: Mahogany tree, scientific name: *Swietenia macrophylla*

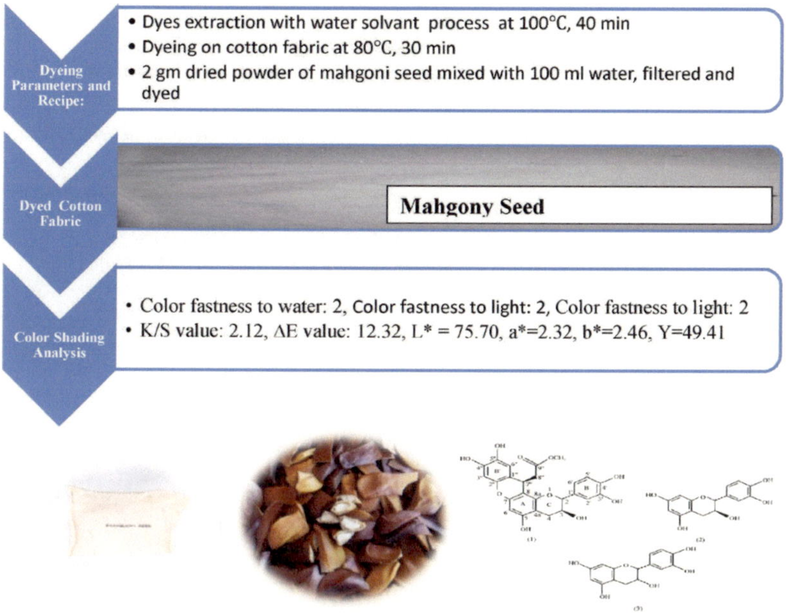

Figure 10.
RMS of Mahogany fruit, seed.

3.11 Natural coloration of cotton fabric with Mahogany fruit (outer peel), types of dyes: natural, source of dyes: Mahogany tree, scientific name: *Swietenia macrophylla*

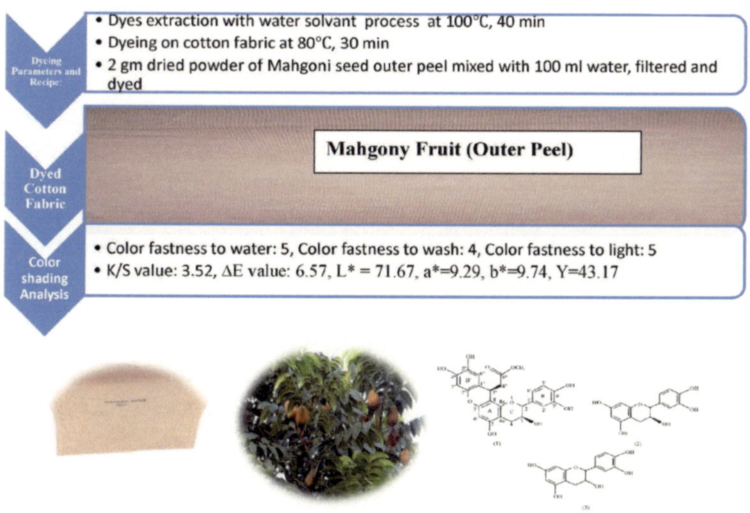

Figure 11.
RMS of Mahogany fruit, outer peel.

3.12 Natural coloration of cotton fabric with Mahogany seed (outer peel), types of dyes: natural, source of dyes: Mahogany tree, scientific name: *Swietenia macrophylla*

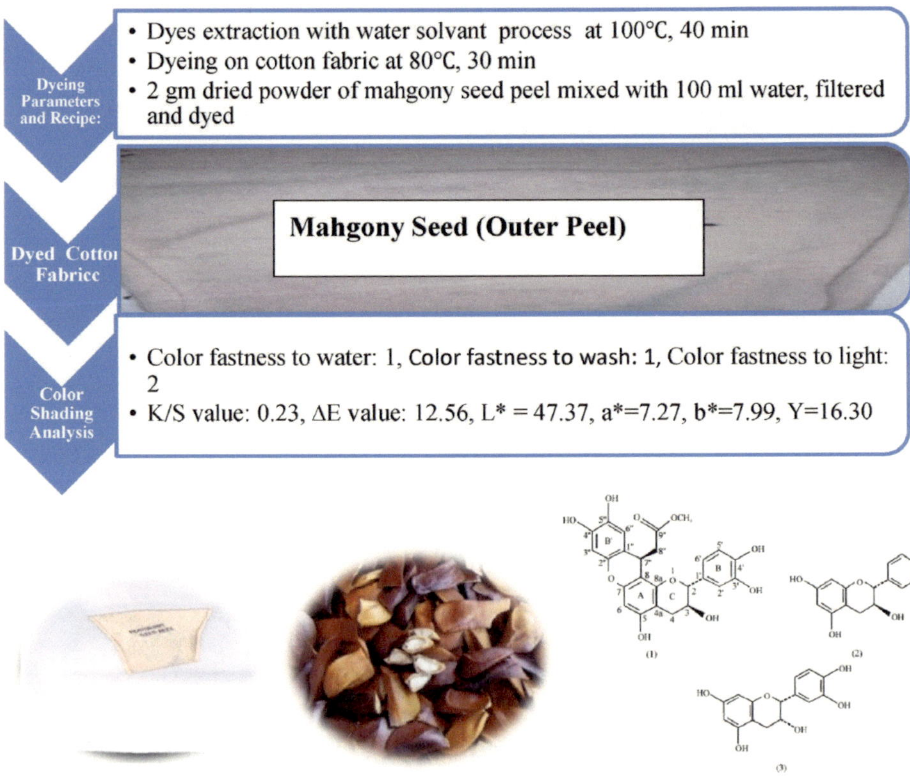

Figure 12.
RMS of Mahogany fruit, outer peel of seed.

OH group and its tannin constituent may be created a natural shading environment on the surface of cotton fiber and its medicinal properties may create the dyed fabric antimicrobial. Flavonoid and falvins constituent of Bohera fruit may have adaptive capabilities of viruses and microorganisms.

RMS of Mahgony Fruit, Seed (Figure 10): Semi-riped Mahogany fruits were plucked from tree, washed, seeds were separated with sharp knife, cut into small pieces and 5 days dried with summer sun light, and made it crispy and fine powdered with grinding machine. Two gram powder was mixed with 100 ml water, heated at boiling temperature for 40 min, and filtered properly. The dyes solution was used for cotton fabric dyeing in open bath medium for 30 min at 80°C.

OH group may be responsible for good dye-absorption and ferrous ion may be responsible for dyeing. Phytochemical constituents and other medicinal properties may have a good source protective clothing like insect repellent, antimicrobial, COVID19, and other viruses.

3.13 Natural coloration of cotton natural coloration of cotton fabric with Haritaki fruit peel, types of dyes: natural, source of dyes: Haritaki tree, scientific name: *Terminalia chebula*

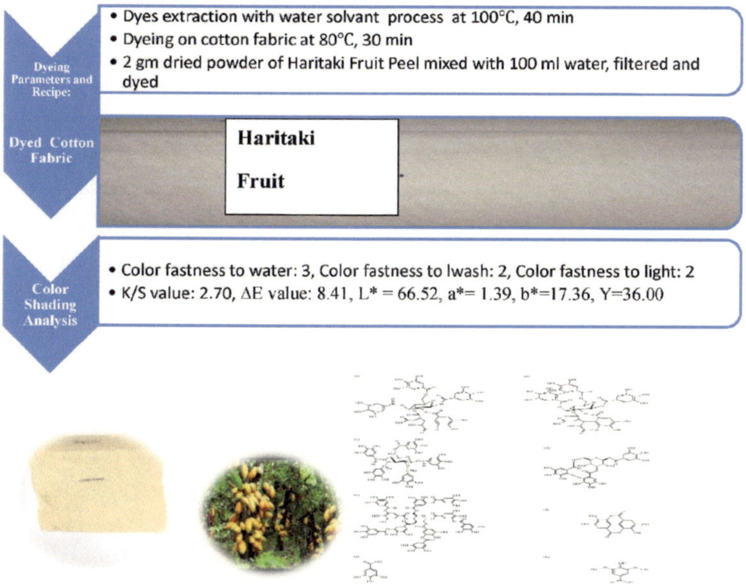

Figure 13.
RMS of Haritaki fruit.

3.14 Natural coloration of cotton fabric with Betel nut, types of dyes: natural, source of dyes: Betel nut tree, scientific name: *Areca catechu*

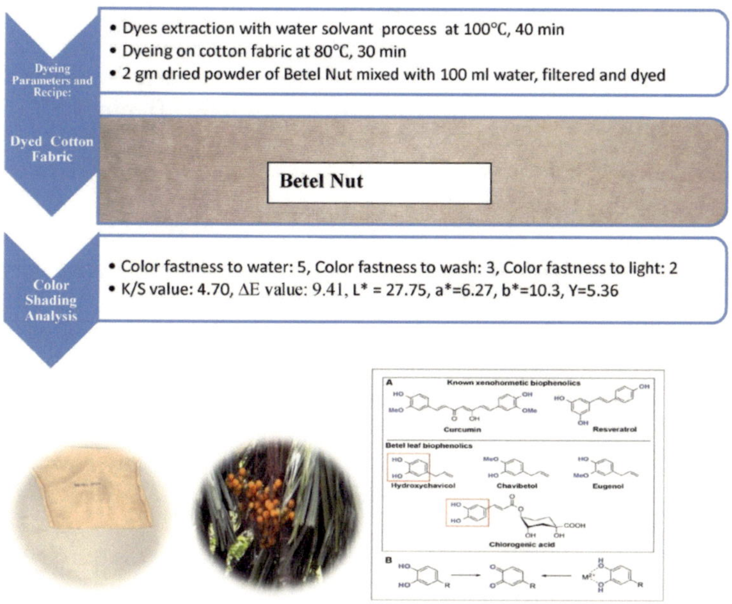

Figure 14.
RMS of Betel nut.

3.15 Natural coloration of cotton fabric with Peach leaves, types of dyes: natural, source of dyes: Peach leaves, scientific name: *Acacia acuminata*

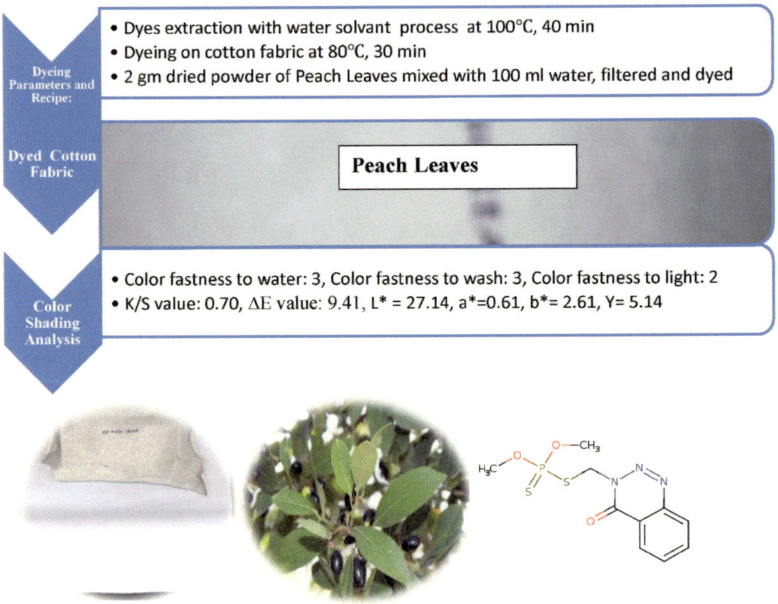

Figure 15.
RMS of Peach leaves.

RMS of Mahgony Fruit, Outer peel (Figure 11): Semi-riped Mahogany fruit were plucked from tree, washed, fruit peels were separated with sharp knife, cut into small pieces and 5 days dried with summer sun light, and made it crispy and fine powdered with grinding machine. Two gram powder was mixed with 100 ml water, heated at boiling temperature for 40 min, and filtered properly. The dyes solution was used for cotton fabric dyeing in open bath medium for 30 min at 80°C.

OH group and ferrous ion may be responsible for good dye-fiber bonding phytochemical constituents and other medicinal properties may have a good source protective clothing like insect repellent, antimicrobial, COVID19, and other viruses.

RMS of Mahgony Fruit, Outer peel of Seed (Figure 12): Semi-riped Mahogany fruit were plucked from tree, washed, seed peels were separated with sharp knife, cut into small pieces and 5 days dried with summer sun light, and made it crispy and fine powdered with grinding machine. Two gram powder was mixed with 100 ml water, heated at boiling temperature for 40 min, and filtered properly. The dyes solution was used for cotton fabric dyeing in open bath medium for 30 min at 80°C.

OH group and ferrous ion may be responsible for good dye-fiber bonding. Phytochemical constituents and other medicinal properties may have a good source protective clothing like insect repellent, antimicrobial, COVID19, and other viruses.

RMS of Haritaki Fruit (Figure 13): Haritaki fruit powder was purchased from herbal medicine shop Tongi, Dhaka, Bangladesh. Two gram powder was mixed with 100 ml water, heated at boiling temperature for 40 min, and filtered properly. The dyes solution was used for cotton fabric dyeing in open bath medium for 30 min at 80°C.

OH group and ferrous ion may be responsible for good dye-fiber bonding. The plant is constituted of Glucoside, Tannins, Gallic acid, Ethyl gallate, and Chebulinic acid where tannin may be responsible for coloring the cotton fiber.

RMS of Betel Nut (Figure 14): Ripe Betel nut were purchased from local village shop, separated the peels, and grinded to make it powder form. Two gram powder was mixed with 100 ml water, heated at boiling temperature for 40 min, and filtered properly. The dyes solution was used for cotton fabric dyeing in open bath medium for 30 min at 80°C.

OH group can create a bond between dye and fiber and remaining curcumin may be responsible for coloration and dyed fabric may be slightly flammable due to having hydroxychavicol in the chemical constituent of betel nut.

RMS of Peach Leaves (Figure 15): Semi-dried Peach leaves were plucked from Peach tree garden. Two gram powder was mixed with 100 ml water, heated at boiling temperature for 40 min, and filtered properly. The dyes solution was used for cotton fabric dyeing in open bath medium for 30 min at 80°C.

Minimum color strength is found but fastness is lightly higher, increasing dye percentage may improve the color on the fiber surface. Remaining ferrous, copper, and zinc may be responsible for coloring the cotton fiber. As per chemical constituent, antimicrobial and anti-inflammatory can be found.

4. Method of natural fixing for mordant free dyeing

Maximum natural dyes sources from natural tree leaves, roots or fruits have ferrous ion, tannin, curcumin, catechin, OH group, and other known and unknown phytochemical constituent. Ferrous ion, tannin, curcumin, and catechin are making various color formations, and OH group is creating affinity with cellulosic fiber as well as the fastness properties may be increased if the dyes have natural pigment which may influence the capability of remaining dyes on the fiber surface after washing. Drying/curing with higher temperature may impact the color making duller or brighter. So specific dye-fiber system of natural fixing for mordant free dyeing is also possible but exact curing/drying temperature should be fixed for getting expected outcome of shading. So expected mechanism of mordant free natural coloration may be proposed as below although dyes and fiber substances may create changing of it.

5. Testing process

5.1 Color fastness to wash method

ISO 105-C06:2010, color fastness to water method: ISO 105-E01 and color fastness to light was tested by AATCC TM16 inQ-SUN XE-2 Xenon Test chamber.

5.2 Machine used for color parameters

Color parameters (L^*, a^*, b^*, Y) were measured by Hunter Lab Spectrophotometer, machine name: ColorFlex EZ, Model 45/0 LAV, Geometry

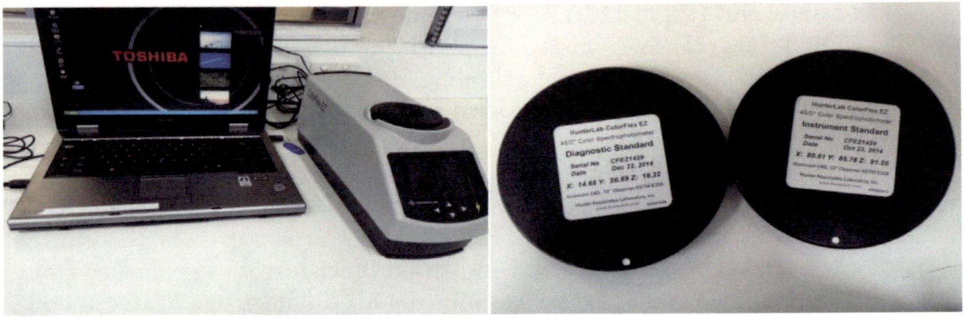

Figure 16.
Color Flex EZ, Spectrophotometer.

45°/0°, viewing area: large, visible spectrum (400–700 nm), testing condition: $D_{65/10°}$, room temperature of testing lab: 18°C (**Figure 16**).

6. Representation of dyed fabric and picture of dyes sources

All dyed shades were scanned with HUAWEI Smart Phone, Camera-13 MP, distance between fabric sample and camera position: 12–16 inch. So there is a possibility of having difference with actual shade. Picture of actual dye-fiber system was also scanned with HP Scanjet 4890 Photo Scanner. All sources of dyes pictures have been mentioned on the basis of grown trees in Asian Countries, specially in Bangladesh and scientific name was used accordingly. All the pictures of chemical structure mentioned here indicate the group of tree, not exactly indicating the chemical structure of specific parts of trees which one is extracted and dyed on cotton fabric.

6.1 Overcoming contradiction of natural dyeing

A huge number of researchers are working on natural dyes, still have a challenging for uneven dyeing, appropriate mordanting and dyes availabilities, shade matching where I can put a logic to protect our environment and human life as well.

6.2 Uneven dyeing

Automation in extraction and dyeing process.

6.3 Appropriate mordanting and dyes availability

Selection of specific dyes for commercial production and/or wastage can be used as dyes sources.

6.4 Shade matching

Customers are not always asking for shade matching when they are shopping although some natural dyes have attractive shading and repeated shading also possible.

7. Conclusion

Mordanting free coloration was practiced establishing the feasibility of environment friendly natural coloration which is the ongoing research of author and a part of his research has been included here. Shade variation of natural dyed fabric may influence the uncontrollable factors for the collection of natural dyes sources on the basis of season to season, region to region and a same source but different parts of same dye sample source. So it is very tough to make a specific declaration of shading behaviors by the textile technologist but specific expertise in natural dyes extraction and natural dyeing may minimize the problem whereas expertise in dyeing with synthetic dyes and dyeing with natural dyes are not same for bulk production.

Following directions for the real shade development mentioned above:

1. Fabric should be well scoured and bleached.

2. Dyes materials should be clean, dried and powdered properly, improper drying and inappropriate powder formation may be responsible for unexpected shading and uneven dyeing.

3. Extraction time, dyes percentage may be maximized or minimized if the actual color is not observed.

4. Improper filtration of dyes may create uneven dyeing.

5. Dyes should not be powdered in wet or sticky condition which may effect the changing of light fastness and color shading.

6. Insect affected dyes source may create the changing of shade.

7. Synthetic mordanting and chemical medium dyes extraction may improve the shade but the concern of chemical and cost.

8. For open bath dyeing, liquor ratio should be higher, and stirred continuously to reduce color mark on the dyed fabric.

Acknowledgements

1. Author acknowledge to Professor (Dr.) AK Samanta, Former Head, Department of Jute and Fiber Technology, University of Calcutta and Professor (Dr.) Md. Nurul Abser, Chairman, Department of Chemistry, Jahangirnagar University for their follow up in my research work on natural coloration.

2. Author also acknowledge to RMIT University and Australian Government for RTP stipend scholarship.

3. Author also thankful to all personnel who provided different sample of natural dyes and testing support for this work.

Author details

Anowar Hossain[1,2,3]

1 School of Fashion and Textiles, RMIT University, Melbourne, Australia

2 Department of Textile Engineering, City University, Dhaka, Bangladesh

3 Department of Chemistry, Jahangirnagar University, Dhaka, Bangladesh

*Address all correspondence to: engr.anowar@yahoo.com

References

[1] BIS (Bureau of Indian Standards). Handbook of Textile Testing. 1st ed. New Delhi: BIS; 1982

[2] Chavan RB. Revival of natural dyes: A word of caution to environmentalists. Colourage. 1995;**42**(4):27

[3] Kant R. Textile dyeing industry an environmental hazard. Natural Science. 2012;**4**:22-26. DOI: 10.4236/ns.2012.41004

[4] Dedhia EM. Natural dyes. Colourage. 1998;**16**(March):45

[5] Gulrajani ML. Present status of natural dyes. Indian Journal of Fibre & Textile Research. 2001;**26**(1-2):191-201

[6] Gupta VK, Sachan RA, Singh VP, Sharma SK, Dye N. Indian Textile Journal. 1988;**108**(4):78-80

[7] Javali UC, Sreenivasa, Radhalakshmi YC. Application of natural dye, an nattoon Mulberry Silk. Colourage. 2009;**56**(3):50-52

[8] Katti MR, Kaur R, Shrihari N. Dyeing of silk with mixture of natural dyes. Colourage. 1996;**43**(12):37-39

[9] Mehra RH, Mehra AR. Natural dyes. Colourage. 1993;**40**(9):58

[10] Mahangade RR, Verma JK, Varadarajan PV, Bosco H. Influence of new mordanting technique on the colour strength and fastness properties of cotton fabric dyed with Kalindrin. Colourage. 2009;**LVI**(1):71

[11] Mathur JP, Gupta NP. Use of natural mordant in dyeing of wool. Indian Journal of Fibre & Textile Research (IJFTR). 2003;**28**:90-93

[12] Hossain A, Samanta AK. Green dyeing on cotton fabric demodulated from *Diospyros malabarica* and *Camellia sinensis* with green mordanting agent. Latest Trends in Textile and Fashion Designing. 2018;**2**(2). DOI: 10.32474/LTTFD.2018.02.000132

[13] Hossain A, Saiful Islam AKM, Samanta AK. Pollution free dyeing on cotton fabric extracted from *Swietenia macrophylla* and *Musa acuminata* as unpolluted dyes and citrus. Limon (L.) as unpolluted mordanting agent. Trends in Textile Engineering & Fashion Technology. 2018;**3**(2). DOI: 10.31031/TTEFT.2018.03.000558

[14] Hossain A, Samanta AK, Bhaumik NS, Vankar PS, Shukla D. Non-toxic coloration of cotton fabric using non-toxic colorant and nontoxic crosslinker. Journal of Textile Science and Engineering. 2018;**8**(5):374. DOI: 10.4172/2165-8064.1000374

[15] Hossain A, Samanta AK, Bhaumik NS, Vankar PS, Shukla D. Organic colouration and antimicrobial finishing of organic cotton fabric by exploiting distillated organic extraction of organic *Tectona grandis* and Azadirachta indica with organic mordanting compare to conventional inorganic mordants. International Journal of Textile Science & Engineering. 2018;**2**(01):IJTSE-113. DOI: 10.29011/ IJTSE-113/100013

[16] Anowar H, Samanta AK, Abser N, Dilruba FA. A review on technological and natural dyeing concepts for natural, dyeing along with natural finishing on natural fibre. International Journal of Textile Science & Engineering. 2019;**3**(01):126. DOI: 10.29011/IJTSE-126/100026

[17] Agarwal A, Paul S, Gupta KC. Effect of mordants on natural dyes. Indian Textile Journal. 1993;**110**(1)

[18] Gohl EPG, Vilensky LD. Textile Science – An Explanation of Fibre

Properties. 2nd ed. India: CBS Publishers and Distributors; 2003. pp. 83-88

[19] Cristea D, Vilarem G. Improving light fastness of natural dyes on cotton yarn. Dyes and Pigments. 2006;**70**(3):238-245

[20] Grierson S, David DF, Sinclair SR. The colour and fastness of natural dyes of the Scottish highlands. Journal of the Society of Dyers and Colourists. 1995;**101**(7/8):220

[21] Deepti G. Fastness properties of natural dyes. Colourage. 1999;**46**(7):35-38

[22] Lee JJ, Lee HH, Eom SI, Kim PJ. UV absorber after treatment to improve lightfastness of natural dyes on proteinic fibres. Coloration Technology. 2001;**117**(3):134

[23] Hirnori O. Improvement of light fastness of natural dyes. Coloration Technology. 2001;**117**(4):204-208

[24] Roshan P, Jayesh VM, Naik SR. Natural dyes: Classification, extraction and fastness properties. Textile Dyer & Printer. 1996;**29**(22):16

[25] Gahlot M, Fatima N, Papnai N. Jatropha flowers: Natural colourant for dyeing of cotton, wool and silk. Colourage. 2008;**55**(2):96-100

[26] Gulrajani ML, Bhaumik S, Oppermann W, Hardtmann G. Dyeing of red sandal wood on wool and nylon. Indian Journal of Fibre and Textile Research. 2003;**28**(2):221-226

[27] Parthasarthy PC, Radhakrishna PV, Rath SS, Venkataraman K. Coloring matters of wood of *Artocarpus heterophyllus* part V. Journal of the Indian Chemical Society. 1969;7(2):101-102

[28] Mahale G, Sakshi G, Sunanda RK. Silk dyed with Acalypha (*Acalypha wilkesiana*) and its fastness. Indian Journal of Fibre and Textile Research. 2003a;**28**(1):86-89

[29] Mahale G, Sakshi RK, Sunanda RK. Arecanut–A natural colourant for silk. Man-Made Textiles in India. 2003b;**46**(4):136-141

[30] Majumdar PM, Das MK, Das S. Buteamonosperma (LAM.) Kuntze-A comprehensive review. International Journal of Pharmaceutical Sciences and Nanotechnology. 2011;**4**(2):1390-1393

[31] Senthikumar S, Umashankar P, Sujatha B. Ultrasonic dyeing of cotton fabrics with Neem leaves. Indian Textile Journal. 2002;**112**(6):15-18

[32] Karnawat R, Mehta A, Mathur JP, Bhandari CS. Use of Neem bark as wool colourant - optimum conditions of wool dyeing. Indian Journal of Fibre & Textile Research (IJFTR). 2003;**28**(1):90-93

[33] Shanker R, Vankar PS. Dyeing wool with Gomphrenaglobosa flower. Colourage. 2005;**52**(4):35-38

[34] Sidhu SP, Grewal JK. Silk dyed with goldendrop (Onosmaechioides) dye. Colourage. 2008;**55**(3):51-56

[35] Vankar PS, Shanker R. Ultrasonic dyeing of cotton and silk with nerium oleander flower. Colourage. 2008;**55**(4):90-94

[36] Srivastava M, Pareek M, Valentina P. A Comparative Study on Dyeing of Wool with Kesula Flowers and Khakre Ka Gond. Colourage. 2006;**LIII**(2):57-59

[37] Siddiqui I, Khaleeque M, Gouse M. Evaluation of colour fastness properties of cotton and silk fabrics dyed with Parijataka (Nyctanthes Arbor) flowers pigment. Colourage. 2008;**55**(10):54-60

[38] Kumar PE, Kulandaivelu AR, Boopathy TS. Eco-friendly natural dyes from Hibiscus Vilifolins and Sesbania

Aegyptiaca for dyeing. Woven Wonder: The Tradition of Indian Textiles. 2005;**LII**(10):53

[39] Vinod KN, Puttaswamy KNN, Gowda N, Sudhakar R. Natural colorant from the bark of macaranga peltata: Kinetic and adsorption studies on silk. Coloration Technology. 2010;**126**(1):48-53

[40] Khan A, Khan M, Srivastava PK, Mohammed F. Extraction of natural dyes from cutch, Ratanjot and madder and their application on wool. Colourage. 2006;**53**(1):61-68

[41] Katyayini VKLT, Jacob M. Dye extraction from Mesta calyx and its assessment. Colourage. 1999;**46**(7):39-42

[42] Gulrajani ML, Gupta D, Agarwal V, Jain M. Some studies on natural dyes: Part II – flavonoids – Kapila/Onion/Tesu. Indian Textile Journal. 1992;**102**(5):78-84

[43] Khan MA, Khan M, Srivastava PK, Mohammed F. Extraction of natural dyes from myrobalan, gallnut and pomegranate and their application on wool. Colourage. 2005;**52**(12):53-60

[44] Saha P, Datta S. Dyeing of textile fibre using Marigold flower as floral dye. Colourage. 2008;**LV**(5):52-56

[45] Mahale G, Sunanda RK, Sabshi. Eco dyeing of animal fibre wool with Marigold. Man-Made Textiles in India. 2006;**49**(9):355

[46] Sudhakar R, Gowda KNN, Padabi NV. Natural dyeing of silk with the nut extract of Areca catechu. Colourage. 2006;**LIII**(7):61

[47] Khanna P, Dedhia EM. Natural colour from Ratanjot and eucalyptus leaves. Colourage. 1999;**46**(10):25

[48] Bhatti IA, Adeel S, Rani S, Gulzar T, Naz S, Mansour HF. Influence of UV radiation on extraction and dyeing of cotton fabrics using Eucalyptus bark powder as natural dye. Research Journal of Textile and Apparel. 2012;**16**(2)

[49] Vankar P, Mishra A, Ghorpade I, Tiwari V. Sonicator dyeing of cotton with eucalyptus bark using copper and chromium mordants, yet eco friendly. Colourage. 2000;**47**(9):25

[50] Patra SK, Nanda BN, Nayak A, Das NB. Yellow dye from debarked Jack fruit' wood. Colourage. 2000;**XLVII**(8):17

[51] Bains S, Kang S, Kaur K. Effect of combination of mordants on colour fastness properties of cotton dyed with peach (Prunus persica) dye. Man-Made Textiles in India. 2003;**46**(6):230-233

[52] Adeel S, Rehman FU, Iqbal MU, Habib N, Kiran S, Zuber M, et al. Ultrasonic assisted sustainable dyeing of mordanted silk fabric using arjun (*Terminalia arjuna*) bark extracts. Environmental Progress & Sustainable Energy. 2019;**38**(s1):S331-S339

[53] Sinha K, Aikat K, Das P, Datta S. Dyeing of modified cotton fiber with natural *Terminalia arjuna* dye: Optimization of dyeing parameters using response surface methodology. Environmental Progress & Sustainable Energy. 2016;**35**(3). DOI: 10.1002/ep

[54] Ibrahim KM et al. Eco-friendly shades on wool using mixed mordants with Acacia catechu (Cutch). Colourage. 2010;**57**(8):81-88

[55] Kawahito M, Urakawa H, Ueda M, Kajiwara K. Colour in cloth dyed with natural indigo and synthetic indigo. Fiber. 2002;**58**(4):122-128

[56] Kasiri Masoud B, Safapour S. Natural dyes and antimicrobials for green treatment of textiles. Environmental Chemistry Letters. 2014;**12**(1):1-13

[57] Shukla SR, Sundar PS. Anti-microbial and moth proofing of textiles. Asian Dyer. 2008;**5**(2):37-39

[58] Samanta AK, Agarwal P. Dyeing of jute and cotton fabrics using jackfruit wood extract: Part I-effects of mordanting and dyeing process variables on colour yield and colour fastness properties. Indian Journal of Fibre and Textile Research. 2007;**32**(4):466-476

[59] Samanta AK, Agarwal P. Application of mixtures of red sandal wood and other natural dyes for dyeing of jute fabric–studies on dye compatibility. International Dyer. 2008a;**193**(2):37-41

[60] Samanta AK, Agarwal P, Datta S. Dyeing of jute with binary mixtures of jackfruit wood and other natural dyes - study on colour performance and dye compatibility. Indian Journal of Fibre and Textile Research. 2008b;**33**(2):171-180

[61] Samanta AK, Agarwal P, Konar A, Datta S. Characterization and application of purified selective dyes on jute. International Dyer. 2008c;**193**(3):25-32

[62] Samanta AK, Agarwal P, Singhee D, Datta S. Application of single and mixtures of red sandalwood and other natural dyes for dyeing of jute fabric: Studies on colour parameters/colour fastness and compatibility. Journal of the Textile Institute. 2009;**100**(7):565-587

[63] Samanta AK, Konar A. Technical Handbook on Natural Dye and Colouration. Kolkata, India: Department of Jute and Fibre Technology, IJT (Calcutta University); 2012a. pp. 97-113

[64] Samanta AK, Konar A. Standardization of methodologies for application of natural dyes on jute. Standards Overview - Bureau of Indian Standards. 2015;**28**(5):22-40

[65] Samanta AK, Konar A, Agarwal P, Datta S. Effects of chemical modifications of jute fabric with EDA and hydrazine on its textile related mechanical properties and dyeability. Journal of Polymer Materials. 2010;**27**(3):203-236

[66] Samanta AK, Konar A, Chakraborti S, Datta S. Dyeing of jute fabric with tesu extract: Part-I: Effects of different mordants and dyeing process variables. Indian Journal of Fibre and Textile Research. 2011;**36**(1):63-73

[67] Samanta AK, Konar A, Datta S. Dyeing of jute fabric with tesu extract: Part-II – Thermodynamic parameters and kinetics of dyeing. Indian Journal of Fibre and Textile Research. 2012b;**37**(2):172-177

[68] Samanta AK, Deepali S, Mitu S. Application of single and mixture of selective natural dyes on cotton fabrics a scientific approach. Colourage. 2003;**50**(10):29-42

[69] Shah HS, Gandhi RS. Instrumental Colour Measurements and Computer Aided Colour Matching for Textiles. Ahmedabad, India: Mahajan Book Distributers; 1990. pp. 76-84

[70] Yoshizumi K, Crews PC. Characteristics of fading of wool cloth dyed with selected natural dyestuffs on the basis of solar radiant energy. Dyes and Pigments. 2003;**58**(3):197-204

Section 4

Colour-Based Natural Dyes and Pigments

Chapter 7

Betanin: A Red-Violet Pigment - Chemistry and Applications

Deepak Devadiga and T.N. Ahipa

Abstract

Nowadays, the demand for eco-friendly/nontoxic natural colorants is growing as an essential alternative to potentially harmful synthetic dyes. Betanin is the chief red pigment of beetroot, and it is the only betalain approved for use in food and pharmaceutical products as a natural red colorant. This chapter is mainly dealing with the betanin pigment, and also the chapter is subdivided into six sections covering the chemistry of betanin, extraction of color using various novel techniques (like microwave- and ultrasonic-assisted extraction) from raw plant material, biosynthesis of betanin followed by chemical synthesis of betanidin, and also the effect of pH, temperature, and light on the stability of betanin followed by its applications.

Keywords: beetroots, betanin, red pigment, extraction, chemical synthesis, stability

1. Introduction

Vegetable beetroot (*Beta vulgaris* L.) has the notable scientific interest, because it is a rich source of nitrate (NO_3^-), a compound with advantageous cardiovascular health effects through the endogen production of nitric oxide (NO) [1, 2]. There are two classes of pigment in plants, i.e., betalains and anthocyanins. Beetroots are the chief sources of betalains which is a water-soluble nitrogen pigment with heterocyclic ring, which can be further subdivided into two classes depending on chemical structure: betaxanthins comprising indicaxanthin; vulgaxanthin I and II, accountable for orange-yellow coloring; and betacyanins, such as betanin, isobetanin, neobetanin, and prebetanin, accountable for red-violet coloring [3, 4]. The most abundant betacyanin is betanin (betanidin 5-O-β-D-glucoside) and is the only pigment which is an approved natural colorant for the use in food products prescribed by the Food and Drug Administration (FDA) in the United States [5, 6]. **Figure 1** represents the chemical structure of betanin.

According to experimental studies, raw beetroot generally contains water (87.1%), carbohydrate (7.6%), protein (1.7%), fat (0.1%), and betanin (0.03–0.06%) [7]. In addition to natural food colorant property, betalain also exhibits antimicrobial, antiviral, and antioxidant activities [8]. Moreover, beetroot dye has nutrient value along with nontoxic nature; therefore, it even finds application in dyeing industry where the health aspect is a foremost criterion. Also, this natural dye extracted from beetroot is eco-friendly in nature and does not cause any environmental problems in contrast to the commercially available synthetic dyes [7].

IntechOpen

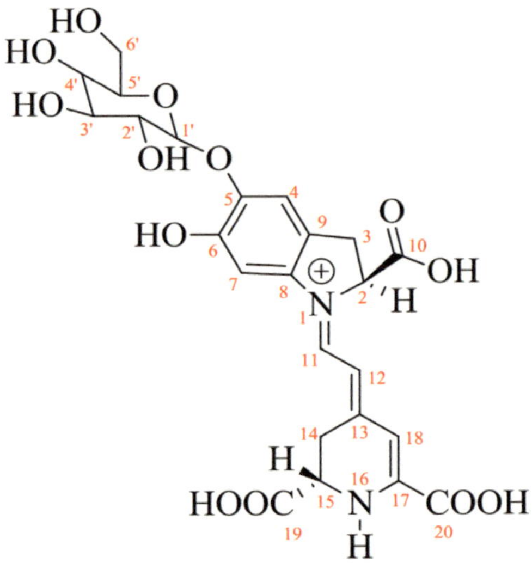

Figure 1.
Chemical structure of betanin.

2. Chemistry of betanin

The first report on the crystallization of betanin was communicated by two independent groups Schmidt and Schonleben [9] and Wyler and Dreiding [10]. These two groups employed an electrophoretic strategy for betanin purification. Wyler and Dreiding [11] recognized three products which were formed by the alkaline degradation of betanidin (**Figure 2**); these were 4-methylpyridine-2,6-dicarboxylic acid, formic acid, and S-cyclodopa(5,6-dihydroxy-2,3-dihydroindole 2-carboxylic acid). When these three products are placed in correct relationship with one another, they form the betanidin's carbon skeleton and also the configuration of the second carbon [12].

2.1 Position of the β-D-glucosyl group

To identify the position of β-D-glucosyl group, betanin was reacted with diazomethane in the presence of air yielded tetramethyl derivative of neobetanidin which was then converted to 6-methoxy-neobetanin-trimethylester by hydrolysis and acetylation [13, 14]. In the alkali condition, neobetanin cleaved to yield 5-hydroxy-6-methoxyindole 2-carboxylic acid. Its methyl ester was prepared from

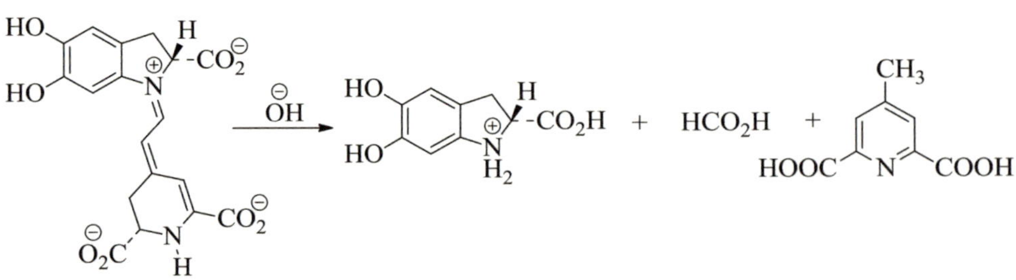

Figure 2.
Alkaline degradation of betanidin.

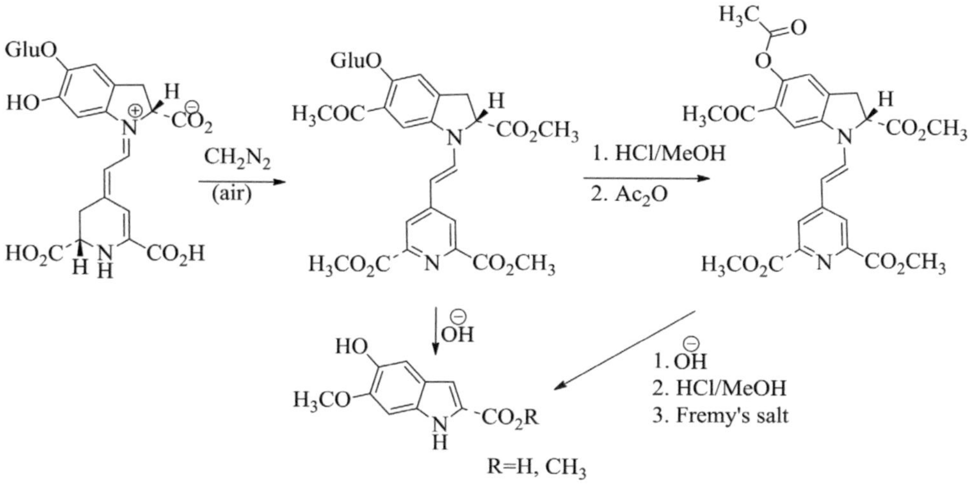

Figure 3.
Reactions that revealed the glucosyl residue position in betanin.

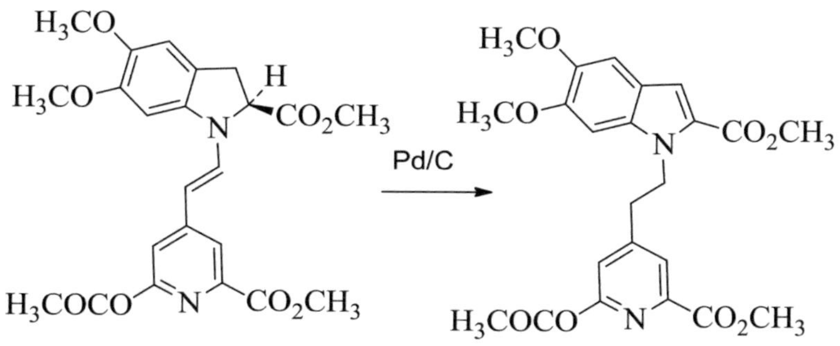

Figure 4.
Reaction which confirmed the presence of vinylene connecting group.

acetylated molecule by degrading in basic condition, and further it was esterified and oxidized. Further, analysis of all these reactions together revealed the glucosyl residue position (**Figure 3**). In addition, NMR of betanin in trifluoroacetic acid and hydrolytic studies of β-glucosidase [15] revealed that betanin was an O-β-D-glucopyranoside.

The yellow-colored compound, i.e., 5,6-di-O-methylneobetanidin trimethyl ester, converted into colorless compound, i.e., 5,6-di-O-methyl-2,3-dehydro-11,12-dihydro-betanidin trimethyl ester, by the palladium-catalyzed disproportionation reaction which confirmed the existence of vinylene connecting group (**Figure 4**) in the derivatives of neobetanidin [12].

Further, betanin can also be interconverted into betanidin. To prepare betanidin, betanin is initially reacted with the excess of L-proline in the presence of dilute ammonia which results in the formation of indicaxanthin which can later be converted into betanidin by reacting it with excess S-cyclodopa [12].

3. Extraction techniques

The drawbacks of conventional approaches, such as time-consuming methods, safety risks with some hazardous solvent systems, contaminated product, and comparatively less yields, have increased the demand for the novel processing methods

Variants	Solvents
V_1	Distilled water
V_2	1% citric acid
V_3	0.5% citric acid
V_4	0.2% citric acid
V_5	0.1% ascorbic acid
V_6	50% ethanol
V_7	20% ethanol
V_8	0.5% citric acid + 0.1% ascorbic acid
V_9	0.2% citric acid + 0.1% ascorbic acid
V_{10}	20% ethanol + 1% citric acid
V_{11}	20% ethanol + 0.5% citric acid

Table 1.
Betanin extraction from beetroot using different solvent systems.

[16, 17]. To improve the betalain extraction, some of the pretreatment methods have been proposed, such as pulsed electric fields [18], gamma irradiation [8], and low-direct current electric fields [19]. But, such methods are quite expensive when compared to the solvent extraction techniques. Further, certain nonthermal techniques, like ultrasound (sonication) processing, and microwave-assisted extractions are also significantly productive in order to enhance the extraction yields of bioactive molecules with minimum degradation [17].

Neagu and Barbu [20] studied the betanin extraction from beetroot using different solvent systems by solid-liquid extraction technique (liquid/solid ratio is 5:1). **Table 1** presents the different extraction solvents used in this study. Results revealed that the highest betanin content of about 20 mg/g of beetroot was obtained with the use of weak acid solution (i.e., V_8, using 0.5% citric acid +0.1% ascorbic acid). Also, they extracted the considerable amount of betanins by using ascorbic acid added solutions. Thus, it is clear that the acidic medium influences positively during the low-temperature extraction process.

In case of ultrasound (sonication) processing, ultrasonic-assisted extraction approach requires the use of ultrasound (with 20–2000 kHz frequencies range) which generally increases the cell wall permeability and generates the cavitations. Because of cavitations, cell membrane breaks down, and internal materials (color and oil) come out [21].

On the other hand, microwave-assisted extractions are also contributed significantly to speed up the sample digestion and extraction of bioactive molecules from matrices. Here, the microwave energy has been utilized and employed in this extraction process. Moreover, this microwave energy induces molecular motion by the rotations of dipoles and migration of ions without varying the structure of the molecules provided the temperature of the system is not too high [21].

4. Biosynthesis of betanin

The biosynthetic pathways for the betanin molecule are depicted in **Figure 5** [22]. Three enzymes such as 4,5-DOPA (dihydroxyphenylalanine)-extradiol-dioxygenase, tyrosinase, and betanidin-glucosyltransferase were involved in the biosynthesis of betalains in the cytoplasm [23]. From arogenic acid, the amino acid L-tyrosine was

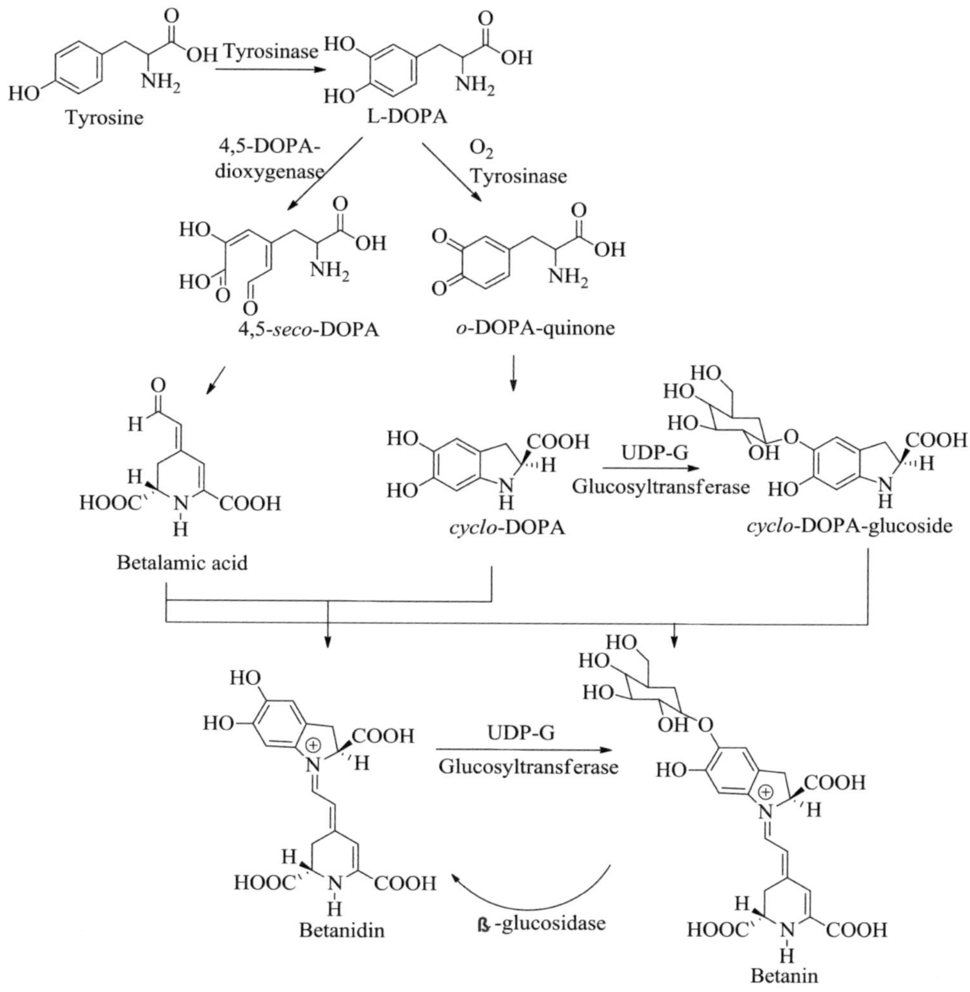

Figure 5.
Biosynthesis of betanin.

formed enzymatically over the shikimate pathway [24], which is the starting material for the biosynthesis of L-DOPA [25]. During this conversion, tyrosinase enzyme helped to convert tyrosine to DOPA through hydroxylation. Further, between carbons 4 and 5 of the L-DOPA cyclic ring was opened by 4,5-DOPA-extradiol dioxygenase enzyme to give 4,5-*seco*-DOPA [26–28] which is then converted into betalamic acid by intramolecular condensation between aldehyde and amine groups [25]. Further, L-DOPA is transformed into *o*-DOPAquinone in the presence of molecular oxygen [29]. Furthermore, it was spontaneously cyclized due to the nucleophilic attack of amino group on the ring system to yield *cyclo*-DOPA [23]. However, *cyclo*-DOPA can also be obtained from the cyclization of L-DOPA in the presence of cytochrome P450 (CYP76AD1) [30].

Finally, betanidin was formed by the formation of imine bond between *cyclo*-DOPA and betalamic acid which is then converted to betanin using betanidin-5-O-glucosyltransferase enzyme by connecting glucose unit of uridine diphosphate-glucose (UDP-G) to the hydroxyl group in position 5 [31]. But this reaction can be reversed in the presence of β-glucosidase [32]. Additionally, it is also concluded that enzyme *cyclo*-DOPA-5-O-glucosyltransferase catalyzes the transport of glucose molecule on *cyclo*-DOPA, by which the *cyclo*-DOPA-glucoside condense with betalamic acid to yield the betanin [33].

5. Chemical synthesis of betanidin

The chemical synthesis of betanidin is illustrated in **Figure 6** [34]. For the synthesis of betanidin, 4-hydroxypyridine-2,6-dicarboxylic acid is used as the starting material which upon hydrogenation and followed by esterification yields all products in *cis* form. Further, Pfitzner-Moffatt oxidation reaction converted the secondary hydroxyl group to a ketone. In the next step, it is converted to semicarbazide using Horner-Wittig reagent. Then, the obtained semicarbazide is further hydrolyzed to give unsaturated ketone, which is converted to betalamic acid by Pfitzner-Moffatt oxidation. Reacting betalamic acid with L-cyclo-DOPA methyl ester yielded betanidin trimethyl ester which is then converted to betanidin through acid hydrolysis by using concentrated hydrochloric acid.

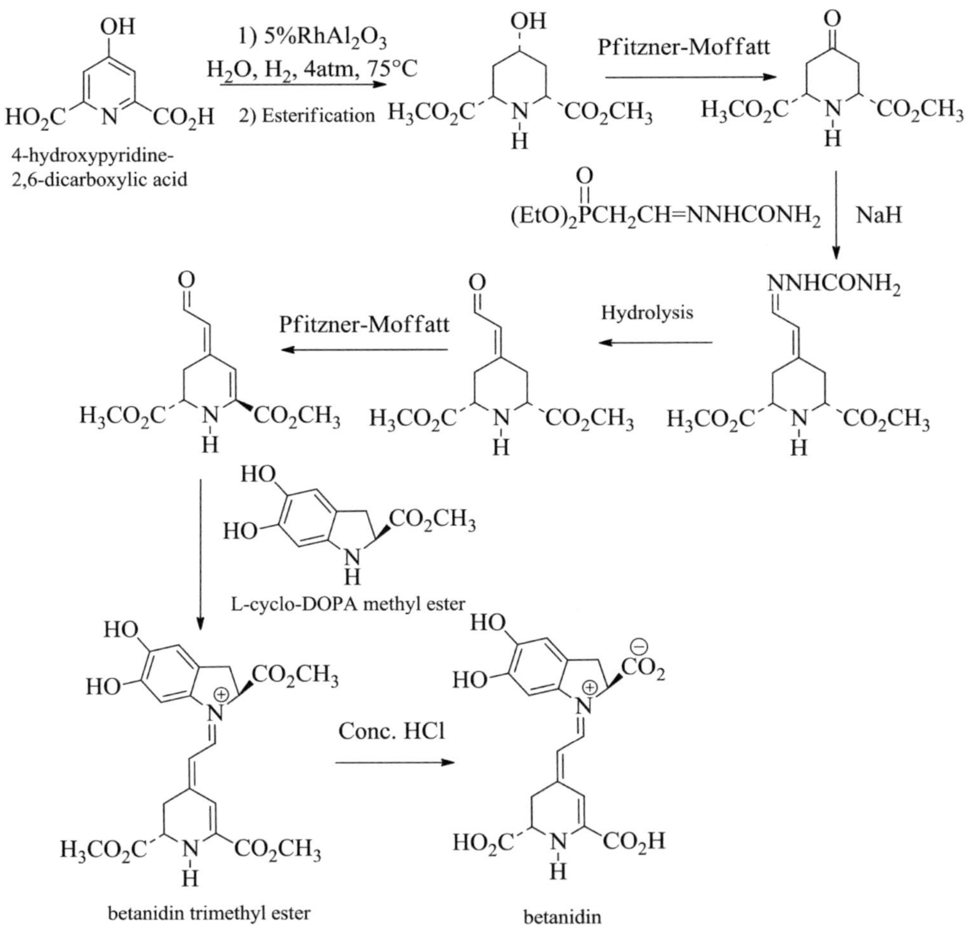

Figure 6.
Chemical synthesis of betanidin.

6. Characterization of betanin

6.1 UV-vis absorption spectra of beetroot extracts

The UV-vis spectra of beetroot extracts in different solvents (ethanol, methanol, and water) are depicted in **Figure 7** [35]. Strong absorption band was observed at around 530 nm in the visible range for the red beet juice which was attributed to the betanin pigment. As the solvent changed from ethanol (532 nm) to water (542 nm) and to methanol (544 nm), absorption maximum shifted towards longer wavelengths. These results showed that methanolic and aqueous extracts mainly contain

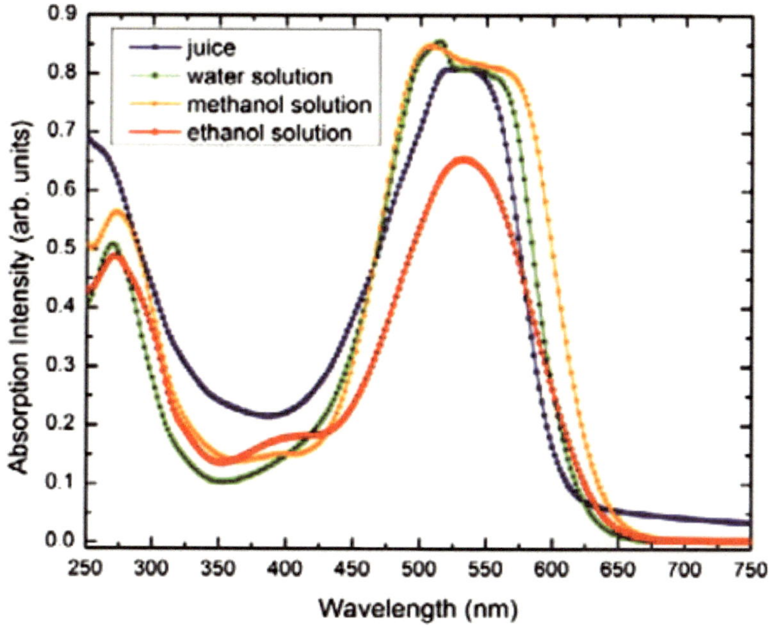

Figure 7.
The UV-vis spectra of beetroot extracts in different solvents (ethanol, methanol, and water).

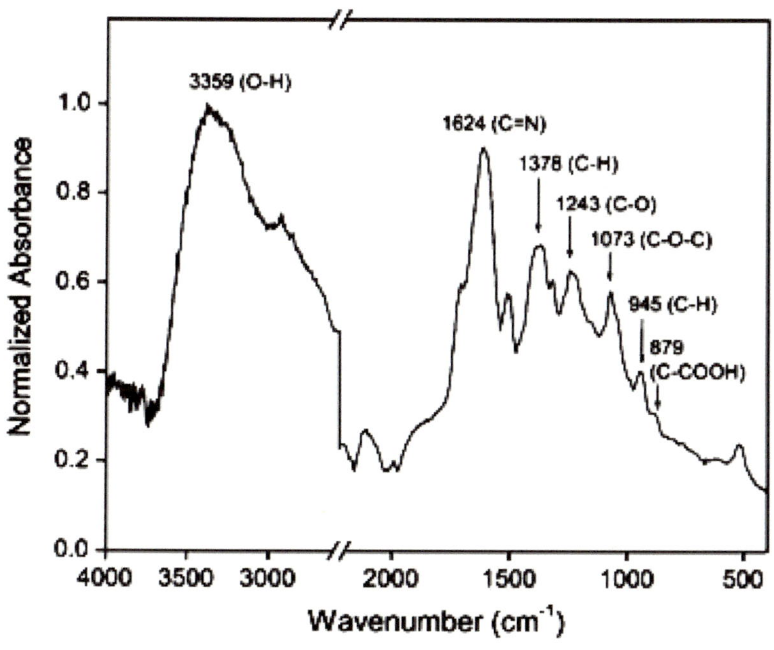

Figure 8.
FT-IR spectrum of betanin (scale range, 400–4000 cm^{-1}).

betanidin, whereas ethanolic extract mainly contains betanin. The intensity of absorption maxima for the aqueous and methanolic extracts is approximately equal, higher than the ethanolic extract.

Further, partially overlapped two absorption bands were observed in case of aqueous and methanolic extracts. For aqueous extract the second band has an absorption maximum at 515 nm and for aqueous extract at 509 nm. These bands are only observed in high concentrated extract, and it vanishes as the solution diluted. Hence, it can be attributed to the formation of supramolecular structure in the concentrated solutions.

6.2 FT-IR data of betanin

Different characteristic absorption bands corresponding to the functional groups of betanins were observed, and its FT-IR spectrum is shown in **Figure 8** [36]. The absorption band around 3359 cm^{-1} was ascribed to the —OH bond stretching vibration [37]; on the other hand, the absorption band around 1624 cm^{-1} was ascribed to the C=N bond stretching vibration [38, 39]. The next absorption band located at 1378 cm^{-1} was ascribed to the C—H bond extension stretching vibration, while the absorption band at 1243 cm^{-1} was ascribed to the C—O bond of the carboxylic acid stretching vibration [38, 39]. Another absorption band centered at 1073 cm^{-1} was ascribed to the C—O—C linked symmetric stretching vibration [40], the absorption band at 945 cm^{-1} was ascribed to the C—H bond deformation, and lastly the absorption band at 879 cm^{-1} was ascribed to the C—COOH bond stretching vibrations [41].

6.3 ^{1}H, ^{13}C, and LC-^{1}H NMR data of betanin

The ^{1}H and ^{13}C NMR data of betanin was obtained by dissolving it in D_2O, and its LC-^{1}H NMR data was also obtained by dissolving it in acetonitrile (MeCN)/D_2O/0.05% trifluoroacetic acid (TFA) using 500 MHz frequency at 25°C. **Figure 9** represents the LC-^{1}H NMR spectrum of betanin and followed by **Table 2** which represents the ^{1}H, ^{13}C, and LC-^{1}H NMR data of betanin [42].

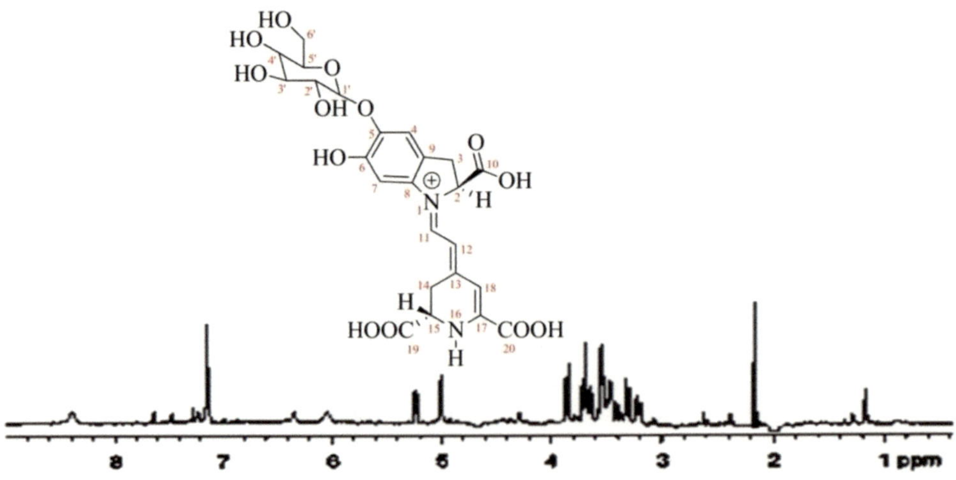

Figure 9.
LC-^{1}H NMR spectrum of betanin.

Atom numbering	^{1}H NMR δ-[ppm], mult, J [Hz]	^{13}C NMR δ-[ppm], mult, J [Hz]	LC-^{1}H NMR δ-[ppm], mult, J [Hz]
2	4.92, *dd*, 3.1, 10.3	65.0	5.23, *dd*, 2.0, 10.2
3a/b	3.53, *dd*, 11.5, 16.9 3.10, *dd*, 4.3, 16.8	32.7	3.66, *dd*, 10.4, 16.7 3.31, *dd*, 2.0, 16.7
4	7.06, *s*	113.9	7.13, *s*
5	—	144.0	—
6	—	146.1	—
7	6.98, *bs*	100.0	7.12, *bs*
8	—	137.4	—
9	—	124.1	—
10	—	175.8	—
11	8.19, *bs* (*d*, 12.6)[b]	144.4	8.38, *bd*, ≈11
12	5.84, *bs* (*d*, 12.6)[b]	106.9	6.04, *bd*, ≈11
13	—	—[a]	—
14a/b	3.20, *bm* 3.12, *bm*	26.5	3.21, *dd*, 7.4, 17.2 3.41–3.59 (overlap)
15	4.40, *bt*, 7.1	53.1	Overlapped by D_2O
17	—	—[a]	—
18	6.22, *bs*	—[a]	6.34, *bs*
19	—	—[a]	—
20	—	—[a]	—
1′	4.98, *d*, 7.4	101.4	5.00, *d*, 6.9
2′	3.55 (overlap)	75.7	3.41–3.59 (overlap)
3′	3.55 (overlap)	73.9	3.41–3.59 (overlap)
4′	3.41 (overlap)	69.3	3.41–3.59 (overlap)
5′	3.52 (overlap)	76.2	3.41–3.59 (overlap)
6′a/b	3.85, *dd*, 1.6, 12.3 3.70, *dd*, 5.3, 12.3	60.6	3.86, *dd*, 1.4, 12.6 3.70, *dd*, 5.5, 12.6

[a]*Chemical shifts were not observable.*
[b]*After acidification (TFA) to pH 2, s = singlet, d = doublet, dd = doublet of doublet, bs = broad singlet, bd = broad doublet, bt = broad triplet, bm = broad multiplet.*

Table 2.
^{1}H, ^{13}C, and LC-^{1}H NMR data of betanin.

6.4 Mass spectrum of betanin

The mass spectrometry of betanin in the positive ionization mode exhibited a molecular ion (*m/z* 551, $[M+H]^+$, 100%). **Figure 10** represents the obtained mass spectrum of betanin molecule [43].

6.5 Thermogravimetric (TG) analysis

The dynamic thermogravimetric (TG) analysis was conducted on fresh betanin and dried betanin (**Figure 11a** and **b**) [44]. Since fresh betanin contains water, immediate mass loss is noted in the temperature range of 40–100°C. Further, the degradation temperature of betanin dye is noted at the temperature of about 204°C.

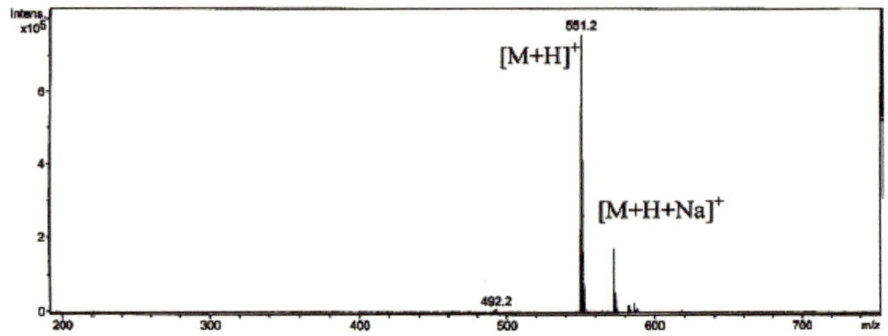

Figure 10.
Mass spectrum of betanin.

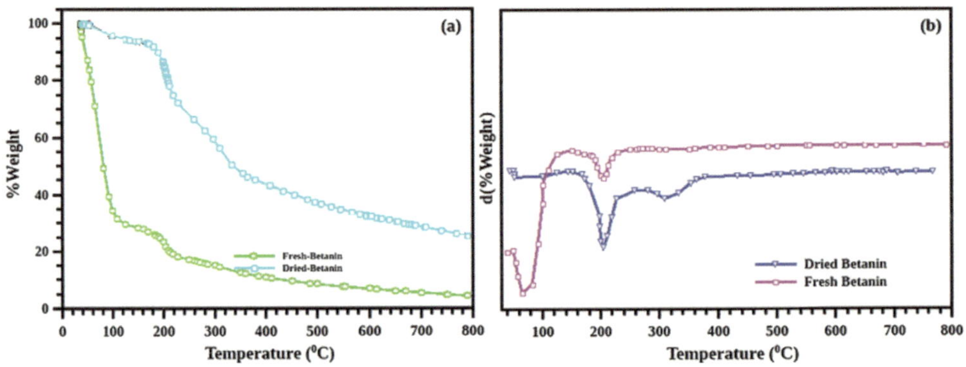

Figure 11.
(a) TG curves and (b) derivative TG curves of fresh and dried betanin.

Furthermore, the dried sample has also exhibited similar behavior except the mass loss at the beginning (in the temperature range of 40–100°C).

7. Factors effecting the stability of betanin

- *pH*: in the buffers of pH 2–9, betanin was stored at 4°C for 7 days and measured the visible spectra at both starting and end of this time span. No shifts in the absorption maxima were noted in between pH 4 and pH 7. A shift of about 2 nm towards a shorter wavelength with a decreased absorbance intensity was observed in case of the buffer which has pH less than 4. In the 575–650 nm region, the spectrum has slightly increased absorbance, and the solution color changed to red-violet from red. While in case of the solutions which has pH value of above 7, i.e., at pH 9, the absorption maximum moved to a longer wavelength region (544 nm) by decreasing the intensity. In the 575–650 nm and 400–450 nm wavelength regions, the absorbance increased to a considerable extent, and the solution color changed to violet from red. These results showed that between pH 3 and pH 7, storage had no effect on betanin solutions, and above and below of these pH values causes the considerable losses of betanin. Visible spectra of betanin compound at pH 2, 5, and 9 are illustrated in **Figure 12** [12, 45].

- *Temperature*: on heating the red color of betanin solutions starts to diminish, and finally it turns to brown color. The color loss was followed by the betanin assay, and the rate indicates that it follows first-order kinetics. The graph indicates that at 100°C, the degradation rate at pH 5 is still less than it is at pH 3 and pH 7. Further, the betanin compound is more stable between pH 4 and 5. However, betanin in beet juice is far more stable at pH 5, which reveals a protective effect by the constituents of juice. The rates of degradation for

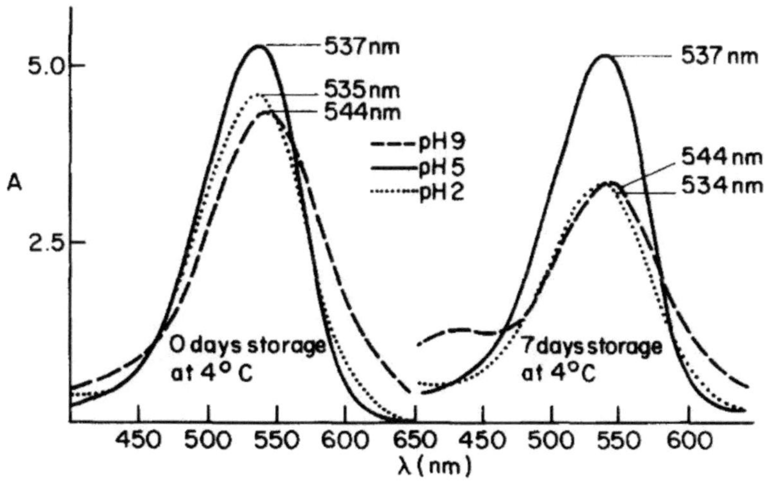

Figure 12.
Visible spectra of betanin compound at pH 2, 5, and 9.

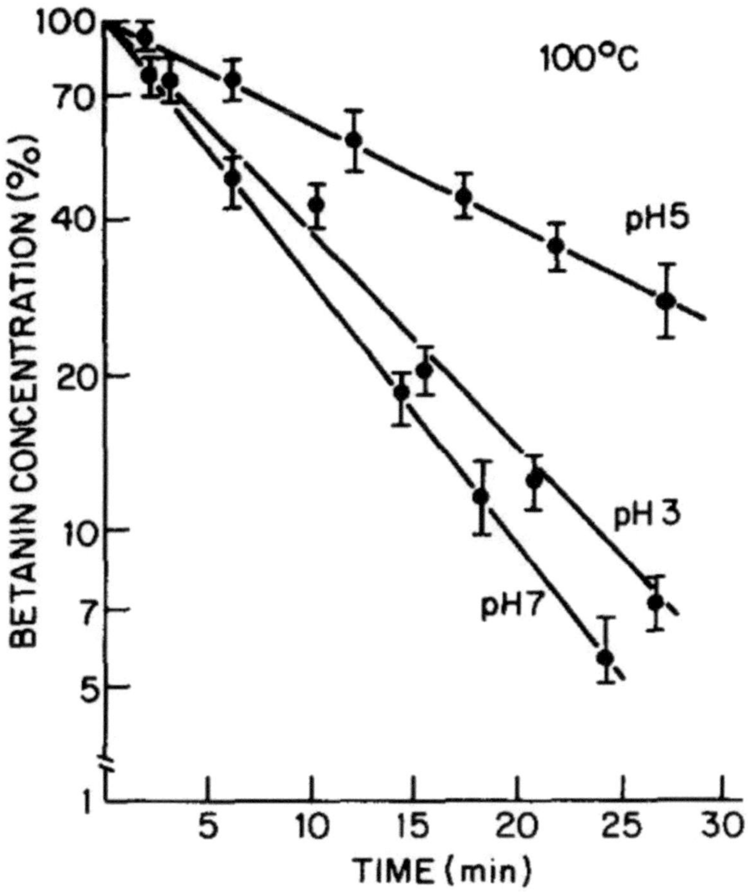

Figure 13.
Rates of degradation for betanin molecule in a system at 100°C at pH 3, 5, and 7.

betanin molecule in a system at 100°C at pH 3, 5, and 7 are depicted in **Figure 13** [12, 45].

- *Light*: at 15°C and pH 7, the presence of light increased the rate of degradation by 15.6% and air (rather than nitrogen) by 14.6%. Both light and air together increased the rate by 28.6% [12, 42].

8. Applicability of betanin

8.1 Dyeing acrylic fabric

Pure betanin dye can compete with synthetic dyes in color depth shade properties and in color fastness properties. Guesmi et al. [43] studied the dyeing of betanin on modified acrylic fabrics and evaluated the effect of dye bath pH, salt concentration, dyeing time, and temperature on dyeing. In a dye bath having sodium chloride (0–15 g/L) and a dye of 30 mg/L concentration with the 40:1 liquor ratio, modified acrylic fabric was dyed using conventional heating.

8.1.1 Effect of pH on dyeing

Over the pH range 1–5, increase in pH increases the adsorption of betanin onto acrylic fabric. Color strength decreases as the pH increased above 5. Generally, amino functional groups of acrylic fibers get protonated as the pH value decreases. Thus, ion-ion forces induced with ionized carboxyl groups in betanin. Betanin may exist in cationized or on monoanion form in a strongly acidic environment which results in the lower depth of dyeing at pH less than 4, and also it is due to the betanin stability loss at low pH [43].

At pH 5 maximal color strength was observed, whereas at pH 4, little decrease in color strength was observed; this is attributed to the increased carboxyl groups in this range and to the high thermal stability of betanin molecule. The number of protonated terminal amino functional groups of fabric decreases at pH > 5, which causes the decreased ionic interaction between the carboxylate anion of the dye and acrylic fibers, thus lowering its dye ability. The structures of betanin molecule as a pH varied are depicted in **Figure 14** [43].

8.1.2 Effect of salt addition

Color strength decreases as the salt concentration increases, hence dyeing without salt addition is the best condition [43].

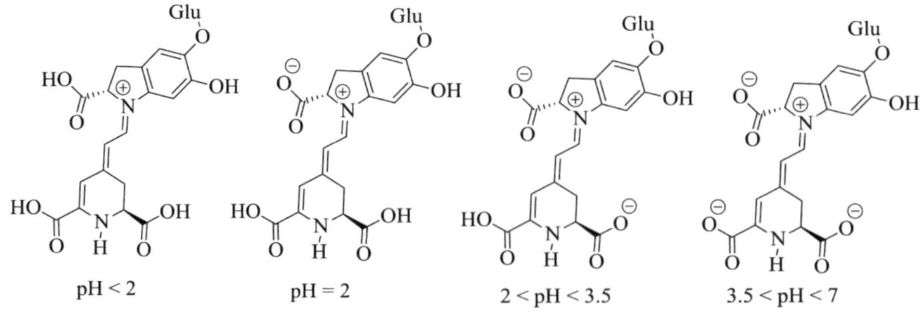

Figure 14.
Structures of betanin molecule as a pH varied [46].

8.1.3 Effect of temperature on dyeing

As the temperature of dyeing increases, the color strength increases up to 50°C, and further by increasing the dyeing temperature, the color strength decreased slowly which is attributed to the decrease in stability of dye at higher temperatures [43].

8.1.4 Effect of time span on dyeing

Color strength increases as the dyeing time increases up to 30 min; from 30 to 45 min, there is no change in color strength, and then it started to decrease as the time increases [43].

8.1.5 Color fastness

The fastness properties of betanin-dyed acrylic fabrics are shown in **Table 3**. The fastness properties of the dyed samples were examined according to ISO standard methods, the specific tests conducted for color fastness to rubbing is as per ISO 105-X12:1987, the color fastness to water is as per ISO 105-E01:1989, the color fastness to washing is as per ISO 105-C02:1989, and the color fastness to light is as per ISO 105-B02:1988 (carbon arc) [43]. It was noted that rubbing, washing, and water fastness of unmordanted acrylic fabrics exhibited significantly good property. But, the light fastness of unmordanted acrylic fabrics was found to be bad. However, light fastness was found to increase from rating 3* to 4* in premordanted fabrics using manganese sulfate and ferrous sulfate, and light fastness was increased from rating 3* to 5* by using cobalt sulfate. Nevertheless, the other mordants did not affect the light fastness of premordanted fabrics. It was found that for the improvements of color strength and light fastness, cobalt sulfate was established as the best mordant [43].

Similarly, Guesmi et al. [46] in the year 2013 studied the dyeing of wool fabric using betanin and chlorophyll-a as biomordant. In a dye bath having sodium chloride (0–5 g/L) and a dye with 40:1 liquor ratio, wool fabric was dyed using conventional heating. Results revealed that the increase in the concentration of biomordant increases the color strength values. They also investigated the effect of variables on the color of dyed fibers and noted that from pH 3.5 to pH 4.5, the color strength considerably increases, the color strength was found to be better without salt than with salt, and the color strength of dyed wool increases as the increase in temperature was up to 40°C and starts to decrease slowly till 50°C. Further increase in temperature, the color strength decreases in pronounced manner. According to the authors, color strength increases with the time span (up to 45 min) of dyeing, and

	Fastness to dry rubbing	**Fastness to wet rubbing**	**Fastness to washing**	**Fastness to water**	**Fastness to light**
None	5*	5*	4*–5*	4*–5*	3*
Manganese sulfate	5*	5*	4*	4*	4*
Ferrous sulfate	4*	4*	3*	3*	3*
Zinc sulfate	5*	5*	4*–5*	4*–5*	4*
Aluminum potassium sulfate	4*	4*	3*	3*	3*
Cobalt sulfate	5*	5*	4*–5*	4*–5*	5*

Table 3.
The fastness properties of betanin-dyed acrylic fabrics represented with the rating scale using star (*).

then it starts to decrease because betanin losses thermal stability, and also it starts to escape from the fiber. Dye exhaustion was examined in both ultrasonic and conventional dyeing approaches. It was exhibited that in a shorter time span of dyeing, sonication increases the dye exhaustion from rating of 30% to rating of 60%. The fastness properties of dyed wool were studied against wet rubbing, light, washing, and dry rubbing. Unmordanted and mordanted samples have good fastness properties too.

8.2 Medicinal application

Antioxidant activity of betanin in biological lipid domain has been exhibited in human macromolecules, like lipoproteins of low density, whole cells, and membranes [2]. Moreover, betanin has attracted researchers because of its anti-inflammatory activities and hepatic safety activities in whole human cells [47]. In cultured endothelia cells, this molecule regulates the redox-mediated signal transduction pathways which is required in responses during inflammation, and betanin also showed antiproliferative effects on tumor cell lines in human [48–50]. In both tumoral and healthy hepatic cell lines in the human body, betanin translocates the antioxidant response element (erythroid 2-related factor 2 (Nrf2)) from the place of cytosol to the place of nuclear domain, which regulates m-RNA and protein levels of antioxidant/detoxifying enzymes, which includes heme oxygenase-1, NAD(P)H quinone dehydrogenase-1, and glutathione S-transferase and, in these cells, bears anticarcinogenic and hepatoprotective effects [51]. Also, it exhibits antidiabetes properties by controlling the activities of liver markers enzymes [52–54].

8.3 Betanin as food colorant

Betanin is the oldest and most abundant red food colorant which has been established in the market, which is noted as E-162 in the European Union and in the United Sates; it is known as 73.40 in the twenty-first chapter of the Code of Federal Regulations (CFR) section of the Food and Drug Earth Administration [2, 5, 6].

Betanins are most commonly used for coloring of ice cream and powdered soft drink beverages. Additionally, betanin is used in some of the sugar confectionery, like sugar coatings, fruit or ice cream fillings, fondants, and sugar strands. At the final part of the processing, it can be added while preparing hot processed candies. Also, it is used in soups as well as bacon and tomato.

9. Summary

In this chapter, the first and second section covered the chemistry of betanin which contains reactions that revealed the glucosyl residue position and presence of vinylene connecting group in betanin. Further, the third section narrated the extraction techniques which mainly included the microwave- and ultrasonic-assisted extraction method. Furthermore, the fourth and fifth sections elucidated the biosynthesis of betanin molecule and chemical synthesis of betanidin molecule, respectively. In addition, different characterization techniques were also explicated in the sixth section which includes UV-Vis, FT-IR, ^{1}H NMR, ^{13}C NMR, LC-^{1}H NMR, mass spectrum, and thermogravimetric analysis of betanin. Also, the factors effecting the stability of betanin were explained in the seventh section which covers the effect of pH, temperature, and light on the stability of betanin. Lastly, the applicability of betanin was taken into account in the eighth section which comprised of dyeing of acrylic fabric, dyeing of wool fabric, and medicinal and food colorant applications of betanin.

DOI: http://dx.doi.org/10.5772/intechopen.88939

Author details

Deepak Devadiga and T.N. Ahipa*
Centre for Nano and Material Sciences, Jain University, Bangalore, India

*Address all correspondence to: tn.ahipa@jainuniversity.ac.in

References

[1] Baião DDS, Silva DVTD, Aguila EMD, Paschoalin VMF. Nutritional, bioactive and physicochemical characteristics of different beetroot formulations. In: Food Additives. InTech; 2017. pp. 1-24

[2] da Silva DVT, dos Santos Baião D, de Oliveira Silva F, Alves G, Perrone D, Del Aguila EM, et al. Betanin, a natural food additive: Stability, bioavailability, antioxidant and preservative ability assessments. Molecules. 2019;**24**(3):458

[3] Azeredo HMC. Betalains: Properties, sources, applications, and stability—A review. International Journal of Food Science and Technology. 2009;**44**(12): 2365-2376

[4] Tumbas Šaponjac V, Čanadanović-Brunet J, Ćetković G, Jakišić M, Djilas S, Vulić J, et al. Encapsulation of beetroot pomace extract: RSM optimization, storage and gastrointestinal stability. Molecules. 2016;**21**(5):584

[5] EFSA ANS Panel (EFSA Panel on Food Additives and Nutrient Sources added to Food). Scientific opinion on the re-evaluation of beetroot red (E 162) as a food additive. EFSA Journal. 2015; **13**(12):4318. Available from: http://doi.wiley.com/10.2903/j.efsa.2015.4318

[6] Nemzer B, Pietrzkowski Z, Spórna A, Stalica P, Thresher W, Michałowski T, et al. Betalainic and nutritional profiles of pigment-enriched red beet root (*Beta vulgaris* L.) dried extracts. Food Chemistry. 2011;**127**(1):42-53

[7] Sivakumar V, Anna JL, Vijayeeswarri J, Swaminathan G. Ultrasound assisted enhancement in natural dye extraction from beetroot for industrial applications and natural dyeing of leather. Ultrasonics Sonochemistry. 2009;**16**(6):782-789

[8] Nayak CA, Chethana S, Rastogi NK, Raghavarao KSMS. Enhanced mass transfer during solid-liquid extraction of gamma-irradiated red beetroot. Radiation Physics and Chemistry. 2006; **75**(1):173-178

[9] Schmidt OT, Schonleben W. Notizen: Zur Kenntnis der Farbstoffe der Roten Rübe (*Beta vulgaris*) II. Zeitschrift fur Naturforsch B. 1957;**12**(4):262-263

[10] Wyler H, Dreiding AS. Kristallisiertes Betanin. Vorlaufige Mitteilung. Helvetica Chimica Acta. 1957;**40**(1):191-192

[11] Wyler H, Dreiding AS. Darstellung und Abbauprodukte des Betanidins. 3. (vorläufige) Mitteilung. Über die Konstitution des Randenfarbstoffes Betanin. Helvetica Chimica Acta. 1959; **42**(5):1699-1702

[12] Harmer RA. Occurrence, chemistry and application of betanin. Food Chemistry. 1980;**5**(1):81-90

[13] Mabry TJ, Wyler H, Sassu G, Mercier M, Parikh I, Dreiding AS. Die Struktur des Neobetanidins. 5. (vorlaufige) Mitteilung. Über die Konstitution des Randenfarbstoffes Betanin. Helvetica Chimica Acta. 1962; **45**(2):640-647

[14] Mabry TJ, Wyler H, Parkih I, Dreiding AS. The conversion of betanidin and betanin to neobetanindin derivatives. Tetrahedron. 1967;**23**(7): 3111-3127

[15] Wilcox ME, Wyler H, Mabry TJ, Dreiding AS. Die Struktur des Betanins. 7. Mittcilung. fjber die Konstitution des Randenfarbstoffes Betanin. Helvetica Chimica Acta. 1965;**48**(1):252-258

[16] Wang L, Weller CL. Recent advances in extraction of nutraceuticals from plants. Trends in Food Science and Technology. 2006;**17**(6):300-312

[17] Bastos EL, Gonçalves LCP. Microwave-assisted extraction of Betalains. In: Water Extraction of Bioactive Compounds: From Plants to Drug Development. Elsevier; 2017. pp. 245-268

[18] López N, Puértolas E, Condón S, Raso J, Alvarez I. Enhancement of the extraction of betanine from red beetroot by pulsed electric fields. Journal of Food Engineering. 2009;**90**(1):60-66

[19] Zvitov R, Schwartz A, Zamski E, Nussinovitch A. Direct current electrical field effects on intact plant organs. Biotechnology Progress. 2003;**19**(3): 965-971

[20] Neagu C, Barbu V. Principal component analysis of the factors involved in the extraction of beetroot betalains. Journal of Agroalimentary Processes and Technologies. 2014; **20**(4):311-318

[21] Amlepatil NM, Miraje SY, Patil PD, Sahoo AK, Mote GV. Natural color extraction from Amaranth and beetroot: A review. Indian Journal of Applied Research. 2015;**5**(5):19-20

[22] Esatbeyoglu T, Wagner A, Schini-Kerth VB, Rimbach G. Betanin—A food colorant with biological activity. Molecular Nutrition & Food Research. 2015;**59**(1):36-47

[23] Gandía-Herrero F, García-Carmona F. Biosynthesis of betalains: Yellow and violet plant pigments. Trends in Plant Science. 2013;**18**(6):334-343

[24] Tzin V, Galili G. New insights into the shikimate and aromatic amino acids biosynthesis pathways in plants. Molecular Plant. 2010;**3**(6):956-972

[25] Fischer N, Dreiding AS. biosynthesis of betalaines. On the cleavage of the aromatic ring during the enzymatic transformation of dopa into betalamic acid. Helvetica Chimica Acta. 1972; **55**(2):649-658

[26] Christinet L, Burdet FX, Zaiko M, Hinz U, Zryd J-P. Characterization and functional identification of a novel plant 4,5-extradiol dioxygenase involved in betalain pigment biosynthesis in *Portulaca grandiflora*. Plant Physiology. 2004;**134**(1):265-274

[27] Takahashi K, Takamura E, Sakuta M. Isolation and expression analysis of two DOPA dioxygenases in *Phytolacca americana*. Zeitschrift fur Naturforsch C. 2009;**64**(7–8):564-573

[28] Gandía-Herrero F, García-Carmona F. Characterization of recombinant *Beta vulgaris* 4,5-DOPA-extradiol-dioxygenase active in the biosynthesis of betalains. Planta. 2012;**236**(1):91-100

[29] Gandía-Herrero F, Escribano J, García-Carmona F. Betaxanthins as substrates for tyrosinase. An approach to the role of tyrosinase in the biosynthetic pathway of betalains. Plant Physiology. 2005;**138**(1):421-432

[30] Hatlestad GJ, Sunnadeniya RM, Akhavan NA, Gonzalez A, Goldman IL, McGrath JM, et al. The beet R locus encodes a new cytochrome P450 required for red betalain production. Nature Genetics. 2012;**44**(7):816-820

[31] Hans J, Brandt W, Vogt T. Site-directed mutagenesis and protein 3D-homology modelling suggest a catalytic mechanism for UDP-glucose-dependent betanidin 5-O-glucosyltransferase from *Dorotheanthus bellidiformis*. The Plant Journal. 2004;**39**(3):319-333

[32] Zakharova NS, Petrova TA. β-Glucosidases from leaves and roots of the common beet, *Beta vulgaris*. Applied Biochemistry and Microbiology. 2000; **36**(4):394-397

[33] Sasaki N, Adachi T, Koda T, Ozeki Y. Detection of UDP-glucose:cyclo-DOPA 5-O-glucosyltransferase activity in four

o'clocks (*Mirabilis jalapa* L.). FEBS Letters. 2004;**568**(1–3):159-162

[34] Hermama K, Dreiding AS. Totafsynthese von betalainen. Helvetica Chimica Acta. 1975;**58**(6):1805-1808

[35] Dumbrava A, Enache I, Oprea CI, Georgescu A, Girtu MA. Toward a more efficient utilisation of betalains as pigments for dye-sensitized solar cells. Digest Journal of Nanomaterials and Biostructures. 2012;**7**(1):339-351

[36] Aztatzi-Rugerio L, Granados-Balbuena SY, Zainos-Cuapio Y, Ocaranza-Sánchez E, Rojas-López M. Analysis of the degradation of betanin obtained from beetroot using Fourier transform infrared spectroscopy. Journal of Food Science and Technology. 2019;**56**(8):3677-3686

[37] Kumar SNA, Ritesh SK, Sharmila G, Muthukumaran C. Extraction optimization and characterization of water soluble red purple pigment from floral bracts of Bougainvillea glabra. Arabian Journal of Chemistry. 2017;**10**: S2145-S2150

[38] Cai Y, Sun M, Wu H, Huang R, Corke H. Characterization and quantification of betacyanin pigments from diverse Amaranthus species. Journal of Agricultural and Food Chemistry. 1998;**46**(6):2063-2070

[39] Molina GA, Hernández-Martínez AR, Cortez-Valadez M, García-Hernández F, Estevez M. Effects of tetraethyl orthosilicate (teos) on the light and temperature stability of a pigment from *Beta vulgaris* and its potential food industry applications. Molecules. 2014;**19**(11):17985-18002

[40] Sengupta D, Mondal B, Mukherjee K. Visible light absorption and photo-sensitizing properties of spinach leaves and beetroot extracted natural dyes. Spectrochimica Acta Part A: Molecular and Biomolecular Spectroscopy. 2015;**148**:85-92

[41] Dong J, Ozaki Y, Nakashima K. Infrared, Raman, and near-infrared spectroscopic evidence for the coexistence of various hydrogen-bond forms in poly(acrylic acid). Macromolecules. 1997;**30**(4):1111-1117

[42] Stintzing FC, Conrad J, Klaiber I, Beifuss U, Carle R. Structural investigations on betacyanin pigments by LC NMR and 2D NMR spectroscopy. Phytochemistry. 2004; **65**(4):415-422

[43] Guesmi A, Ladhari N, Ben Hamadi N, Sakli F. Isolation, identification and dyeing studies of betanin on modified acrylic fabrics. Industrial Crops and Products. 2012; **37**(1):342-346

[44] Susha N, Nandakumar K, Nair SS. Enhanced photoconductivity in CdS/ betanin composite nanostructures. RSC Advances. 2018;**8**(21):11330-11337

[45] Von Elbe JH, Young Maing I, Amundson CH. Color stability os betanin. Journal of Food Science. 1974; **39**(2):334-337

[46] Guesmi A, Ladhari N, Ben Hamadi N, Msaddek M, Sakli F. First application of chlorophyll-a as biomordant: Sonicator dyeing of wool with betanin dye. Journal of Cleaner Production. 2013;**39**:97-104

[47] Khan MI. Plant betalains: Safety, antioxidant activity, clinical efficacy, and bioavailability. Comprehensive Reviews in Food Science and Food Safety. 2016;**15**(2):316-330

[48] Gentile C, Tesoriere L, Allegra M, Livrea MA, D'Alessio P. Antioxidant betalains from cactus pear (*Opuntia ficus-indica*) inhibit endothelial ICAM-1 expression. Annals of the New York Academy of Sciences. 2004;**1028**:481-486

[49] Kapadia GJ, Azuine MA, Rao GS, Arai T, Iida A, Tokuda H. Cytotoxic

effect of the red beetroot (*Beta vulgaris* L.) extract compared to doxorubicin (Adriamycin) in the human prostate (PC-3) and breast (MCF-7) cancer cell lines. Anti-Cancer Agents in Medicinal Chemistry. 2011;**11**(3):280-284

[50] Reddy MK, Alexander-Lindo RL, Nair MG. Relative inhibition of lipid peroxidation, cyclooxygenase enzymes, and human tumor cell proliferation by natural food colors. Journal of Agricultural and Food Chemistry. 2005; **53**(23):9268-9273

[51] Krajka-Kuźniak V, Paluszczak J, Szaefer H, Baer-Dubowska W. Betanin, a beetroot component, induces nuclear factor erythroid-2-related factor 2-mediated expression of detoxifying/antioxidant enzymes in human liver cell lines. The British Journal of Nutrition. 2013;**110**(12):2138-2149

[52] Dhananjayan I, Kathiroli S, Subramani S, Veerasamy V. Ameliorating effect of betanin, a natural chromoalkaloid by modulating hepatic carbohydrate metabolic enzyme activities and glycogen content in streptozotocin—Nicotinamide induced experimental rats. Biomedicine & Pharmacotherapy. 2017;**88**:1069-1079

[53] Sutariya B, Saraf M. Betanin, isolated from fruits of *Opuntia elatior* Mill attenuates renal fibrosis in diabetic rats through regulating oxidative stress and TGF-β pathway. Journal of Ethnopharmacology. 2017;**198**:432-443

[54] Amjadi S, Ghorbani M, Hamishehkar H, Roufegarinejad L. Improvement in the stability of betanin by liposomal nanocarriers: Its application in gummy candy as a food model. Food Chemistry. 2018;**256**: 156-162

Section 5

Printing with Natural Dyes and Pigments

Chapter 8

Printing of Textiles Using Natural Dyes: A Global Sustainable Approach

M. Kavyashree

Abstract

Globalization has completely changed the fashion industry and its approach toward sustainability. Fast changing trends are majorly focused on synthetic dyes and fabrics. And these products have failed building the bridge between sustainability and environment-friendly designs. Therefore, this chapter is aimed at the approach toward design development and development of sustainable eco-friendly approach for screen printing by using natural dyes on fabrics like silk and cotton. The chapter is based on sustainable design approach which gives much provision for using natural dyes for everyday change in fashion. The chapter also focuses on color fastness properties of the used dyes. These results were evaluated by carrying out color fastness tests for light, wash, rubbing, and perspiration using standard testing methods. Design methodology used in the study also has the potential for skill development programs for entrepreneurs and also contributes to rural development programs by creating sustainable growth.

Keywords: natural dyes, indigo, sappanwood, madder, screen printing, eco-friendliness, sustainable approach, silk and cotton textiles

1. Introduction

The use of natural dyes has been in India for thousands of years and is part of India's cultural identity. In ancient times, dyed textiles were symbols of status and, because of their cost, were reserved only for the wealthiest classes. Dye sources were earlier extracted from plants, animals, and minerals. Madder and indigo were the major dyes used in India since 2500 BC. With the intervention of synthetic dyes, the use of natural dyes for textile dyeing has been reduced tremendously. Fast fashion demands low cost, simple, and reproducible dyeing or printing process. Both the synthetic dyes and natural dyes play a key factor for coloration and ornamentation of textile materials like silk and other fibers. Synthetic dyes are playing important and predominant role for coloring the majority of silk materials, while natural dyes are being specially used for developing uncommon soothing shades on silk yarns or fabrics in small-scale sector or rural sector, which is gradually gaining popularity due to its nontoxicity and eco-friendly character against possible unsafe ecotoxicity criteria of synthetic dyes [1, 2].

Textile printing in India has been practiced over centuries. Screen printing using natural dyes in India has helped in creating bridge between conventional printing techniques and mass production of textiles in an efficient way. In this chapter natural dyes such as madder, sappanwood, and indigo are used under optimized temperature, pH, and duration for extraction. The cotton and silk substrates are prepared for screen printing by scouring the fabrics. Screens are developed based on the patterns obtained by conventional shibori dyeing techniques. During printing process, madder, sappanwood, and indigo dye extract were used to provide color. More often it is assumed that color will change depending on printing methods and the substrate used. But in natural dyes, this expectation is somewhat high [3], and especially in printing processes, the effect of mordant type is of great importance for the shade of the color as printing is totally a different approach compared with dyeing of textiles. As a sustainable approach toward screen printing, mordants and binding agents used in this chapter are natural. 20% alum is used with madder and sappanwood extraction as metamordanting agent. Tamarind kernel powder is used as thickening agent for madder and sappanwood. Cornstarch is used as thickening agent for indigo.

2. Material and methodologies

2.1 Fabrics

Most of the studies available for dyeing with natural dyes relate to woolen textiles, and such studies on silk are still insufficient and that study on cellulose or lignocellulose textiles is scant. Hence, there is ample scope of undertaking an integrated study of the surface appearance, durability, dye intake, and screen printing of silk and cotton with selective natural dyes.

Silk and cotton fabrics have been considered for experimentation in this chapter. As silk fabrics are best available for commercial use, this delicate filament fiber is well known for its sheen, texture, water absorbency, dye affinity, thermal tolerances, and insulation properties [4]; its dyeability is one of the prime properties needed to analyze the screen printing of shibori designs using natural dyes.

2.1.1 Pretreatment for fabrics

Natural fabrics must be treated to ensure good dye absorbency before dyeing or printing to remove any natural or added impurities while manufacturing and handling the textile materials. Silk and cotton fabrics used in this experiment are treated for scouring, desizing, and degumming. Like in any other dyeing and printing processes, it was found that pretreatment process influenced the printability of cotton with natural dyes [5].

2.2 Dyes and mordants

Commercially available natural dyes like madder, sappanwood, and indigo are used for the experiments, which are sourced from KMA Exports, Tindivanam, Tamil Nadu. Dye sources used for this study are easily available, and the effluents are not hazardous for the human health and nontoxic to the ecosystem. Myrobalan and alum were used as metamordants while preparing the printing paste of madder and sappanwood. Alum is used to treat the fabrics for its low environmental toxicity [6].

2.2.1 Thickening agents

The thickening agents are tamarind kernel seeds and cornstarch.

2.2.2 Reduction of indigo

According to discovery of European wood vat reduction recipe from medieval manuscripts [7], it was found that a hitherto unknown bacterium,

Clostridium isatidis, is responsible for indigo reduction; this bacterium reduces the indigo particles to soluble leucoindigo. As per the traditional indigo fermentation process, vat routinely requires the addition of red dye madder, which contains high concentration of anthraquinones; these quinines significantly increase the rate of indigo reduction by *C. isatidis*. As quoted by J N Liles in his book *The Art and Craft of Natural Dyeing*, in 1794, Bancroft suggested that the use of glucose alone will reduce the indigo and if reduction did not adequately occur, the addition of oxides of tin can be used. Hence, in this experiment instead of sodium dithionite, jaggery will be used along with sodium hydroxide to keep the vat alkaline.

Reduction with jaggery and madder: indigo cake was powdered, and 10 g was taken in a vat; 1 l of water was kept over water bath at temperature of 60°C, to which 20 g of madder and jaggery was added. About 4 g of sodium hydroxide was then added to the solution, and the solution is maintained at pH 12 by adding sodium hydroxide.

2.2.3 Dye extraction with madder and sappanwood

Dye extraction was carried out at acidic and neutral medium for madder and sappanwood at the temperature of 70°C for 60 min, where 15 g of madder and 15 g of sappanwood were used with two separate vessels containing 150 ml of water. When the extracted solution cools down to room temperature, 7.5 g of alum is added as metamordanting agent for both madder and sappanwood dye extract. Additionally, 15 g of myrobalan was added to madder dye extract solution as mordanting agent. Then the extracted solution of madder and sappanwood was steeped for 12 h, which was later stirred and filtered.

2.3 Printing paste preparation

To make printing paste of indigo, 100 ml water was taken in a beaker and 20 g of cornstarch is added and boiled, to which 100 ml of fermented indigo solution was added and left for further fermentation for 12 h.

To make printing paste of madder and sappanwood, 15 g of tamarind kernel seed powder was taken for each solution and mixed with constant stirring using an electric blender.

2.4 Printing technique

The printing process involves the printing pastes, fabrics, and the screen developed using shibori designs. Fabric is laid on a waxed wooden surface to ensure no movement of fabric while printing. The print paste is guided through the screen mesh or the open areas of the design on the fabrics by using a flexible rubber squeegee. A constant pressure is maintained while drawing the paste from one end to another and during each printing. The printed fabrics are then allowed to air-dry for 24 h.

2.5 Washing and fixing

Printed fabrics were washed with plain water to remove the cornstarch and tamarind seed kernel powder. Indigo samples were washed with cold water followed by hot water with constant stirring. Printed samples of madder and sappanwood were treated with dhawadi (*Woodfordia floribunda*) as postmordanting for fixing the color on to the substrates. Fixation was done at 60°C for 40 min followed by cold water rinse.

3. Testing and evaluation methods

3.1 Color parameters and color fastness properties

Dried samples were then analyzed further for color parameters and fastness according to the specified standards. **Figures 1** and **2** shows printed cotton and silk using indigo, **Figures 3** and **4** printed cotton and silk using madder, and **Figures 5** and **6** printed cotton and silk using sappanwood.

Figure 1.
Sappanwood on cotton fabric.

Figure 2.
Sappanwood on silk fabric.

Figure 3.
Madder on cotton fabric.

Figure 4.
Madder on silk fabric.

Figure 5.
Indigo on cotton fabric.

Figure 6.
Indigo on silk fabric.

3.1.1 Evaluation of CIE lab coordinates

Color value of the sample was analyzed based on L*a*b* values using reflectance spectra through spectrophotometer. The L* value is a measure of lightness and darkness of the color to define the color on a two-dimensional chromatic space of green-red axis and blue-yellow axis. Negative a* value indicates greenness in the screen printed samples and negative b* value indicates blueness in the screen printed samples, while positive a* value indicates redness and positive b* value indicates yellowness in the screen printed samples. L*, a*, and b* values were evaluated for five samples, and the average readings are recorded in **Tables 1–3**.

3.1.2 Measurement of color fastness properties

Color fastness to washing of dyed samples was determined by launder-o-meter in accordance with method prescribed in ISO 105-C02. Color fastness to rubbing was determined by crockmeter in accordance with method prescribed in AATCC crockmeter method [8]. Color fastness to light of dyed cotton and silk fabrics was assessed on IS 2454 1985. Light fastness of printed fabric is influenced by chemical and physical state, concentration of dye, nature of the fibers, and mordant type [9]. Color fastness to perspiration of dyed cotton and silk fabrics were determined by perspi-o-meter in accordance with method prescribed in ISO 105-E04:2013 test method. The data obtained are recorded in **Tables 4–6**.

Fabrics	Mordants	CIE color coordinates			Description
		L*	a*	b*	
Cotton	Simultaneous mordanting (alum)	59.44	30.14	−9.60	Lighter, more red, and more blue
	Postmordant (*Woodfordia floribunda*)	60.88	27.30	−11.49	Lighter, more red, and more blue
Silk	Simultaneous mordanting (alum)	56.53	40.13	−17.91	Lighter, more red, and more blue
	Postmordant (*Woodfordia floribunda*)	60.09	37.45	−8.23	Lighter, more red, and more blue

Table 1.
Color characteristics of sappanwood with mordanting.

Fabrics	Mordants	CIE color coordinates			Description
		L*	a*	b*	
Cotton	Simultaneous mordanting (alum)	61.21	13.84	19.14	Lighter, more red, and more yellow
	Postmordant (*Woodfordia floribunda*)	61.32	13.07	19.11	Lighter, more red, and more yellow
Silk	Simultaneous mordanting (alum)	67.69	13.53	11.29	Lighter, more red, and more yellow
	Postmordant (*Woodfordia floribunda*)	61.56	16.40	17.01	Lighter, more red, and more yellow

Table 2.
Color characteristics of madder with mordanting.

Fabrics	Reducing agents	CIE color coordinates			Description
		L*	a*	b*	
Cotton	Jaggery	38.90	−2.94	−22.10	Lighter, more green, and more blue
Silk	Jaggery	51.33	−1.17	−25.74	Lighter, more green, and more blue

Table 3.
Color characteristics of indigo with honey and jaggery.

Fabric	Mordants used	Fastness properties									
		Washing		Light	Perspiration				Rubbing	
		Change in color (cc)	Staining on cotton (cs)		Alkaline		Acidic		Dry	Wet
					cc	cs	cc	cs		
Cotton	Simultaneous mordanting (alum)	1/2	5	1	3/4	4	1/2	4	4/5	4
	Postmordant (*Woodfordia floribunda*)	1/2	5	1/2	3/4	4	1/2	4	4/5	4
Silk	Simultaneous mordanting (alum)	4	5	1/2	3/4	4	2	4/5	5	4/5
	Postmordant (*Woodfordia floribunda*)	3	5	1/2	3	4/5	2	4/5	5	4/5

Table 4.
Color fastness for sappanwood.

Fabric	Mordants used	Fastness properties									
		Washing		Light	Perspiration				Rubbing	
		Change in color (cc)	Staining on cotton (cs)		Alkaline		Acidic		Dry	Wet
					cc	cs	cc	cs		
Cotton	Simultaneous mordanting (alum)	4	5	2/3	4/5	5	4/5	5	5	4
	Postmordant (*Woodfordia floribunda*)	3/4	5	2/3	5	5	4	5	5	4
Silk	Simultaneous mordanting (alum)	4	5	2	4/5	5	5	5	4/5	3/4
	Postmordant (*Woodfordia floribunda*)	4	5	2	4/5	5	5	5	4/5	4

Table 5.
Color fastness for madder.

Fabric	Reducing agents	Fastness properties									
		Washing		Light	Perspiration				Rubbing	
		Change in color (cc)	Staining on cotton (cs)		Alkaline		Acidic		Dry	Wet
					cc	cs	cc	cs		
Cotton	Jaggery	4/5	4/5	3	5	5	5	5	4	3
Silk	Jaggery	4	5	2/3	4/5	5	4/5	5	4/5	3

Table 6.
Color fastness for indigo reduced with jaggery.

4. Results and discussions

It is indicated by CIE color coordinate results. The a* value indicates redness or greenness, and b* value indicates yellowness and blueness. From **Table 1**, it can be seen that a* values are positive which indicates redness and b* values are negative which indicates blueness in printed samples with both metamordanting and post-mordanting agents. From **Table 2**, it can be seen that a* values are positive which indicates redness and b* values are positive which indicates yellowness in printed samples with both metamordanting and postmordanting agents. From **Table 3**, it can be seen that a* values are negative which indicates greenness and b* values are negative which indicates blueness in printed samples with both metamordanting and postmordanting agents.

The light fastness properties of printed cotton and silk fabrics dyed in sappanwood with alum and dhawadi (*Woodfordia floribunda*) as mordants (see **Figures 1** and **2**) are found poor which shows that dyes are unstable to photodegradation. Screen printed samples were evaluated for rubbing fastness. Rubbing fastness grades of dry rubbing are found very good to excellent, but for wet rubbing fastness, it is good. Washing fastness grades for color change for both mordants were found poor for cotton and average for silk, and color staining was observed to be good for all the samples. Washing fastness grades clearly reveal that color change was observed in both the mordants and very slight color staining was seen on cotton and silk fabrics. Screen printed samples showed moderate to good performance during alkaline perspiration test and poor performance during acidic perspiration test (see **Table 4**).

The light fastness properties of printed cotton and silk fabrics dyed in madder with alum and dhawadi (*Woodfordia floribunda*) as mordants (see **Figures 3** and **4**) are moderate which shows that dyes are slightly stable to photodegradation and printing paste interaction is good. Screen printed samples were evaluated for rubbing fastness. Rubbing fastness grades of dry rubbing are found very good to excellent, but for wet rubbing fastness, it is good. Washing fastness grades for color change for both mordants were found good for cotton and very good for silk, and color staining was observed 5 for all the samples. Almost all screen printed samples showed good to excellent performance during alkaline and acidic perspiration tests (see **Table 5**).

Data for color fastness to washing, rubbing, light, and perspiration of cotton and silk screen printed samples (see **Figures 5** and **6**) with natural indigo reduced using natural ingredients has found out during the study by following standard testing methods. The use of natural auxiliaries has resulted in good fastness rating on cotton and silk screen printed samples with indigo. Based on the result, the printed fabrics have good color fastness properties in terms of wash, light, and perspiration. Rubbing was found moderate to poor (see **Table 6**).

5. Conclusions

In the current world of fast fashion, there is an increased concern globally toward the use of hazardous and carcinogenic synthetic dyes like azo and benzidine; these dyes have irrevocable effects on nature and mankind. The growing awareness about sustainability and environment-friendly dyes has created an essential platform for young researchers to revive and experiment with traditional approach of textile dyeing and printing. However, the textile dyed from natural dyes does need value addition to reach the desired market acceptance; printing different motifs using natural dyes can help overcome the requirements of value addition for textiles. Natural dyes have always been preferred for its soothing colors; with the experiment and results obtained in this chapter, it can be seen that the screen printing on cotton and silk fabrics using indigo, madder, and sappanwood has resulted in promising colors and also can be considered as the recommendable alternative to harmful synthetic dyes. This chapter was focused on the application of conventional shibori patterns for screen printing using natural dyes. From the history of fashion, it is known that the comeback of green clothing is in high fashion. In addition, design process of textile design is also served as eco-route in this chapter. There is a huge scope to textile printing using natural dyes; new sources of natural dyes and new trends in fashion applications can create a global niche market for textiles and for natural dyes. This research can be an innovative method for textile design with screen printing. The major significance of this chapter was to add a value on textile by improving the esthetic functions and commercial values of fabric. The application of these techniques can serve as a new and complete method to create innovative fabric design.

Acknowledgements

Sincere thanks are expressed by the author to Dr. K.N. Ninge Gowda, for valuable guidance and support including use of facilities of the Department of Apparel Technology and Management, Bangalore University, during my PhD research. Sincere thanks are also expressed to my family Shri Vasanth Kumar, my husband; Smt H.V. Vimala, my mother; and Lavanya, my sister for constant support and encouragement.

Author details

M. Kavyashree
Department of Apparel Technology and Management, Bangalore University, Bangalore, India

*Address all correspondence to: vassa.kavya@gmail.com

References

[1] Shahid M, Ahmad A, Yusuf M, Khan MI, Khan SA, Manzoor N, et al. Dyes and Pigments. 2012;**95**(1):53-61. DOI: 10.1016/j.dyepig.2012.03.029

[2] Shahid-ul-Islam M, Shahid FM. Journal of Cleaner Production. 2013;**57**:2-18. DOI: 10.1016/j.jclepro.2013.06.004

[3] Samanta AK, Agarwal P. Application of natural dyes on textiles. Indian Journal of Fibre & Textile Research. 2009;**34**:384-399

[4] Sheikh RK, Farouqui FI, Modak PR, Hoque A, Yasmin Z, et al. Dyeing of Rajshahi silk with basic dyes: Effect of modification on dyeing properties. Journal of the Textile Institute. 2006;**97**:295-300

[5] Bahtiyari MI, Benli H, Yavas A, Akca C. Journal of Textile & Apparel/Tekstil ve Konfeksiyon. 2017;**27**(3):259-265

[6] Ezatollah M, Bijan M. Alum mineral and the importance for textile dyeing. Current Trends in Fashion Technology & Textile Engineering. 2018;**3**(4):555619. DOI: 10.19080/CTFTTE.2018.03.555619

[7] Padden AN, Dillon VM, John P, Edmonds J, Collins MD, Alvarez N. Clostridium used in medieval dyeing. Nature. 1998;**396**:225. DOI: 10.1038/24290

[8] AATCC. TM6, TM8, TM15, TM16, Technical Manual. Vol. 85. USA: American Association of Textile Chemist. 2010. pp. 18-25

[9] Cristea D, Vilarem G. Improving light fastness of natural dyes on cotton yarn. Dyes and Pigments, Elsevier. 2006;**70**:238-245

Section 6

Functional Properties and Functional Finishing with Natural Dyes and Pigments

Chapter 9

Functional Properties of Natural Dyed Textiles

Deepti Pargai, Shahnaz Jahan and Manisha Gahlot

Abstract

Recently, due to harmful effect of climate change and other environmental concerns, the interest towards natural dyes has gained momentum. Presently in this high technological era, only aesthetic appeal of natural dyes could not get full attention of consumers. A consumer is attracted towards a type of product which is not only aesthetic but also have some functional properties. Hence considering this functional aspect, various researches are being carried out to find out the new dimensions of natural dyes. Colours and other functional properties can be achieved by various synthetic agents, but the exploration of new functional properties of natural dyes would be beneficial and relevant for the present as well as for the future. This chapter compiles all functional aspects of natural dyes which are related to adding functional properties such as antimicrobial, UV protection, insect repellent, etc. to textiles which can protect the human being from various kinds of harmful effects of surroundings such as UV radiation, microbes, bacteria and other harmful insects. This chapter will not only deal with available methods for analysis of these functional properties of natural dyes but also explore the reason for functional properties of natural dyes.

Keywords: functional properties, natural dyes, UV protection textiles, antimicrobial property

1. Introduction

Natural dyes are obtained from different natural sources such as plants, insects, animals and minerals. These are used for coloration of textiles and food since ancient time [1–3]. With the advent of synthetic dyes, natural dyes had been subdued for a while. Recently environmental consciousness of consumer increases due to several harmful impacts of synthetic dyes; thus, interest towards natural dyes has again gained momentum [4, 5]. Natural dyes are biodegradable as well as nontoxic. These dyes are safer for environment in terms of safe discharge [6]. In this era of climate change and different lifestyle of human being, the environment around human being becomes more harmful than the previous era. Microbes, moth insects and UV rays are causing various kinds of harmful effects to human being [7–9]. Presently the human does not wear clothes only for modesty. Hygiene, cleanliness and protection become major issues, and thus the demand for functional clothes has also increased. Textile fabric majorly covers the human skin, creating a microclimate. Presently as the climate changes, the requirement for functional clothing arises. Besides providing colour, natural dyes have inherent functional properties such as resistance for bacteria, fungus and moth, UV protection, etc. [10–12]. Fabric as a second skin covers the major part of the body and hence can be used as a preventive measure from near environment. At present, the researches on utilisation of natural dyes in functional finishing of textiles have increased because of the efficiency of natural dyes which provides protection against various harmful agents as well as provides greater comfort. Dyed fabric remains fresh and odour-free in use [13, 14]. Natural dyes obtained from traditional dyes giving plants contain a variety of compounds such as curcumin, crocin, bixin, carthamin, punicalagin, nimbin, etc. known to have therapeutic properties [15]. Many plants contain secondary antimicrobial which gives protection against microbial attack.

Applications of natural dyed textile can be extended to a diverse field such as sportswear and medicinal field due to its various functional properties such as antimicrobial as well as UV protection [16–18].

2. Functional properties of natural dyed textiles

Natural dyes having functional groups such as $-OH$, $-NH_2$, $-COOH$, etc. groups and the textile fibres with active sites ($-OH$, $-SO_3H$, $-COOH$, $-C_6H_5OH$, $-NH_2$) can make a complex with or without mordant. Because of the different interactions of dye and fabric, many functional properties with fastness can be achieved [19]. Besides functional group of natural dye such as tannins, flavonoids, anthraquinone, etc., functional groups of respective fabrics are also responsible for functional properties in the fabric [20]. The following functional properties are obtained through natural dyeing of textiles.

2.1 Antimicrobial/antibacterial property

The surrounding of human includes a variety of microorganisms like bacteria and other microbes, which are invisible for the naked eyes. Bacterium is a unicellular organism. It is divided in Gram-positive (*Staphylococcus aureus*) and Gram-negative (*E. coli*) on the basis of chemical and physical properties of their cell walls [21]. Examples of common Gram-positive bacteria are *Staphylococcus aureus, Streptococcus epidermidis and Bacillus cereus*, and the examples of Gram-negative bacteria are *Escherichia coli, Klebsiella pneumonia, Shigella flexneri and Proteus vulgaris* [22]. Clothing majorly covers the human skin or nearby area of direct contact from the human skin. Thus the growth of these microorganisms on fabric affects the wearer as clothes on the human body provide a favourable environment for microbial growth; it provides warmth, oxygen and water as well as nutrients from perspiration. Textile fabrics itself can also be affected by the growth of these microorganisms [23]. Cotton is one of the most affected fabrics by these microbes [24]. Naturally dyed textiles can provide protection against these microorganisms. Antimicrobial properties of natural dyes are due to the presence of various compounds such as anthraquinones, flavonoids, tannins, naphthoquinones, etc. [25]. Various antimicrobial agents are present in natural dye for common human pathogen, but very few are reported in the case of textiles with respect to human pathogenic strain, as testing method is different in the case of testing against textiles. Many plants which have been traditionally used for dyeing are also considered having medicinal properties which provide protection against these harmful microorganisms. For instance, natural dye which is obtained from *Punica granatum* peels having hydrolysable tannins exhibited a notable antimicrobial activity when applied on tencel fabric [26]. Henna leaves having lawsone compound were found to be very effective against *Candida glabrata* in solution as well as after application on wool substrate [27]. It is reported that antibacterial and antifungal properties of natural dye are due to its phenolic content. Phenolic compounds attach on the surface of textiles by forming a complex. When the fabric comes in contact with microbes, these attached phenolic compounds hamper the enzyme production in microbes; thus, further cell reaction would not take place, and at the end cell dies. Natural dyes obtained from *M. composita* leaves having alkaloids, are found to be effective against Gram-positive and Gram-negative bacterial strains [28]. Madder (alizarine functional group) and safflower having carthamin with alum, Zn-sulphate and tannic acid also imparted antibacterial properties, to the dyed poly-amide- 6 fabric [29]. *Rumex maritimus (*golden dock*) and Quercus infectoria (indigo) dyes are found to be the most effective against common microbes like Escherichia coli*, *Bacillus subtilis*, *Klebsiella pneumoniae*, *Proteus vulgaris* and *Pseudomonas aeruginosa* [12]. Berberine compound as a cationic dye having quaternary ammonium structure can act as an antibacterial agent. The dyed wool represented a high level of antibacterial activity [30]. Application of natural dyes peony, pomegranate, clove, *Coptis chinensis* (Chinese goldthread) and gallnut extracts on cotton silk and wool fabric provides excellent antibacterial activity against *Staphylococcus aureus*. The possible reported reason for antimicrobial activity of natural dyes obtained from pomegranate peels, *Coptis chinensis*, peony and clove is due to ellagic acid, berberine and eugenol, respectively [31] (**Table 1**).

Antimicrobial testing is a valuable aid for textile production, distribution and consumption. To measure the antimicrobial efficacy of natural dyed textiles, test methods are performed under controlled conditions. The antimicrobial activities are generally tested both qualitatively and quantitatively through standard tests. Some of the available and majorly used AATCC standards for textile are as follows:

One prominent quantitative standard for antimicrobial testing of textiles is AATCC 100-2004 (bacterial reduction method) [32]. Under this test, the test microorganism of standardised concentration is grown in liquid culture. This prepared culture is diluted in a sterilised nutritive solution. Control and natural dyed fabric swatches are inoculated with microorganisms. Inoculated control and test fabrics are allowed to incubate for 24 hours at 37°C, in sealed jars. After incubation, shake for 1 minute; then concentrations of microbes are observed. Finally it is calculated how much microorganism reduces as compared to initial concentration.

$$\text{Percent reduction of bacteria } R = 100\ (B - A)/B \tag{1}$$

where A is the number of bacteria recovered from the inoculated treated test specimen swatches in the jar incubated over the desired contact period while B is the number of bacteria recovered from the inoculated treated test specimen swatches in the jar immediately after inoculation (at "0" contact time).

Another standard method for antimicrobial testing of textiles is parallel streak method (AATTC Test Method 147-2004). The agar surface is streaked with an inoculum of test bacterium. The samples treated with natural dye and the undyed sample (control sample) are placed in close contact with this agar surface. This is incubated for 37°C for 18–24 hours.

The following equation is used to calculate the width of a zone of inhibition along a streak on either.

$$W = (T - D)/2 \tag{2}$$

Name of the natural dye	Botanical name	Responsible component for antimicrobial properties
Pomegranate	*Punica granatum*	Tannins
Henna	*Lawsonia inermis*	Lawsone
Neem	*Melia azedarach*	Phenolic compounds and flavonoids
Madder	*Rubia tinctorum*	di- and trihydroxyanthraquinones
Golden dock	*Rumex maritimus*	Tannins
Oak galls	*Quercus infectoria*	Tannins
Peony	*Paeonia officinalis*	Paenol/paenoside/paeonolide/paenoniflorin
Clove	*Syzygium aromaticum*	Eugenol (2-methoxy-4 allyl-phenol)
Goldthread/Canker root/Huang Lian	*Coptis chinensis*	Alkaloid berberine
Turmeric	*Curcuma longa*	Curcumin or diferuloylmethane with chemical formula of (1,7-bis(4-hydroxy-3-methoxyphenyl)-1,6-heptadiene-3,5-dione) and curcuminoids
Safflower	*Carthamus tinctorius*	Flavonoids

Table 1.
Natural dyes with responsible component for antimicrobial properties.

where W is width of clear zone of inhibition in mm, T is total diameter test specimen and clear zone in mm and D is diameter of the test specimen in mm [33].

It has been reported that dyed fabric is able to retain almost half of its initial antibacterial properties after few washings. Application of mordant like tannin or other cross-linking agent such as citric acid or any other surface modifications with cationisation or by applying biopolymer is also responsible for wash stability of functional properties of natural dyed textiles.

2.2 UV protection

Presently as the climate change are showing its harmful effects, the need for protection against UV rays has been increased in order to avoid incidences of UV-induced skin damages. Various sunscreen and synthetic UV absorbers for textile fabric are currently available in the market, but researches are carried out to search out eco-friendly alternative such as natural dyes to enhance the aesthetic as well as UV protection property of the fabric. UV protection properties of dyed fabric is analysed using UPF of the fabric. Ultraviolet protection factor (UPF) indicates the UV protection properties of the fabric.

Application of natural dyes on fabric significantly enhances the UPF of the fabric. UPF of fabric is affected by the absorption characteristics of natural dyes [34]. Tannins in plant act as a chemical protector against the UV radiation [35]. Tannin as a phenolic compound absorbs UV radiation due to resonance in structure and hence after the process of attachment on textile surface. These attached molecules absorb the UV radiation and thus protect the skin from absorbing UV rays. Tannin-based natural dyes such as *R. maritimus, M. philippinensis, K. lacca, A. catechu and A. nilotica* have good UV protection properties [36].

Various kinds of natural dyes provide protection against microbes as well as UV rays on different kinds of fabric such as wool, cotton and silk. Natural dyes from eucalyptus leaf extract with ferrous sulphate mordant can provide UV protection properties with antimicrobial properties to the dyed silk fabric [37]. The cotton fabric is dyed with ***Xylocarpus granatum*** (*cedar mangrove*) bark extract. In this experiment, the UPF values of all dyed samples were achieved above the range of 50. It was also reported that it is only possible because of the use of metal mordants as it helped in the formation of tannate complex [38]. Woollen fabric dyed with the blossoms of broom (***Cytisus scoparius***) and dandelion (***Taraxacum officinale***) also provides effective UV protection [39]. *Flos Caryophyllata (clove)* and *acutissima* shell with $FeSO_4$ mordant provide anti-UV properties to the silk fabric and thus the wearer [40]. Weld, woad, logwood lipstick tree, madder, brazilwood and cochineal as natural dyes could provide good UV protection on hemp and flax fabric [41]. Cotton and silk fabrics dyed with gromwell roots absorb most of the UV rays [42]. Cationised cotton fabric dyed with henna dye extract exhibited outstanding enhancement in UV protection [43]. Dyeing of polyester fabric with chitosan and turmeric dye enhances the UPF of the fabric [44]. It was reported in a study that simultaneous dyeing and finishing of silk and wool fabrics with the use of cochineal and weld natural extracts enhanced the UPF and fastness of the fabric. As a result, excellent UPF range with high fastness was obtained [45]. Application of gallnuts, areca nuts and pomegranate peels dyed on silk fabric imparted the UV protection with deodorization and antimicrobial properties [46]. Antibacterial and UV protection properties were also obtained by cotton fabric dyed with banana peel [47]. Mordant with natural dyes such as cutch and madder also positively affects the UPF of the cellulosic nettle fabric. In several cases, the use of mordant improves the UPF value, whereas in most of the cases, mordant lowers the UPF value in comparison to blank [48]. Addition of mordant lowers the UPF value severely. This was most likely due to the complex formation ability of dye. Due to coordination bond formation between the dye and mordant, the structure of the dye changes and so does its light absorption properties. The absorption shifts to higher wavelengths of visible region, thus giving deeper colour but less UPF [49]. Very good UPF values were obtained when the pre-mordanting of jute fabric was done before the actual dyeing with *babool, ratanjot*, annatto and *manjistha.* Both bio and chemical mordant were used for this [50]. Chitosan mordanting of green tea dyed cotton fabrics enhanced the UV protection property [51]. Ferrous sulphate as a mordant establishes ternary complex with both fibre and the dye, and the remaining coordination sites of Fe metal can absorb UV radiation by converting electronic excitation energy into thermal energy [47]. Although natural dyes improve the UV protection properties of the textiles, some limitations are associated with it. For example, the amount of UV protection of dyed fabrics also tends to decrease due to exposure to sunlight as well as in laundry process. Therefore, there is a need to do research with the aim of enhancing the durability of natural dye. Traditionally mordants have been used since a long time to improve the stability and durability of natural dyes, but presently various kinds of surface modification techniques are being used to enhance the durability [52] (**Table 2**).

Standard testing of specific functional property helps to maintain a quality of textile products [53]. (AATCC-183: 2004) test method is used for analysis of UPF [54]. UPF of the treated fabric samples was determined by using “Lab-sphere’s UV transmittance analyser”. This method is used to determine the UV radiation blocked and transmitted by the textile fabric. The UPF is computed as the ratio of the erythemally weighted ultraviolet radiation (UV-R) irradiance at the detector with no specimen to the erythemally weighted UV-R irradiance at the detector with a specimen present. The formula which is used for calculation of UPF is as follows:

$$UPF = \frac{\Sigma E_\lambda \times S_\lambda \times \Delta\lambda}{\Sigma E_\lambda \times S_\lambda \times T_\lambda \times \Delta\lambda} \quad (3)$$

Name of the natural dye/mordant used	Botanical name	Main component for UV protection properties	Fabric	UPF range
Jamun leaves	*Syzygium cumini*	Flavonoids catechin	Cotton	Excellent
Eucalyptus	*Flavonoid*	Tannin	Cotton, wool	Excellent
Madder	*Rubia cordifolia*	Anthraquinone	Cotton, Himalayan nettle	Very good
Cutch	*Acacia catechu*	Catechin (condensed tannin)	Cotton, Himalayan nettle	Excellent
Pomegranate peels	*Punica granatum*	Tannin	Cotton	Excellent
Banana peels	*Musa paradisiaca*	Luteolin	Giza cotton	Excellent

Table 2.
Natural dyes with responsible component for UV protection properties.

UPF rating	UV-R protection	Effective UV-R transmission (%)	UPF labelling
15–24	Good protection	6.7–4.2	15, 20
25–39	Very good protection	4.1–2.6	25, 30, 35
40–50, 50+	Excellent protection	≤2.5	40, 45, 50, 50

Table 3.
UPF classification system (AS/NZS 4399:1996) [55].

where
E_λ = relative erythemal spectral effectiveness.
S_λ = solar spectral irradiance.
T_λ = average spectral transmittance of the specimen (measured).
$\Delta\lambda$ = measured wavelength interval (nm).
The UV protection category was determined by the UPF values described by the Australian Standards/New Zealand AS/NZS 4399 (1996) given in **Table 3.**

2.3 Deodorising

Odour in textiles is caused by bacterial colonies due to favourable conditions such as perspiration [56]. Various studies reported about the deodorising performance of natural colourants. A study in the deodorising efficiency of gardenia, *Cassia tora* L., coffee sludge and pomegranate rind dyed fabric was observed. Maximum deodorising capacity was found in pomegranate followed by *coffee sludge (Coffea arabica), Cassia tora* and gardenia [57].

Natural dyeing of cotton, wool and silk fabric using gallnut also provides a better deodorising function against ammonia, trimethylamine and acetaldehyde and showed bacterial resistance against *Staphylococcus aureus* and *Klebsiella pneumonia*. The main component in the gallnut extract was found to be gallotannin which is the reason for these functional properties [58].

Natural dyeing of cellulosic and silk fabric with peony, clove and pomegranate (*Punica granatum*) also provides deodorising functionalization [59].

Application of fresh dye of indigo plant provided antimicrobial, sterilising or deodorising effect and treatment effect of atopic dermatitis [60].

Gas detecting tube method was used for deodorising test. It measures the concentration of ammonia gas. Natural dyed fabric is placed in the tube and the concentration of ammonia is observed. The reduction in concentration of ammonia signifies the deodorising capacity of dyed fabric. In blank (reference) tube, the concentration of ammonia is about 500 ppm.

The deodorising capacity of the dyed fabric is calculated using the following formulae:

$$\text{Deodorization performance } (\%) = (\text{Cb} - \text{Cs})/\text{Cb} \times 100 \tag{4}$$

where Cb is the gas concentration (ppm) of test tube without fabric (blank state) and Cs is the concentration of tube with fabrics [61].

2.4 Moth proof and insect repellent

Moth larvae which usually remain concealed cause the great losses to woollen textiles [62]. It has been estimated that 92.5 pounds wool fibre are eaten in 1 year due to the presence of protein in the wool [63]. Various product ranges such as carpet, blanket, namda (felted carpet), shawl and knit wears are majorly produced using wool fibre. Woollen carpet and handicraft play a major role in Indian export [64]. Dark humid conditions with 25–35° temperature are favourable conditions for moth larvae attack. Clothes moth (*Tineola bisselliella*) and carpet beetle (*Anthrenus verbasci*) are distributed in all the areas. Anti-moth finishing agents are DDT, permethrin, permethrin/hexahydro-pyrimidine derivative, cyhalothrin, etc. Some of these chemicals have been banned, while permethrin pyrimidine chlorine-based compounds are used widely as an anti-moth finishing agent for textiles, but these are becoming less effective on beetle larvae. The demand for replacing the permethrin and other synthetic anti-moth agent is increasing due to ecotoxicity of these anti-moth agents. The demand for natural anti-moth agent is increased due to eco-consciousness of consumers. In the case of natural dyes, its chemical structure also plays an important role in determining the anti-moth properties. Very few info are available for natural dyed anti-moth properties. Saffron flower waste, onion skin, henna, myrobolan, silver oak leaf, madder, wall nut, dholkanali and yellow roots were observed to impart anti-moth properties to wool, depending on the amount of tannin in their chemical composition. Natural dyes having higher amount of tannin repel the moths more effectively. In various experiments it has been reported that the natural dye having more than about 40% tannin is effective as an anti-moth agent, for instance, the dye extracted from silver oak, walnut husk and pomegranate rind having 47.87, 44.31 and 45.23% tannin, respectively, works as effective anti-moth agent [65].

For testing anti-moth properties, natural dyed and undyed woollen fabric samples were kept in petri dishes. Ten alive carpet beetles were put on each petri dish. Petri dish were kept in incubator (time: 15 days, temp: 30–35, RH 50–60%). Weight loss of the fabric due to moth attack is observed. Visual examination of the damaged fabric and number of alive moths were also done. For the comparative analysis, Eulon (a synthetic anti-moth) is also used [66].

2.5 Mosquito repellent

Dangerous diseases due to mosquito bite like dengue, malaria and chikungunya hit over 1.13 million people in the country last year [67]. Global warming also increases the growth of mosquitoes and thus the growth in disease like malaria, yellow fever and dengue fever [68]. Pomegranate peel dyed cotton fabric using different conc. of polyvinyl alcohol provides 80 percent mosquito repellency [69].

Mosquito repellency test is performed using a prepared cage of 40 cm × 40 cm. Fifty mosquitos are collected. Perforated transparent plastic is used to cover two opposite sides, while the other side of the cage is covered with carton [70].

3. Chemical compound structure and related functional properties of natural dyes

Indigoid, pyridine, carotenoid, quinonoid, flavonoids, betalains, anthocyanin, anthraquinone and tannins are the major chemical compounds found in natural dyes. These chemical structures are also one basis for the classification of natural dyes [3]. These functional groups provide a specific functional property to the textiles.

3.1 Indigoid

The colouring matter in indigo plant leaves is a light yellow substance called indican (1H-indol-3yl b-D-glucoside) (**Figure 1**) [71]. Natural indigoid dyes are mainly obtained from woad (*Isatis tinctoria L. Brassicaceae*, also known as dyer's woad) and the indigo plant (*Indigofera tinctoria L*) in temperate climates [72].

The cotton fabric dyed with fermented indigo leaves exhibited excellent UV protection as well as antimicrobial activity against *Staphylococcus aureus*. But dyed sample was observed to be relatively inactive against *Klebsiella pneumonia*, while in the case of application of silk the negligible protection was observed. Deodorisation capacity of these dyed fabrics was found to be low [73]. Indigo dyed samples with ferrous sulphate as a metal mordant were observed to fall under good UV protection category [74].

3.2 Anthraquinone

These dyes have anthraquinone (**Figure 2**) as a main colouring agent. These are generally in red colour also called mordant dyes. Madder, lacs, kermes and cochineal natural dyes have anthraquinone chemical structure [75].

The madder dye having anthraquinone structure improved both the UV protection performance and the antibacterial activity (against *Staphylococcus aureus* (*S. aureus*) and *Escherichia coli* (*E. coli*)) of the PET fabric. UV protection factor increased up to 106 and antibacterial activity up to 86% against both types of bacteria tested [76]. Excellent UV protection properties were achieved with the application of cochineal dye on silk and wool fabrics [45].

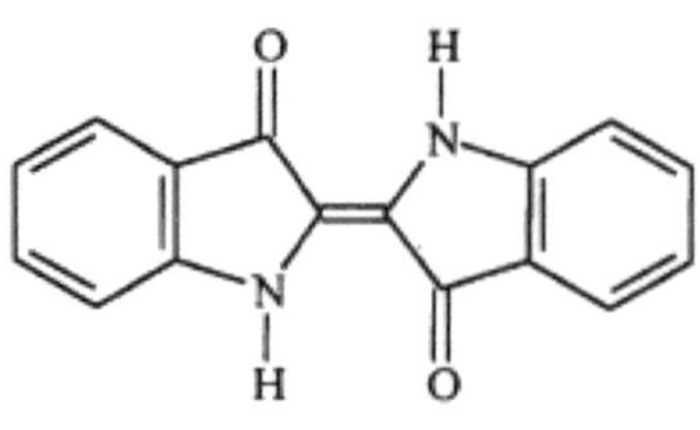

Figure 1.
Indigo.

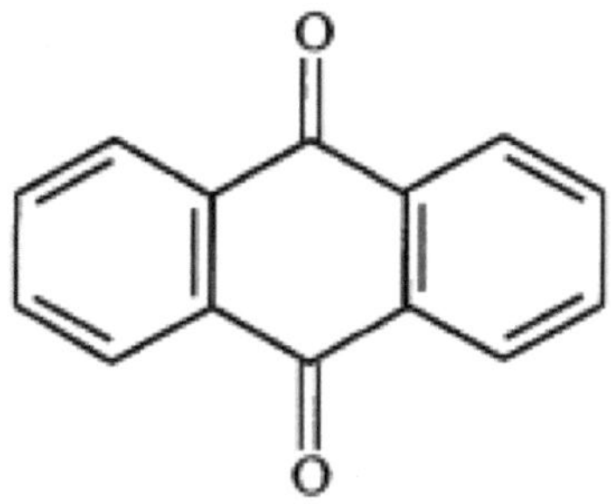

Figure 2.
Anthraquinone.

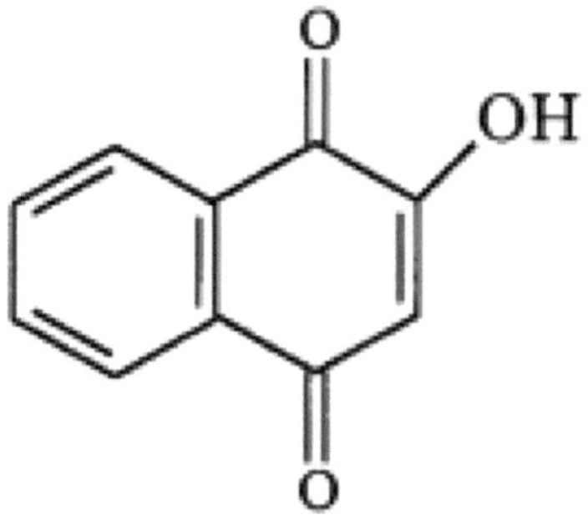

Figure 3.
Alpha naphthoquinone.

The ultraviolet protection factor (UPF) values of lac (having anthraquinone structure) dyed silk fabric with and without metal mordants also ranged between very good and excellent for the silk fabric [77].

3.3 Alpha naphthoquinones

Lawsone, henna, juglone and shells of unripe walnut natural dyes contain alpha naphthoquinone (**Figure 3**); these dyes are similar to disperse dyes [78].

It has been reported in a study that application of henna extract on cantonised cotton fabric showed outstanding enhancement in UV protection with minimal impact on the tensile strength against harmful UV radiation due to alpha-napthaquinone [79].

3.4 Flavonoids

The basic structural feature of flavonoid compounds is the 2-phenyl-benzo pyrane or flavan nucleus, which consists of two benzene rings linked through a heterocyclic pyrane ring (**Figure 4**). Mostly all yellow-coloured natural dyes are derivatives of hydroxyl and methoxy-substituted flavones and isoflavones. A common example of flavonoid containing dye is weld (*Reseda luteola)* (containing luteolin pigment). Other plant sources are *Allium cepa* (onion), *Artocarpus heterophyllus/Artocarpus integrifolia* (jackfruit), *Myrica esculenta* (Kaiphal), *Datisca cannabina* (Hemp), *Delphinium zalil* (Yellow larkspur), *Gossypium herbaceum*, *Sophora japonica/Styphnolobium japonicum*, *Butea monosperma/Butea frondosa* (flame of the forest/palas), *Mallotus philippinensis* (Kamala), *Bignonia chica/Arrabidaea chica* (Carajuru/Puca), *Commelina communis and Pterocarpus santalinus* (red sandalwood) [80].

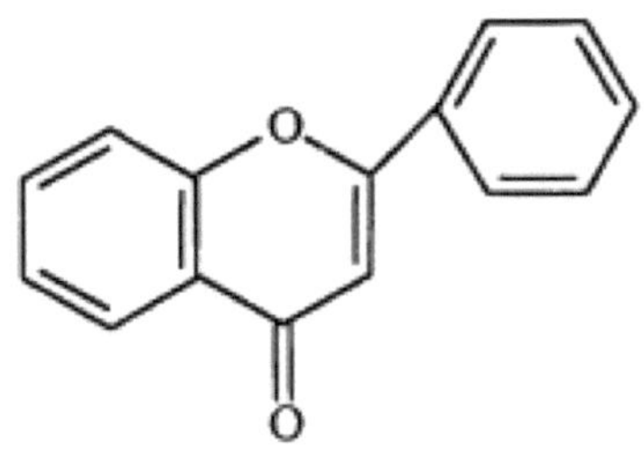

Figure 4.
Flavones (a class of flavonoids).

It has been reported that quercetin, a flavonoid, imparted better antioxidant, antibacterial and UV protection performance to silk fabric [81]. Flavonoids and anthocyanin from red onion skin (*Allium cepa* L.) are natural dyes with antibacterial activity and UV protection after application on wool and cotton fabric [82].

3.5 Di-hydropyrans

These are closely related to flavones in terms of chemical structure [83]. Logwood, brazilwood and sappanwood are common example of di-hydropyrans (**Figure 5**) which generally provide darker shades on silk wool and cotton [84].

Dyeing with logwood dye having di-hydropyrans showed maximum zone of inhibition against *F. solani* and *P. decumbens* even without any mordanting [85].

3.6 Anthocyanidins

Anthocyanins (**Figure 6**) are water-soluble natural pigments belonging to the phenolic family [86]. Carajurin, a direct orange colour for silk coloration, is obtained from the leaves of *Bignonia chica*. These are commonly found in red, purple, blue-coloured flowers and fruits.

It was reported that red reddish dyed silk fabric also provides the antioxidant, antimicrobial activities and UV protection property due to anthocyanin [87].

3.7 Carotenoids

Carrots are the main source of carotene presence due to double bond conjugation.

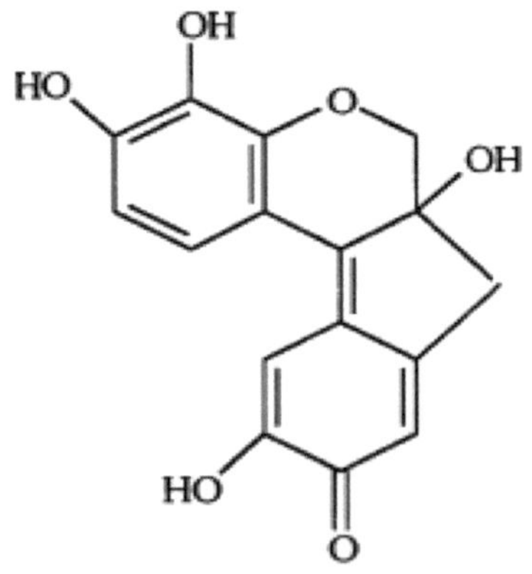

Figure 5.
Di-hydropyrans.

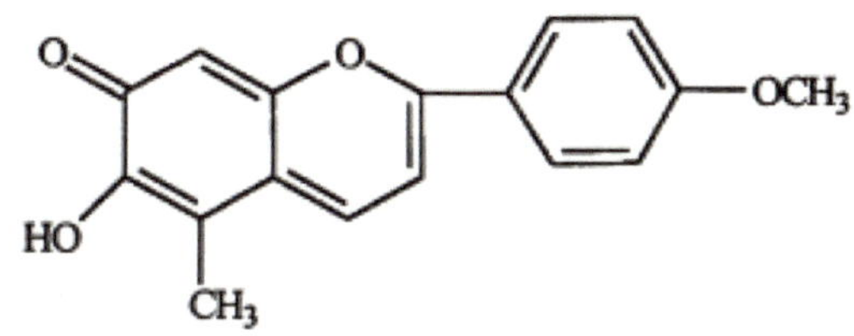

Figure 6.
Anthocyanin.

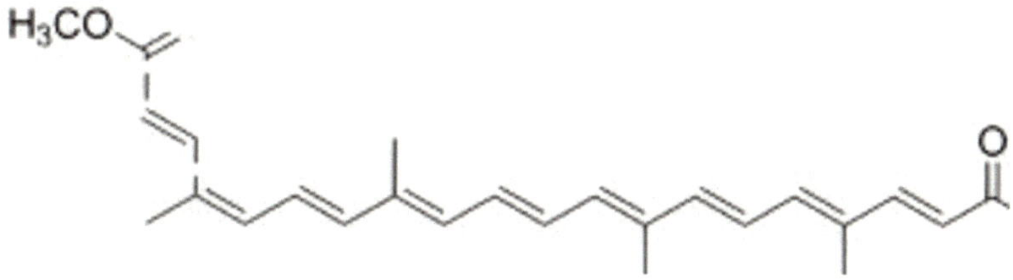

Figure 7.
Carotenoids.

Annatto and saffron come under the carotenoid (**Figure 7**) family. It has been reported that carotenoids have anti-radiation property [88].

3.8 Tannins

Tannins are higher-molecular-weight phenolic compounds. The molecular weight of tannins ranges between 500 and 3000. It is found in a wide range of natural flora such as fruit, pods, plant galls, leaves, bark, wood and roots. Tannins are generally divided into two groups (**Figures 8** and **9**):

1. Condensed tannins (proanthocyanidins)

2. Hydrolysable (pyrogallol) [89]

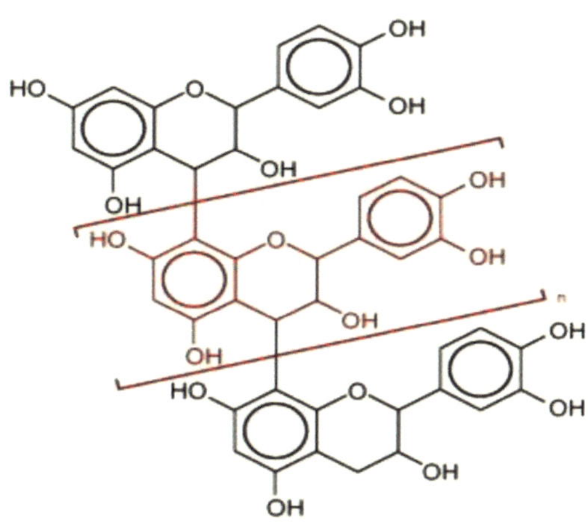

Figure 8.
Condensed tannins.

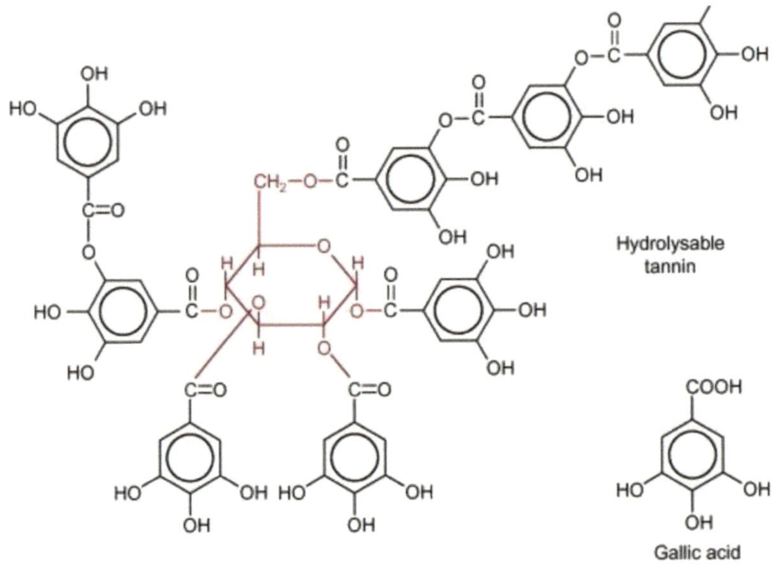

Figure 9.
Hydrolysable tannins.

Tannin is considered as a protective agent against microbes and UV rays. Antimicrobial and antioxidant properties of dyed wool fabric were observed to have improved after application of tannin [90].

4. Conclusion

Presently the consumers become aware of hygienic healthy and protective lifestyle; hence there is a necessity of clothing with functional properties as it covers the major part of the body. The various changes in climate and environment require more protection against UV rays, microbes and insects protecting human from vector-borne diseases, and clothes do not only cover the modesty or basic requirement but also are more functional in terms of protection. Aesthetic as well as medicinal properties of natural dyes has been exploring since ancient time. The glory of natural dyes somewhat subdued after the invention of synthetic dye and presently with the increase of eco-consciousness. Various researches have been conducted on functional aspect of natural dyes, but still there is a need to find out the possible sources for more prominent protection.

Although the natural dyed textiles are promising to provide functional properties such as antimicrobial, UV protection and mosquito as well as moth repellence, still the stability-related issues with natural dyes also need to be significantly addressed. Various researches are being organised with this aspect. This stability-related issue can be enhanced with proper knowledge of interaction of fabric natural dye and mordant future performance of natural dyes. Proper combination of dye, fabric and mordants help to enhance the wash stability. Different techniques like surface modification (plasma treatment and UV irradiation, etc.) and microencapsulation can be used to enhance the stability of the functional properties of dyes. This would further lead to a more stable functional property. Extraction- and application-related issues of natural dyes should also be sorted out. The whole life cycle of natural dyes requires different areas of science; thus, collaborative efforts are required for more prominent results in terms of colour as well as providing functional properties.

Author details

Deepti Pargai*, Shahnaz Jahan and Manisha Gahlot
G.B. Pant University of Agriculture and Technology, Pantnagar, Uttarakhand, India

*Address all correspondence to: pargai.deepti16@gmail.com

References

[1] Sen T, Barrow CJ, Deshmukh SK. Microbial pigments in the food industry—Challenges and the way forward. Frontiers in Nutrition. 2019;**6**:7. DOI: 10.3389/fnut.2019.00007

[2] Gokhale SB, Tatiya AU, Bakliwal SR, Fursule RA. Natural dye yielding plants in India. Natural Product Radiance. 2004;**3**(4):228-234

[3] Saxena S, Raja ASM. Natural dyes: Sources, chemistry, application and sustainability issues. In: Muthu S, editor. Roadmap to Sustainable Textiles and Clothing. Textile Science and Clothing Technology. Singapore: Springer; 2014. DOI: 10.1007/978-981-287-065-0

[4] Hassaan MA, Nemr AE. Health and environmental impacts of dyes: Mini review. American Journal of Environmental Science and Engineering. 2017;**1**(3):64-67. DOI: 10.11648/j.ajese.20170103.11

[5] Chavan RB. Health and environment hazards of synthetic dyes. Textile Review Magazine. May 2013:1-6

[6] Kumar A, Chowdhury AKR. Eco-friendly dyes and dyeing. Advanced Materials and Technologies for Environmental Sciences. 2018;**2**(1): 145-176

[7] United States Environmental Protection Agency. A Guide to UV Index. 2004. Available from: https://www.epa.gov/sites/production/files/documents/uviguide.pdf

[8] Jose S, Nachimuthu S, Das S, Kumar A. Moth proofing of wool fabric using nano kaolinite. The Journal of Textile Institute. 2017;**109**(2):225-231. DOI: 10.1080/00405000.2017.1336857

[9] Liu Q, Meng X, Li Y, Zhao CN, Tang GY, Li HB. Antibacterial and antifungal activities of spices. The International Journal of Molecular Sciences. 2017;**18**(6):1283. DOI: 10.3390/ijms18061283

[10] Mariselvam R, Singh R, Kaliraja K. Antifungal activity of different natural dyes against traditional products affected fungal pathogens. Asian Pacific Journal of Tropical Biomedicine. 2012; **2**(3):S1461-S1465. DOI: 10.1016/S2221-1691(12)60438-9

[11] Przewoźna KS, Zimniewska M. Natural dyeing plants as a source of compounds protecting against UV radiation. Herba Polonica. 2009;**55**(3): 311-318

[12] Singh R, Jain A, Panwar S, Gupta D, Khare SK. Antimicrobial activity of some natural dyes. Dyes and Pigments. 2005;**66**(2):99-102

[13] Gupta D, Khare SK, laha A. Antimicrobial properties of natural dyes against gram negative *bacteria*. Coloration Technology. 2004;**120**(4):167-171. DOI: 10.1111/j.1478-4408.2004.tb00224.x

[14] Hassan MM. In: Muthu S, editor. Handbook of Sustainable Luxury Textiles and Fashion. 1st ed. 211p. Singapore: Springer; 2015. DOI: 10.1007/978-981-287-633-1

[15] Chengaiah B, Rao KM, Kumar M, Alagusundaram M, Chetty CM. Medicinal importance of natural dye. International Journal of Pharm Tech Research. 2002;**2**(1):144-154

[16] Rani J, Avinashilingam PC. Eco Friendly Dyeing and Finishing on Bamboo Cotton Knitted Fabrics for Sportswear [thesis]. Coimbatore: Avinashilingam Institute for Home Science and Higher Education for Women; 2014

[17] Jayalakshmi I, Soudanyaa D. Effect of natural dyeing in bamboo material. Man-Made Textiles in India. 2015: 465-472

[18] Wells K. Wearing Naturally Dyed Clothes may Boost Health. 2019. Available from: http://timesofindia.indiatimes.com

[19] Vankar P. Chemistry of natural dyes. Resonance. October 2000:73-80. Available from: https://www.ias.ac.in/article/fulltext/reso/005/10/0073-0080

[20] Malomo D, Abimbade AD, Oluwaseun AK, Eghareba O. Likely mechanism of dye adhesion on fabrics. In: Proceedings of 62nd ISERD International Conference; Boston, USA; 2017

[21] Anonymous. Types of Microorganism. 2019. Available from: https://courses.lumenlearning.com

[22] Lowy F. Bacterial classification, structure and function. 2009;**44**(12): 977-983. Available from: http://www.columbia.edu/itc/hs/medical/pathophys/id/2009/introNotes.pdf

[23] Boryo DEA. The effect of microbes on textiles material: A review on the way out so far. The International Journal of Engineering and Science (IIJES). 2013;**2**(8):09-13

[24] Seventekin N, Ucarci O. The damage caused by micro-organisms to cotton fabrics. Journal of Textile Institute. 1993;**84**(3):304-313

[25] Yusuf M, Shabbir M, Mohammad F. Natural colorants: Historical, processing and sustainable prospects. Natural Products and Bioprospecting. 2017;7(1): 123-145. DOI: 10.1007/s13659-017-0119-9. Epub 2017 Jan 16

[26] Rehman F, Sanbhal N, Naveed T, Farooq A, Wang Y, Wei W. Antibacterial performance of Tencel fabric dyed with pomegranate peel extracted via ultrasonic method. Cellulose. 2018;**25**(7):4251-4260. DOI: 10.1007/s10570-018-1864-6

[27] Yusufa M, Shahida M, Khan MI, Khan SA, Khan MA, Mohammed F. Dyeing studies with henna and madder: A research on effect of tin (II) chloride mordant. Applied and Environmental Microbiology. 2014;**80**(21):6611-6619

[28] Pal A, Tripathi YC, Kumar R, Upadhyay L. Antibacterial efficacy of natural dye from Melia composita leaves and its application in sanitized and protective. Textiles Journal of Pharmacy Research. 2016;**10**(4):154-159

[29] Ibrahim NA, El-Zairy WM, El-Zairy MR, Ghazal HA. Enhancing the UV-protection and antibacterial properties of polyamide-6 fabric by natural dyeing. Textiles and Light Industrial Science and Technology (TLIST). 2013;**2**(1): 31-46

[30] Haji A. Antibacterial dyeing of wool with natural cationic dye using metal mordants. Materials Science (MEDŽIAGOTYRA). 2012;**18**(3): 267-270

[31] Lee YH, Hwang EK, Baek YM, Kim HD. Colorimetric assay and antibacterial activity of cotton, silk, and wool fabrics dyed with peony, pomegranate, clove, Coptis chinensis and gallnut extract materials. Fibers and Polymers. 2009;**17**(4):560-568

[32] AATCC 100-Assessment of Antibacterial Finishes on Textile Materials. 2004

[33] AATCC 147. Parallel Streak Method. Antibacterial Activity Assessment of Textile Materials. 2004

[34] Feng XX, Zhang LL, Chen JY, Zhang JC. New insights into solar UV protective properties of natural dye. Journal of Cleaner Production. 2007;**15**: 366-372

[35] Svobodová A, Psotová J, Walterová D. Natural phenolics in the prevention of UV-induced skin damage: A review. Biomedical Papers of the Medical Faculty of the University

Palacky, Olomouc, Czech Republic. 2003;**147**(2):137-145

[36] Gupta D, Jain A, Panwar S. Anti-UV and anti-microbial properties of some natural dyes on cotton. Indian Journal of Fibre and Textile Research. 2005;**30**: 190-195

[37] Mongkholrattanasit R, Cholachatpinyo A, Tubtimthai N, Nattadon R, Mai C. An evaluation of UV protection imparted by wool fabric dyed with natural dye from eucalyptus leaf. Chiang Mai Journal of Science. 2011;**41** (5.2):1208-1219

[38] Pisitsak P, Hutakamol J, Jeenapak S, Wanmanee P, Nuammaiphum J, Thongcharoen R. Natural dyeing of cotton with *Xylocarpus granatum* bark extract: Dyeing, fastness, and ultraviolet protection properties. Fibers and Polymers. 2016;**17**(4):560-568

[39] Křížová H, Wiener J. Comparison of UV protective properties of woollen fabrics dyed with yellow natural dyes from different plant sources. Environmental Sciences. 2016;**2**(7): 2454-9916

[40] Cao HM, Yang JF, Zhu GY. The performance of *Flos Caryophyllata* extract on black dyeing and UV resistance. Advanced Materials Research. 2013;**821**:564-568

[41] Griffoni D, Bacci L, Zipoli G, Carreras G, Baronti S, Sabatini F. Laboratory and outdoor assessment of UV protection offered by flax and hemp fabrics dyed with natural dyes. Photochemistry and Photobiology. 2009;**85**:313-320

[42] Hong K, Bae JH, Jin SR, Yang JS. Preparation and properties of multi-functionalized cotton fabrics treated by extracts of gromwell and gallnut. Cellulose. 2011;**19**(2):507-515. DOI: 10.1007/s10570-011-9613-0

[43] Alebeida OK, Taoa Z, Seedahmedc A. New approach for dyeing and UV protection properties of cotton fabric using natural dye extracted from henna leaves. Fibres and Textiles in Eastern Europe. 2015;**23**(5):61-65

[44] Vorabodee S, Chotima S, Jantip S, Potjanart S, Porntip SB. Effect of chitosan and turmeric dye on ultraviolet protection properties of polyester fabric. Applied Mechanics and Materials Academic Journal. 2014;**535**:658

[45] Gawish SM, Helmy HM, Ramadan AN, Farouk R, Mashaly HM. Eco-friendly multifunctional properties of cochineal and weld for simultaneous dyeing and finishing of proteinic fabrics. International Journal of Engineering and Technology. 2016;**8**(5): 2246-2253

[46] Jung JS. Study of fastness, UV protection, deodorization and antimicrobial properties of silk fabrics dyed with the liquids extracted from the gallnuts, areca nuts, and pomegranate peels. MATEC Web of Conferences. 2016;**49**:1-6

[47] Salah M. Antibacterial activity and UV protection property of some Egyptian cotton fabrics treated with aqueous extract from banana peel. International Journal of Clothing Science. 2012;**1**(1):1-6

[48] Pargai D, Gahlot M, Rani A. Ultraviolet protection properties of nettle fabric dyes with natural dyes. Indian Journal of Fibre and Textile Research. 2016;**41**(4):418-425

[49] Gupta D, Ruchi UPF. Characteristics of natural dyes and textiles dyed with them. Colourage. 2007;**54**(4):75-80

[50] Chattopadhyay SN, Pan NC, ROY AK, Saxena S, Khan A. Development of natural dyed jute fabric with improved colour yield and UV protection

characteristics. The Journal of the Textile Institute. 2013;**104**(8):808-818

[51] Kim SH. Dyeing characteristics and UV protection property of green tea dyed cotton fabrics. Focusing on the effect of chitosan mordanting condition. Fibers and Polymers. 2006; 7(3):255-261

[52] Pargai D, Jahan S. Application of *Vitis vinifera* microcapsules on cotton fabric: A potential to prevent UV-induced skin problems. Journal of Natural Fibers. July 2018:1-15. DOI: 10.1080/15440478.2018.1500334

[53] NPTEL. Objective of Testing. 2019. Available from: https://nptel.ac.in/courses/116102029/2

[54] AATCC-183:2004. Transmittance or blocking of erythemally weighted ultraviolet radiation through fabrics. In: Technical Manual of American Association of Textile Chemist and Colourist. USA; 2010;**85**:318-321

[55] Australia/New Zealand Standard (AS/NZS 4399:1996). AS/NZS 4399: 1996. Australian/New Zealand Standard® Sun protective clothing—Evaluation and classification. Published Jointly By: Standards Australia: The Crescent, Homebush Australia and Standards New Zealand: Level Wellington New Zealand

[56] Callewaert C, Maeseneire ED, Kerckhof FM, Verliefde A, Wiele TVD, Boon N, et al. Microbial odor profile of polyester and cotton clothes after a fitness session. Applied and Environmental Microbiology. 2014:**80**(21). DOI: 10.1128/AEM.01422-14

[57] Lee YH. Dyeing, fastness, and deodorizing properties of cotton, silk, and wool fabrics dyed with coffee sludge (*Coffea arabica L.*) extract. Journal of Applied Polymer Science. 2007;**103**(1):251-257

[58] Lee YH. Deodorizing function and antibacterial activity of fabrics dyed with gallnut (*Galla Chinensis*) extract. Textile Research Journal. 2014;**85**(10): 1045-1054

[59] Lee YH, Kyung HE, Do KM. Colorimetric assay and antibacterial activity of cotton, silk, and wool fabrics dyed with peony, pomegranate, clove, *Coptis chinensis* and gallnut extract materials. 2009;**2**(1):10-21

[60] Anonymous. Medicinal Properties of 6 Natural Dyes. 2019. Available from: https://nptel.ac.in/courses/116104046/12.pdf

[61] Hwang EK, Lee YH, Kim HD. Dyeing and deodorizing properties of cotton, silk, and wool fabrics dyed with various natural colorants. Journal of the Korean Society of Dyers and Finishers. **19**(6):12

[62] Anonymous. Agriculture and food. Identifying and controlling clothes moths, carpet beetles and silverfish. 2018. Available from: https://www.agric.wa.gov.au/pest-insects/identifying-and-controlling-clothes-moths-carpet-beetles-and-silver%EF%AC%81sh

[63] Basuk M, Behera J. A review on woollen cloth's moth and its remedies. Textile Today. March 2018. Available from: https://www.textiletoday.com.bd/review-woolen-cloths-moth-remedies/

[64] Carpet Industry and Exports. 2019. Available from: https://www.ibef.org/exports/carpet-industry-in-india.aspx

[65] Shakyawa DB, Raja ASM, Kumar A, Pareek PK. Antimoth finishing treatment for woollens using tannin containing natural dyes. Indian Journal of Fibre and Textile Research. 2015;**40**:200-202

[66] Kumar A, Pareek PK, Kadam VV, Shakyawar DB. Antimoth efficacy of

neem on woollen fabric. Indian Journal of Fibre and Textile Research. 2016; **15**(2):272-277

[67] WHO. Mosquito Borne Diseases. 2015. Available from: https://www.who.int/neglected_diseases/vector_ecology/mosquito-borne-diseases/en/

[68] Niang EHA, Bassene H, Fenollar F, Mediannikov O. Biological control of mosquito-borne diseases: The potential of Wolbachia-based interventions in an IVM. Framework Journal of Tropical Medicine. 2018;**2018**:15. Article ID: 1470459

[69] Rimpi SA. Protection and application of natural mosquito repellents cotton fabric through dyeing. International Journal for Innovative Research In Multidisciplinary Field. 2017;**3**(8):873-878

[70] Jajpura L, Saini M, Rangi A, Chhichholia K. A review on mosquito repellent finish for textiles using herbal extracts. International Journal of Engineering Sciences and Management Research. 2015;**2**(8):17-24

[71] Wahyuningsih S, Ramelan AH, Wardani DK, Aini FN, Sari PL, Tamtama BPN, et al. Indigo dye derived from *Indigofera tinctoria* as natural food colorant. IOP Conference Series: Materials Science and Engineering. 2017;**193**(1):012048. DOI: 10.1088/1757-899X/193/1/012048

[72] Olusola A, Emmanuel SE, Funmilola S. Extraction of indigo dye (powdered, form) from the leaf of *Indigofera tinctoria*. Dye and Textile Technology. 2012;**6**(1):137-143

[73] Jung JS. Dyeing properties and functionality of cotton and silk dyed with fermented indigo and fresh indigo. International Journal of Advances in Science Engineering and Technology. 2017;**5**(3):38-41

[74] Mongkhorattanasit R, Klaichoi C, Orawan M, Punrattanasin N, Sasivatchutikool N, Rungruangkitkrai N. Effect of ferrous sulfate to improve UV-protection property of cotton fabric dyed with natural indigo. In: Advanced Material Research. 2014; Vol. 1030–1032. pp. 418-421

[75] Samanta AK, Konar A. Dyeing of textiles with natural dyes. In: Kumbasar EK, editor. Natural Dyes. Crotia: Intechopen; 2011. 33p

[76] Tove A, Zhou Y, Guan J, Chen G, Perwuelz BN, Nierstrasz V. Bioactive and multifunctional textile using plant-based madder dye: Characterization of UV protection ability and antibacterial activity. Fibres and Polymers. 2017; **18**(11):2170-2175

[77] Mongkholrattanasit R, Rungruangkitkrai N, Tubtimthai N, Sasivatchutikool N. UV protection property of colorant from lac for silk fabric dyeing by cold pad-batch: The influence of metal mordants concentration. Advanced Materials Research. 2014;**884-885**:257-260. DOI: 10.4028/www.scientific.net/AMR.884-885.257

[78] Vankar PS, Shukla D. New Trends in Natural Dyes for Textiles. India: Woodhead Publishing; 2017. 368p

[79] Alebeida OK, Taoa Z, Seedahmed AI. New approach for dyeing and UV protection properties of cotton fabric using natural dye extracted from henna leaves. 2015;**5**(113):60-65. DOI: 10.5604/12303666.1161758

[80] Panche AN, Diwan AD, Chandra SR. Flavonoids: An overview. Journal of Nutritional Science. 2016;**5**: e47. DOI: 10.1017/jns.2016.41

[81] Zhou Y, Tang RC. Natural flavonoid-functionalized silk fiber presenting antibacterial, antioxidant, and UV protection performance. ACS

Sustainable Chemistry and Engineering. 2017;**5**(11):0518-10526

[82] Mothaghi Z. An economical dyeing process for cotton and wool fabrics and improvement their antibacterial properties and UV. 2018; **15**(6):777-788

[83] Parthiban M, Shrikrishnan MR, Kandhavadivu P. Sustainability in Fashion and Apparel. India: Woodhead Publishing; 2017. 226 p

[84] ParthibanM, Srikrishnan MR, Kandhavadivu P. Sustainability in Fashion Apparel Challenge and Solutions. India: Woodhead Publishing; 2017. 226 p

[85] Sahab AF, Waly A, Marie M, Elzoeerey WR, Elela RA. Studies on the antifungal activity of natural dyes and their application on textile materials. Der Pharmacia Lettre. 2016;**8**(17):66-72

[86] Khoo HE, Azlan A, Tang ST, Lim SM. Anthocyanidins and anthocyanins: Colored pigments as food, pharmaceutical ingredients, and the potential health benefits. Food and Nutrition Research. 2017;**61**(1):1-7

[87] Grifoni D, Bacci L, Lonardo SD, Romani A. UV protective properties of cotton and flax fabrics dyed with multifunctional plant extracts. Dyes and Pigments. 2014;**105**:89-96. DOI: 10.1016/j.dyepig.2014.01.027

[88] Natália M, Ferreira SRS. Carotenoids functionality, sources, and processing by supercritical technology: A review. Journal of Chemistry. 2016; **2016**:16. Article ID: 3164312

[89] Hagerman AE, Riedl KM, Jones GA, Sovik KN, Ritchard NT, Hartzfeld PW, et al. High molecular weight plant polyphenolics (tannins) as biological antioxidants. Journal of Agricultural and Food Chemistry. 1998;**46**(5):1887-1890. DOI: 10.1021/jf970975b

[90] Hong KH. Effects of tannin mordanting on coloring and functionalities of wool fabrics dyed with spent coffee grounds. Hong Fash Text. 2018;**5**(33):1-11. DOI: 10.1186/s40691-018-0151-3

Section 7

Eco Safe Synthtic Dyes and Chemicals

Chapter 10

Dyeing Chemicals

Shekh Md. Mamun Kabir and Joonseok Koh

Abstract

Dyeing auxiliaries play an important role in the determination of the final dyeing quality. The formation of additional complexes with dyes and auxiliary agents enhances the exhaustion of dyes on textile substrates. For aqueous-based dyeing, dye auxiliaries such as chelating agents, dispersing agents, leveling agents, electrolyte, pH control agents, and surfactants form complexes with the dye on natural and synthetic fibers. A growing awareness of the impact of industrial pollution on the environment became crucial for the dyeing industry in the closing decades of the twentieth century. These chapters discuss the characteristics of dyeing chemicals and how auxiliary substances can assist in achieving outstanding dyeing performance.

Keywords: chelating agent, dispersing agent, leveling agent, electrolyte, pH control agent, surfactant, antifoaming agents

1. Introduction

The textile dyeing industry is constantly increasing because of the growing consumption of fabrics and garments; moreover, till the next decade, a billion consumers will be added to the global market [1]. The processing of such a large volume of fabrics and garments is conducted through dyes and chemicals. A variety of factors can influence the quality of dyeing and its complex mechanisms in batch reproduction. The degree of levelness and reproducibility of dyeing depends on not only the dyes and chemicals used but also the control of temperatures and pH conditions in the dye bath [2]. To achieve any progress in such studies, it is necessary to control the parameters in order to determine their effect on the dyeing system. To study the performance of dyeing in an aqueous dye bath system, it is essential to alter the dye concentration while maintaining, and all other conditions constant so that the changes in the chemical structure of the solvent and the nature of the dye species can be analyzed [3].

Natural and synthetic dyes play an important role in the process of dyeing textile fabrics and garments. Different classes of dyes are used for coloring different textile materials with the aid of auxiliaries, which facilitate the homogenization of the mixtures [4]. The method of determining the equilibrium constant of the dye–auxiliary complex can be constructed.

$$\text{Dye} + \text{Auxiliary Agent} \rightleftharpoons \text{Addition complex}$$
$$D + AA \rightleftharpoons AC$$

The association between the dye and the auxiliary agent may proceed as far as the colloidal particles are dispersed in dyes. Apart from this, there exists an equilibrium between the dyes and the textile auxiliaries, which have been added to the dye bath [5]. The auxiliaries exist in a state of association equilibrium [6]. A new era of dyeing chemical research began in 1930 when soap was replaced by synthetic surfactants [7]. Since many researchers have described the use of various textile auxiliaries in dyeing [8]. Thermodynamic studies of dyeing make useful contributions to the general theories of intermolecular forces while diffusion processes are also influenced by the dye concentration, temperature, nature of dyeing auxiliaries, and polymer structure [9]. A major challenge for textile wet processors is addressing the increasing cost and demand associated with the use of different auxiliaries in industrialized countries. However, in recent years, there has been significant demonstration of a lack of knowledge about the usage of dyeing chemicals. In this study, the potential for dyeing auxiliaries was explained.

2. Dyeing chemicals

This chapter demonstrates the characteristics of different auxiliary agents, which tend to form colloidal flocculation and dye–auxiliary complexes with dyes. A chelating agent is used to remove the hardness of water by bonding with calcium and magnesium ions and other heavy metal ions in hard water, by forming a stable complex compound that does not decompose over a prolonged processing period [10]. Dispersing agents, pH, and temperatures affect the changes in shade and fastness of disperse dyes [2, 11]. A greater degree of levelness is linked with greater retarding effect, which means a longer dyeing time [4]. In the presence of electrolytes, dyeing of cotton fabric with anionic dyes is to suppress negative charge at the fiber surface and to promote increased dye exhaustion [12]. If the dye bath pH is adjusted prior to the dyeing process, it may affect the absorption of dye on the fiber [13]. Synthetic surfactants have been extensively studied by many researchers [14]. After World War II, fabric softeners emerged through the introduction of the synthetic surfactant [15].

2.1 Chelating agent

The word "chelate" is obtained from the Greek word "chel," which means crab's claw. Chemically, a chelate is an organic compound that can form a ring structure by bonding with metal ions. Chelating agents are mostly used in dye baths, as they remove the hardness of water by bonding with heavy metal ion. Chelating agents form a stable complex compound that does not decompose over a prolonged processing period [10]. The coordination of water with metal ions enhances the acidity of the dye solution, which is dependent on the physical and chemical characteristics of the metal ions [16]. Inorganic chelating agents are also used as detergent by suspending and dispersing agents. They require less than the stoichiometric quantity predicted to keep ions in solution (threshold effect). They will dissociate to sodium phosphate in water over time losing their ability to chelate, especially in hot water. Polyphosphates are derivatives of phosphoric acid and are made by reacting phosphorous pentoxide with phosphoric acid. Important polyphosphates are tetra sodium pyrophosphate ($Na_4P_2O_7$), sodium phosphate ($Na_5P_3O_{10}$), and sodium hexametaphosphate ($Na_6P_6O_{18}$) [17] (**Figure 1**).

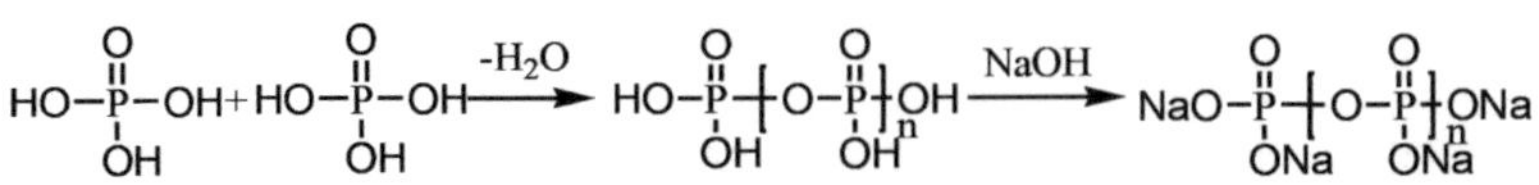

Figure 1.
Formation of sodium polyphosphates.

DOI: http://dx.doi.org/10.5772/intechopen.81438

Organophosphonic acids also act as chelating agents and aid detergency by dispersing and suspending soil. They are more stable than inorganic polyphosphates in hot water and exhibit threshold effect (**Figure 2**).

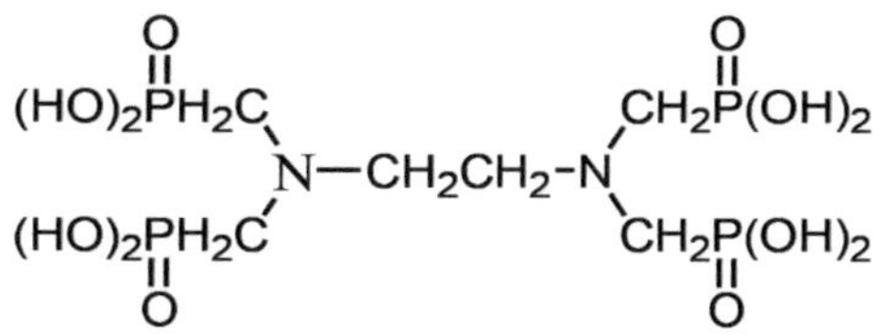

Figure 2.
Organophosphonic acids.

Amino carboxylic acids form very stable complexes, when reacting with metal ions. They react stoichiometrically and can be formed quantitatively to determine calcium and magnesium by titration. Disodium-ethylenediaminetetraacetic acid (EDTA) and nitrilotriacetic acid (NTA) are categorized in these groups [17] (**Figure 3**).

Several hydroxyl groups containing organic compounds precipitate bi- and trivalent metal cations in alkaline medium and also acts as effective sequestering agents; however, they are not effective for calcium and magnesium. Some of the well-known products in this category are glycolic acid, gluconic acid, and citric acid (**Figure 4**).

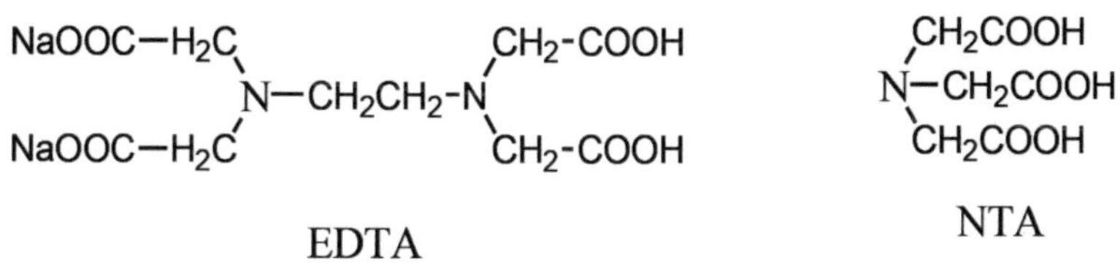

Figure 3.
Disodium-ethylenediaminetetraacetic acids and nitrilotriacetic acid.

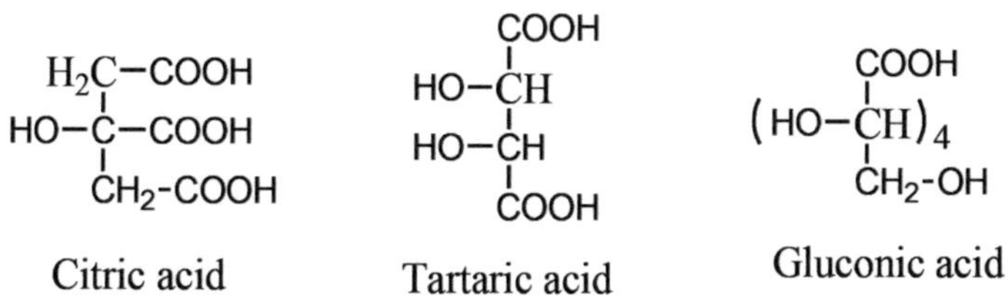

Figure 4.
Hydroxy acids sequestering agents.

Ethylenediaminetetraacetic acid also prevents or removes scales by forming water-soluble complex compounds, which vigorously react with metal ions [18]. Gluconic acid is obtained from fruit, honey, and wine. In an aqueous solution at neutral pH, gluconic acid forms gluconate ion, which is used in cleaning products in which it dissolves mineral deposits through the formation of ligand with metal compounds [19]. Citric acid also strongly interacts with active metal components. This interaction helps metal dispersion and decreases the specific surface area and pore volume of the metal components. The addition of citric acid increases the amount of metal ion adsorption. Increasing the adsorption depends on the formation of more complex structures (**Figure 5**).

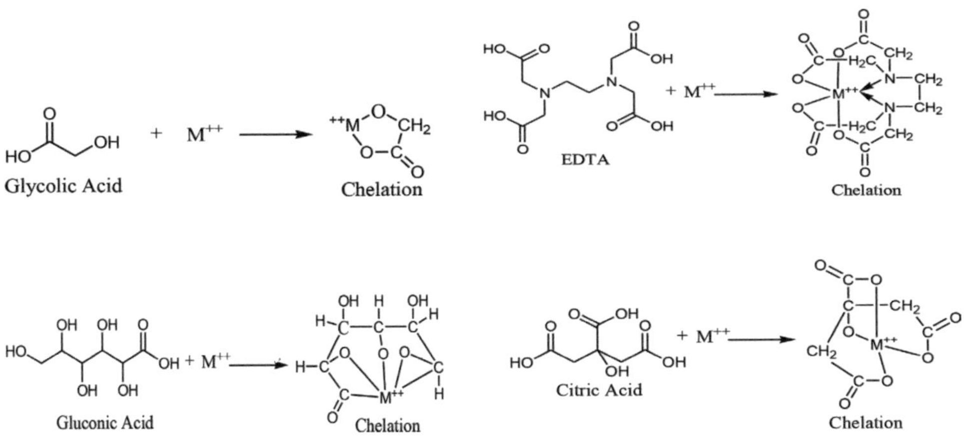

Figure 5.
Coordination of metal ions with different chelating agents.

Glycolic acid exhibits better chelation power in all metal ions because of the lower molecular weight, which helps to remove all metal ions. Glycolic acid, which is a naturally occurring organic agent, exhibits more penetration of metal ions. Natural chelates are biodegradable and non-toxic to the environment. Under natural conditions, EDTA is converted into ethylenediaminetriacetic acid, and then forms a ring structure with other metals, creating persistent organic pollutants [20]. Kabir and Koh [21] have demonstrated that glycolic acid provides effective chelation efficiency for a moderate acid donor. Consequently, glycolic acid exhibited a higher percentage of dyebath exhaustions and better dye ability than other organic chelating agents.

2.2 Dispersing agent

Disperse dyes are nonionic chemicals that are barely soluble in water and often crystallize with varying particle size [22]. These characteristics are inadequate for dispersing dyes of water and cause unleveled dyeing. To achieve the required particle size and distribution, the disperse dye is milled, usually in the presence of a dispersing agent [23]. Generally, the dispersing agents are anionic, e.g., ligninsulfonates, or polycondensates of arylsulfonic acids with formaldehyde, which facilitate milling. Dispersing agents have shown a dual function role: breaking down aggregated dye particles and dispersing dyes in the dye liquor. Dispersing agents consists of high-molecular weight or polymeric compounds in which polar or ionizing groups alternate with nonpolar groups along the chain. The backbone of a dispersing agent is nonpolar, while the polar or ionizing groups are located in the side chains. Johnson [24] demonstrated that the adsorption of dispersing agent is oriented parallel to the surface, so that the nonpolar groups adjoin to the surface and the polar groups turn outward.

The application of disperse dyes on polyester fabric mainly occurs via the inclusion of dispersing agents. The hydrophobic tails of the dispersing-agent molecules are oriented toward the center of the dye micelle, which facilitates micellar solubilization of the disperse dye molecules, thereby conferring higher dye solubility. The disperse dyeing mechanism has been categorized in four stages: (i) dissolution of dye in water by the formation of the dye micelle with dispersing agents; (ii) transference of dye molecules from the solution to the surface of the fiber; (iii) replenishment of dyebath by the dissolution of solid material from the dispersion; and (iv) diffusion of dye into the fiber (**Figure 6**).

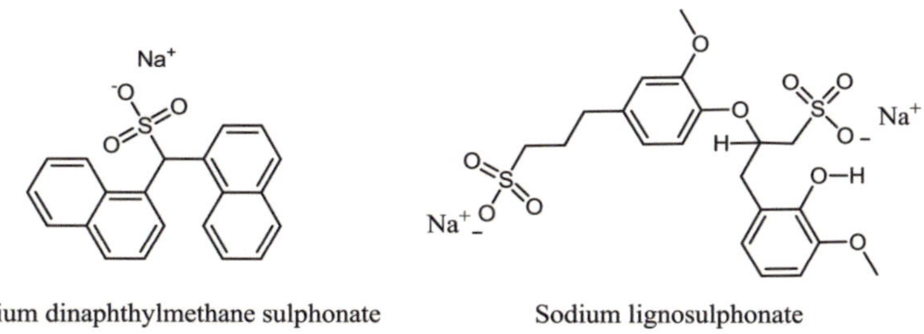

Figure 6.
Sodium dinaphthylmethane sulfonate and sodium lignosulfonate.

Murray and Mortimer [25] mentioned sodium dinaphthylmethane sulfonate and lignosulfonates as agents used for disperse dyes. A combination of dispersing agents and water-soluble polymers, for example, poly (vinyl alcohol), sodium polyacrylate, and the maleic anhydride-styrene copolymer, in admixture with anionic surfactants, would be useful in processing the pigments. Wolf and Bauer [26] suggested the use of dissolved disperse dyes in dimethylformamide and formed dispersions by pouring these solutions into aqueous solutions of the dispersing agents. Dispersing agents formed the dispersion by varying particle size of disperse dyes (**Figure** 7).

Heimanns [27] explained that a high-concentration of dispersing agent acts as a barrier to diffusion and reduces the dye yield in thermo-fixation. For better color yield, liquid brands of disperse dyes contain a smaller amount of dispersing agent. To ensure stability, the amount of dispersing agents must be maintained in the dye bath [8].

2.3 Leveling agents

To achieve uniform dyeing on fabric, it is essential to add a suitable leveling agent in the dye baths. However, it is quite difficult to explain the functions and actions of leveling agents. Werner [4] explained that the effects of textile auxiliaries, such as leveling agents, would appear at first sight to contradict the very nature of physical methods.

Figure 8 shows that equilibrium in dyebath-fiber systems is caused by a large number of processes that play an important part in producing good dyeing. Leveling agents enhance the force in a state of association equilibrium in the dye bath. Werner [4] investigated that the formation of dyes and leveling agents represents bonding similar to what occurs in the dyeing of fibers. The effects of leveling agents are: (i) leveling agents tend to decrease the absorption of dyes by forming a dye-auxiliary

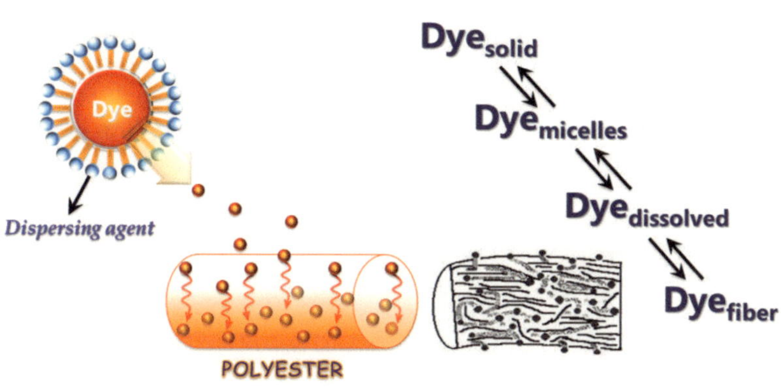

Figure 7.
Mechanism for dyeing of polyester with disperse dye and dispersing agents.

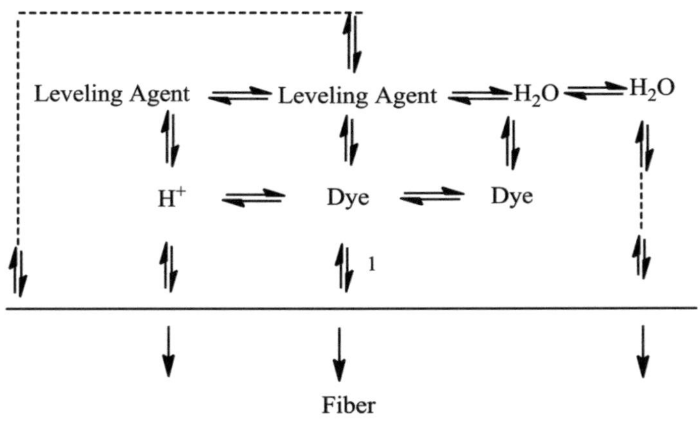

Figure 8.
Equilibrium in dyebath-fiber systems containing leveling agents.

complex with free dyes maintaining equilibrium; (ii) leveling agents act as retarding agents; and (iii) dye migration proceeds slowly, leads better leading to improved leveling of dye on the fiber. A cationic polyethoxylated amine can perform strong leveling action. The greater cationic character a strong complex formation, pronounced retardation of dyeing, and higher risk of precipitation (**Figure 9**).

The polyethoxylated chain is longer (n > 50), and the dye auxiliary complex is dispersed by the cationic leveling agents. Amphoteric leveling agents have both anionic and cationic groups, so its activation depends on the dye bath pH [20].

2.4 Electrolyte

During dyeing in the dyebath, electrolyte serves as three important roles—driving dye into textiles causing, maximum exhaustion of dye molecules through the presence of salt, and fixing dyestuff to the cellulose material [28]. The dyeing mechanism of reactive dye can be classified into two phases, exhaustion and fixation. The process is lengthy because considerable time is spent on the controlled heating of the dyebath and the portion-wise addition of salt and alkali to avoid unleveled dyeing and maximize the exhaustion and fixation [20]. A colorless crystalline solid NaCl composed of inorganic compound of sodium and chloride, a salt in which ionic bonds hold the two components together in the familiar water-soluble white crystals, has a key role in the textile dyeing process in maintaining the electrolytic balance in the textile dyeing process. Glauber's salt is the common name for sodium sulfate dehydrates $Na_2SO_4 . 10H_2O$; it occurs as white or colorless monoclinic crystals, which are most commonly used in the dyeing industry [29]. Vacuum salt is manufactured by recrystallization of purified brine solution. In the vacuum crystallization process, raw salt is dissolved in water to make a saturated solution and clarify the impurities from the bottom. A vacuum is generated by using a suitable vacuum pump [30].

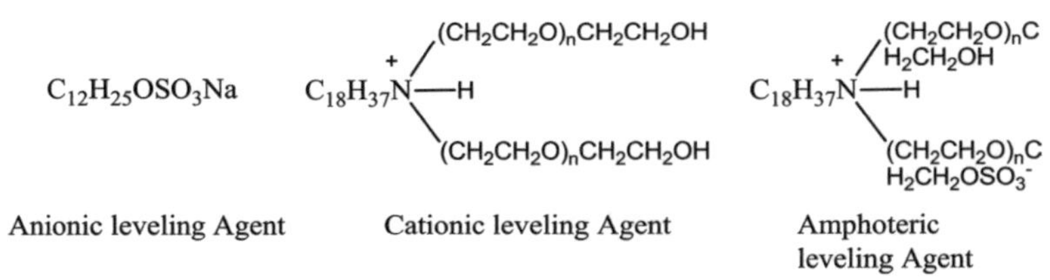

Figure 9.
Different leveling agents.

When cotton/bast fibers are immersed in water, its surface due to the hydroxyl ions also becomes anionic; hence, the dye particles and the cellulosic fiber tend to repel each other. So, the level of substantivity is reduced. The addition of salt creates an electrical positive double layer, which hides the negative electrostatic charge (Donnan Potential) of the cellulose surface. This allows the dye to approach the fiber, allowing better interaction of Van der Waals forces as well; this improves the substantivity. Neale et al. [31] proposed the existence of the Donnan equilibrium. The idea of the Donnan equilibrium was employed for the modern electrochemical theories of the dyeing of cellulose from aqueous solutions of direct dyes [32].

NaCl enhances the diffusion of dye and its adsorption onto fiber. Bicarbonate and carbonate increase the dye bath pH and perform dye fixation through the formation of covalent bonds. "Bolaform" electrolytes contain anionic and cationic organic groups, which are separated by large distances. The word 'bola' means a long cord with heavy balls. The alkyl chains whose lengths are sufficiently long act as electrolyte to become surface-active compounds. Hamada et al. [33] extensively studied various physical properties of bolaform electrolytes or amphiphilies. The bolaform electrolytes or amphiphilies contain a single positive or negative site (**Figure 10**).

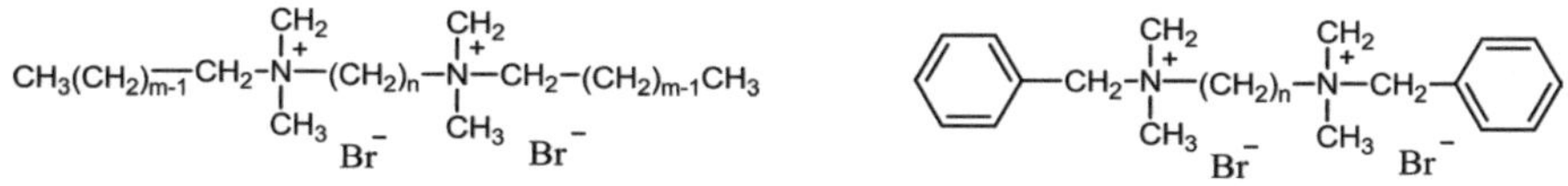

Figure 10.
Bolaform electrolyte.

2.5 pH control agents

Dyebath pH is adjusted prior to the dyeing process; otherwise, it may affect various factors, such as the absorption of dye into the fiber, by increasing or reducing alkalinity [34]. The pH controlling agents play three important roles: (a) maintain a high degree of acidity; (b) control the pH within narrow tolerances; and (c) slide the pH in acidic conditions [35]. The pH controlling agents are usually based on two chemicals, a weak acid and its salt, with a stronger base such as acetic acid, sodium acetate, or phosphoric acid-sodium phosphate. When the dyeing temperature increases, pH control agents release more acidic compounds. Ammonium sulfate decomposes gradually, proucing ammonia and sulfuric acid, which is a strong acid that subsequently lowers the pH when ammonia escapes, at boiling temperatures. However, enclosed or partially enclosed machine, such as a winch, it is not very efficient because ammonia is prevented from escaping into the dye bath (**Figure 11**).

Organic esters are also used as an alternative method for obtaining a pH that slides in the direction of acidity under the conditions of processing. In 1953, Brotherton Co. Ltd. introduced the Estrocon process, which is based on the addition

$$(NH_4)_2SO_4 + 2H_2O \rightleftharpoons H_2SO_4 + 2NH_4OH$$

$$NH_4OH \rightleftharpoons NH_3 + H_2O$$

$$H_2SO_4 \rightleftharpoons 2H^+ + SO_4^{2-}$$

Figure 11.
Dissociation of ammonium salt.

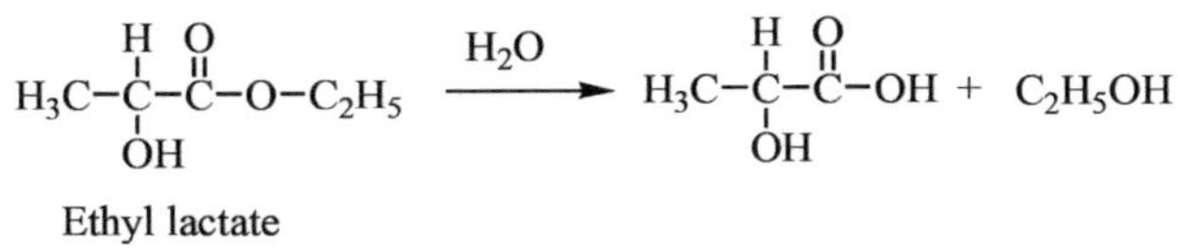

Figure 12.
Action of ethyl lactate as pH sliding agents.

of diethyl tartrate or ethyl lactate for dyeing wool with acid milling dyes and chrome dyes by the single-bath method [36] (**Figure 12**).

Almost 35 years ago, Sandoz introduced the Sandacid V process, in which γ-butyrolactone undergoes hydrolysis to produce butyric acid [37] (**Figure 13**).

Koh et al. [37] found that hydrolysable organic esters achieved a relatively wider range of pH sliding than ammonium sulfate or sodium di hydrogen phosphate, which can produce good exhaustion with satisfactory dyeing in the closed dyeing process.

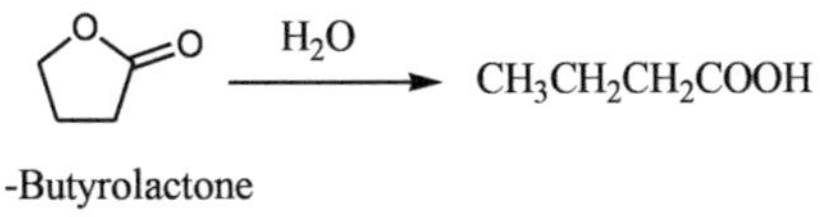

Figure 13.
Dissociation of γ-butyrolactone.

2.6 Surfactants

Surfactant molecules should have surface active properties with the chemical structure of a hydrophilic (water loving) and hydrophobic (having little attraction for organic molecules) balance [38] (**Figure 14**).

These molecules get preferentially oriented at the interface between air and water, which lowers the surface tension of water substantially when dissolved in the concentration range of 0.1–10 g/l. The driving force for the activation of surface active agents is the formation of micelles, in which the lyophobic tails are associated with themselves, and the hydrophilic heads are surrounded by water molecules. Surfactants are classified into different ways such as use, ionic charge, and chemical structure. According to the ionic structure, surfactants can be classified into four categories: anionic, cationic, amphoteric, and nonionic [39]. Anionic surfactants are surfactants that are ionized into anions and cations but the anion is the dominating ion in the solution. Examples are alkyl benzene sulfonate and sodium lauryl sulfate (**Figure 15**).

An ionic surface active agent that produces cation as the dominating ion when dissolved in water is called a cationic surfactant. For example, dodecyl dimethyl ammonium chloride and distearyl dimethyl ammonium chloride (**Figure 16**).

Nonionic surfactants that does not show any ionic behavior when dissolved in water are called nonionic surfactants. Examples are Lanolin ethoxylate and glycerol monostearate (**Figure 17**).

Figure 14.
Schematic of a surfactant molecule.

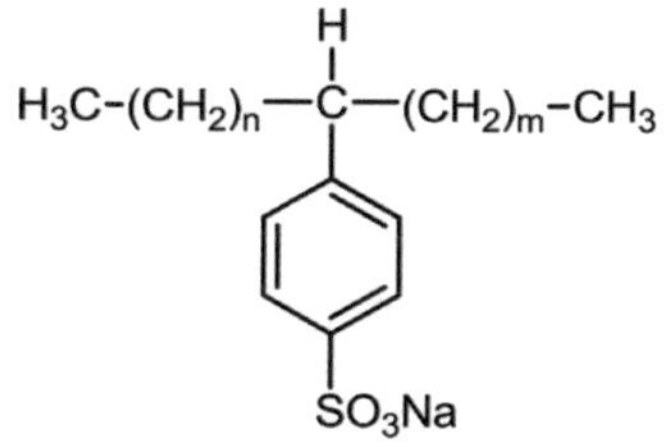

Alkylbenzene sulfonate

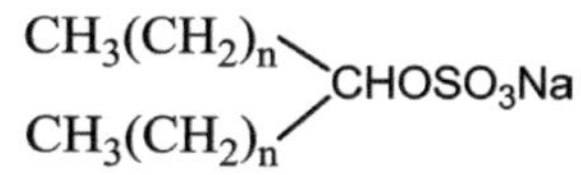

Sodium lauryl sulfate

Figure 15.
Structure of anionic surfactants.

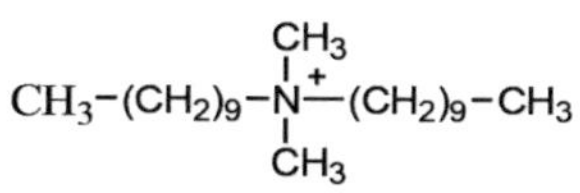

Dodecyldimethyl ammonium chloride

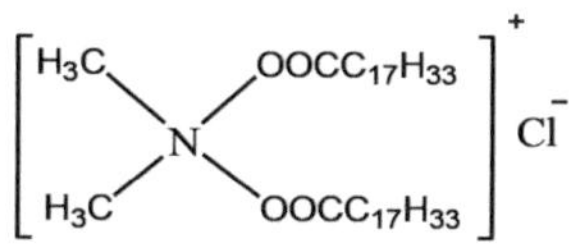

Disteryl dimethyl ammonium chloride

Figure 16.
Structure of cationic surfactants.

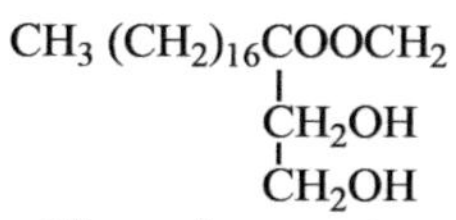

Glycerol mono stearate

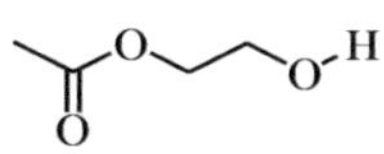

Lanolin ethoxylate

Figure 17.
Structure of nonionic surfactants.

Surfactants that ionize and produce large segments (these segments are called Zwitter ions) carrying both anionic and cationic ions when dissolved in water are called amphoteric surfactants. Lauryl dimethyl betaine and cocamidopropyl betaine are the most commonly used (**Figure 18**).

Surfactants are widely used as wetting agents, emulsifiers, detergent, lubricants, or dispersing agents [40].

2.7 Anti-foaming agents

The dyeing process can lead to the formation of macro and micro foams during circulation of dyes and auxiliaries in the dye bath. A macro foam is displayed with large bubbles, which are visible on the surface of the system and produce cosmetic imperfections [41]. A typical list of anti-foaming agents is as follows: 2-ethyl hexanol (EH), tributyl phosphate (TBP), poly (dimethyl siloxane) (PDMS) amides, mineral oil, fatty acids, and their derivatives [42] (**Figure 19**).

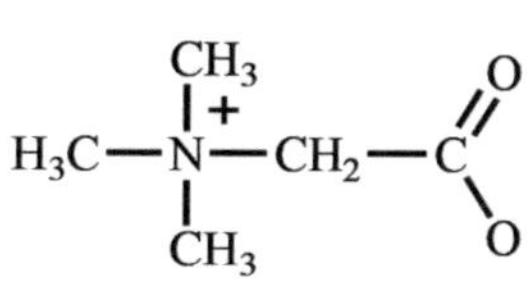

Cocamidopropyl betaine

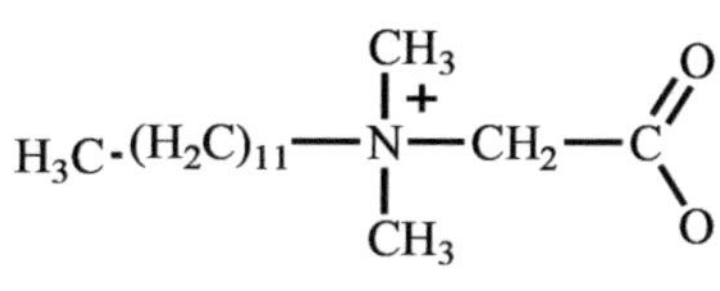

Lauryl dimethyl betaine

Figure 18.
Structure of amphoteric surfactants.

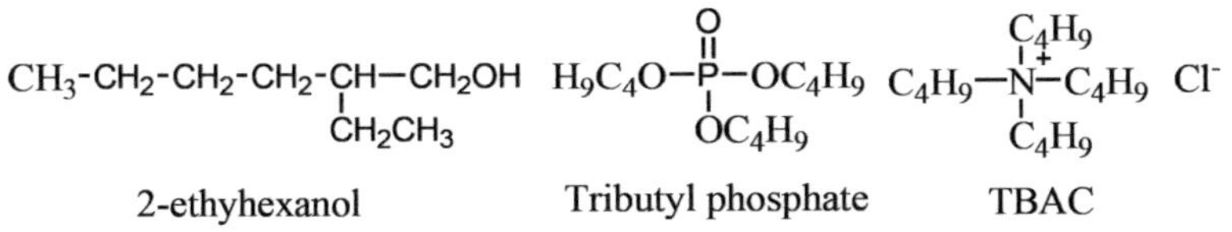

Figure 19.
Anti-foaming agents.

Silicone-based anti-foaming agents consist of small polar groups and hydrophobic polymer chain. The hydrophobic polymer chain is typically a permethylated siloxane. The PDMS chain gains its hydrophilic character by modification with poly (ethylene glycol) (PEG) and poly (propylene glycol) (PPG). PEG and PPG are water-soluble polymers that allow materials to retain good water swellability [43]. A decrease in foaming occurs with an increase in the hydrophilicity of the co-polymers. When the hydrophilicity of the copolymers increased, the micro foams were removed more effectively. Kekevi et al. [43] demonstrate that the uniqueness of these anti-foaming agents is caused by two properties—flexibility, which enhances the acquired conformations that result in efficient packing at various interfaces and lower cohesive energy, which is derived from the cross-sectional area of the siloxane at the interface (**Figure 20**).

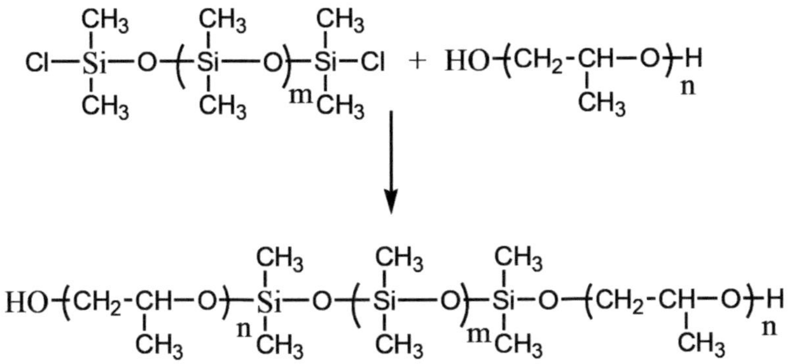

Figure 20.
Silicone-based anti-foaming agents.

2.8 Reducing agents

Beginning of the nineteenth century, sodium dithionate ($Na_2S_2O_4$) was introduced as a reducing agent for the vat dyeing process [8]. Effective sodium dithionate, also called hydrose, provides an effective reduction in indigo dyes [44]. The amount of $Na_2S_2O_4$ used in the reduction of indigo dyeing, formed a large amount of by products and sulfite (SO_3^{2-}) and sulfate (SO_4^{2-}) ions- (**Figure 21**).

$$Na_2S_2O_4 + 3\,H_2O \longrightarrow NaHSO_4 + NaHSO_3 + 4\,H^+$$

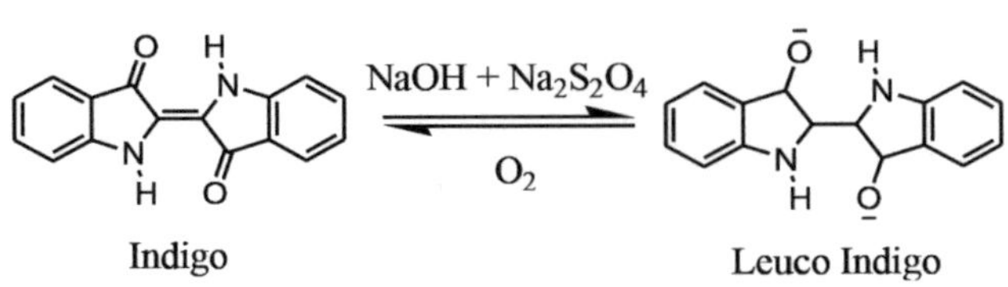

Figure 21.
Oxidation-reduction of indigo.

Stripping methods, mostly used in textile finishing, can remove dye from colored fabric. The process is named as 'back stripping' or 'destructive stripping'. The depth of shade is changed by back stripping; on the other hand, dyes are chemically altered by destructive stripping. Fono and Montclair [45] showed that dyes containing azo groups that can be chemically reduced to an almost colorless amine by using chemical reducing agents. The mechanism of reductive stripping depends on the structure of dyes and fibers and the chemical nature of reducing agents. Chavan [46] explained that different chemical combinations of reducing agents and stripping assistants, which are being used to strip the dye from fabric (**Figure 22**).

$$—N{=}N— \xrightarrow{NaOH + Na_2S_2O_4} —NH_2 + —NH_2$$

Figure 22.
Stripping process of azo dyes.

Sodium formaldehyde sulfoxylate (CH_3NaO_3S) (Rongalite C) and Rongalite FD are also used as reducing agents in the dyeing industry (**Figure 23**).

$$N(CH_2SO_2Na)_2 + NH(CH_2SO_2Na)_2 \longrightarrow \text{Rongalite FD}$$

Figure 23.
Rongalite FD salts.

As sodium dithionate is highly toxic, many researchers [47] have investigated to replacing it with ecological reducing agents such as glucose, fructose, inverted sugar, and molasses (**Figure 24**).

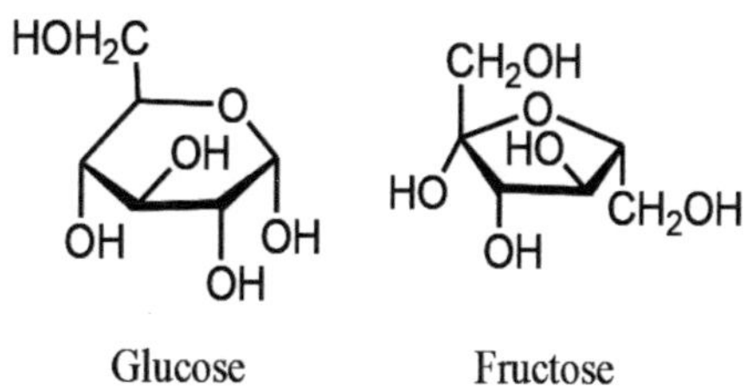

Figure 24.
Ecological reducing agents.

Many studies have been devoted to catalysts that accelerate the reduction of vat dyes when added to certain reducing agents. BASF-marketed a catalyst, bis-(dimethylglyoximato-diamminocobaltinitrite) for use as a reducing agent.

2.9 Softeners

A softener is a chemical used to make the touch of fabric more pleasing. Softened fabrics are fluffier and have better drape ability. In addition to esthetics, softeners improve abrasion resistance, improve tearing strength, and reduce needle cutting when the garments are sewn. Softeners are divided into three major chemical categories: anionic, cationic, and nonionic. Anionic softeners have a negative charge on the molecule, which comes from the carboxylate group (**—**COO**—**), sulfate group (**—**OSO_3**—**), or phosphate group (**—**PO_4^-). Fatty alcohol sulfates are made by the reaction of a hydrophobe with sulfuric acid [17] (**Figure 25**).

$$R{-}OH + HOSO_3H \longrightarrow R{-}O{-}SO_3H + H_2O$$

Figure 25.
Fatty alcohol sulfates formation.

Fewer sulfate groups result in better softeners. Lightly sulfonated oils are sometimes called self-emulsifying because they form turbid water solutions (**Figure 26**).

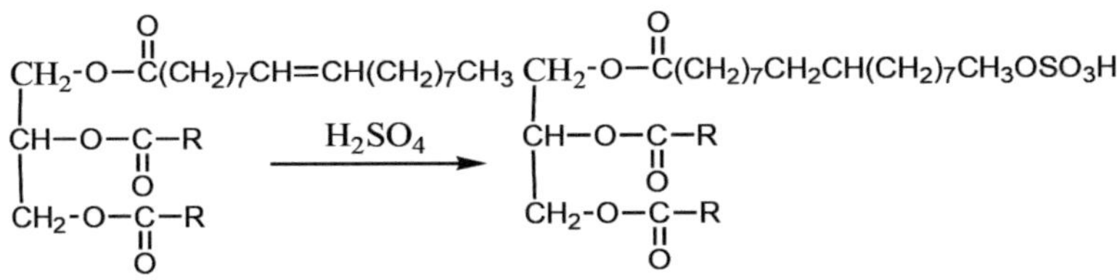

Figure 26.
Formation of sulfated triglycerides.

Cationic softeners have a positive charge on the large part of the molecule. The amine becomes functionalized under a pH of 7. Quaternary ammonium salts are activated at all pH levels. The ionic interaction causes complete exhaustion from baths and orientation on the fiber surfaces, which cause good slipperiness and reduction in the static charge on the fabric surface. There are several cationic softeners that are usually used in the textile industry, such as fatty amines, fatty amino esters, fatty amino amides, and quaternary ammonium salts (**Figure 27**).

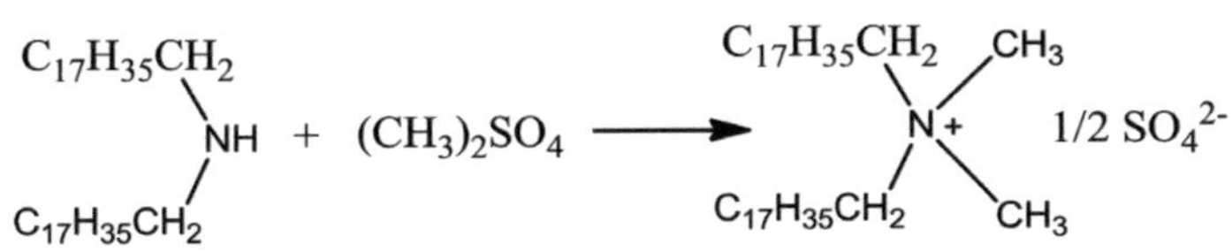

Figure 27.
Synthesis of quaternary ammonium salts.

The nonionic softener has three subcategories: ethylene oxide derivatives, silicones, and hydrocarbon waxes based on paraffin or polyethylene. Polyethylene emulsions are hard, waxy films, which serve to reduce the coefficient of friction. Silicones are polysiloxane polymers. Silicon resembles carbon in which it is tetravalent and forms a covalent bond with other elements (**Figure 28**).

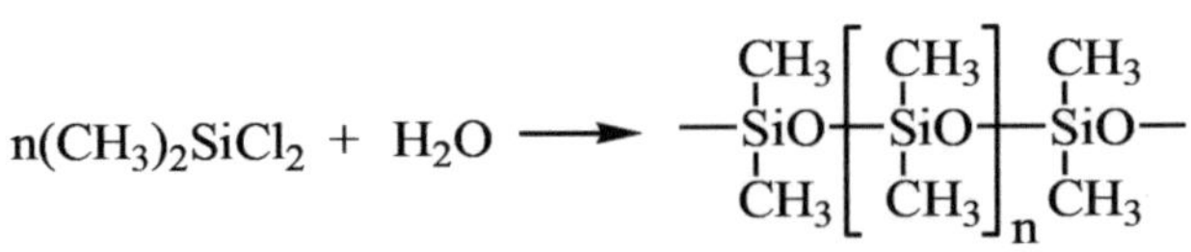

Figure 28.
Reaction of dimethylpolysiloxane.

2.10 Anti-creasing agents

Crease is formed in cotton fiber because of the intermolecular hydrogen bonding of primary and secondary hydroxyl groups of the polymer chains. The amorphous regions of the hydrogen bonds can easily break down by folding. In the folding state, the broken hydrogen bonds stabilize. Different durable press finishes have been used for many years for use in cotton fabric. N-methylol compounds such as dihydroxyethylene urea react readily with the hydroxyl groups of cellulose chains [48] (**Figure 29**).

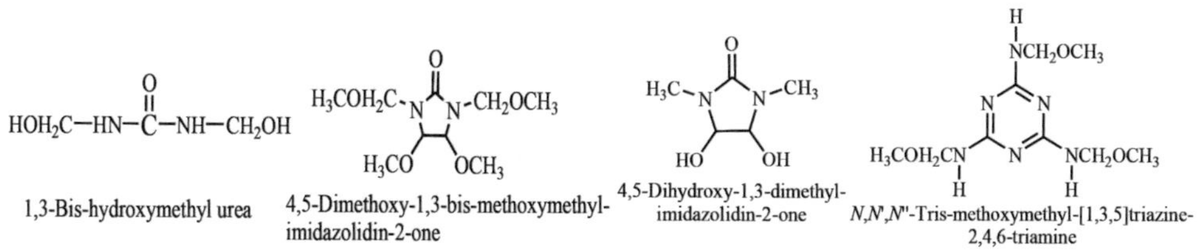

Figure 29.
Cross-linking agents based on N-hydroxyethyl and N-alkoxymethyl compounds.

N-Hydroxyethyl acrylamide is also used as a cross-linking agent for cellulosic fibers (**Figure 30**).

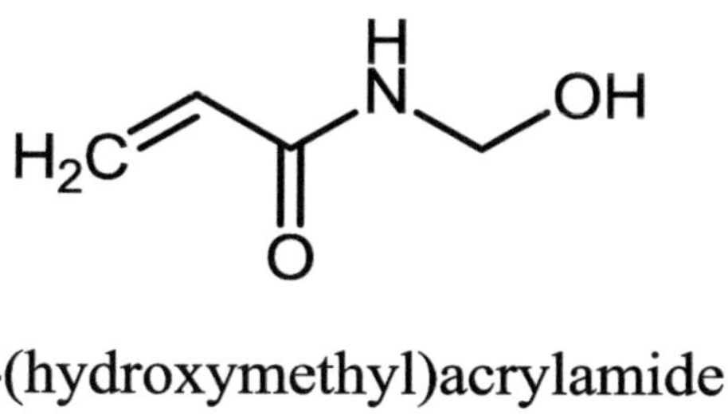

Figure 30.
Cross-linking agents based on reactive monomers.

Activated vinyl groups or divinyl sulfone and N, N-methylene-bis-acrylamide are also used cross-linking agents. Divinyl sulfone precursors are used for the cross-linking reaction with cellulose. Cross-linking reactions with epoxides such as epichlorohydrin are mostly used on cellulosic fibers. In addition, aziridine derivatives such as tetramethylene-bis-(N, N-ethylene urea) and trisaziridinyltriazine are effective cross-linking agents [49] (**Figure 31**).

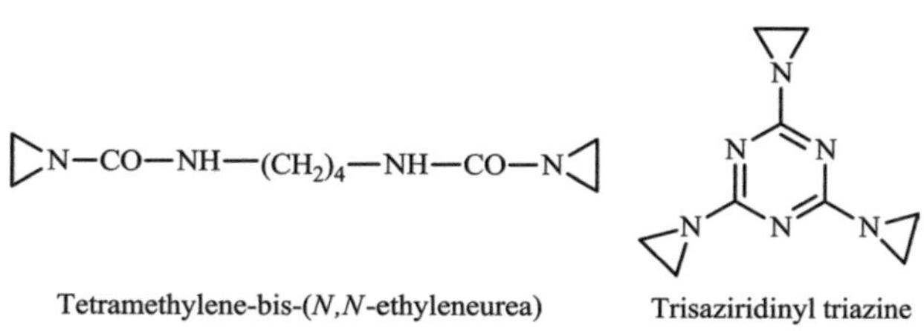

Figure 31.
Cross-linking agents based on aziridine derivatives.

3. Conclusion

Although considerable research has been conducted regarding dyeing auxiliaries, chemical and other related industries have faced many problems with respect to environmental issues and cost effectiveness for consumers. Although thermodynamic studies of dyeing can make useful contributions to the general theories of intermolecular forces, diffusion processes, and the influence of different parameters such as dye concentration, temperature, dyeing time, and polymer structure remain largely unexplained. When equilibrium studies are carried out and assessed in conjunction with the growing amount of information on the structure of dyeing chemicals and aqueous solutions, their relevance to the dyeing theory should not be underestimated. Due to the lack of chemical knowledge of dyers, quality dyeing is deteriorating with a significant effluent load. This study clearly shows that the different dyeing chemicals used in the dye houses also need investigation to determine the procedures in which the chemicals are used to dye natural and synthetic textiles. This may be a good first step in presenting different chemicals used in the coloration industry.

Author details

Shekh Md. Mamun Kabir[1*] and Joonseok Koh[2]

1 Department of Wet Process Engineering, Bangladesh University of Textiles, Dhaka, Bangladesh

2 Division of Chemical Engineering, Konkuk University, Seoul, South Korea

*Address all correspondence to: mamunkabir.butex@gmail.com

References

[1] Woodall P. The new titans: A survey of the world economy. The Economist. 16 September 2006

[2] Aspland JR. A series on dyeing, Chapter 8: Disperse dyes and their application to polyester. Textile Chemist and Colorist. 1993;**25**(1):21-25. ISSN 0040-490X

[3] Holme I. Fibre physics and chemistry in relation to coloration. Coloration Technology. 1967;**1**(1):31-43. ISSN: 1478-4408

[4] Werner L. The mode of action of non-ionic levelling agents. J.S.D.C. 1958;**74**(4):221-233. ISSN: 1478-4408

[5] Derbyshire AN, Peters RH. Interaction between Chlorazol Sky Blue FF and Chrysophenine G in aqueous solution. J.S.D.C. 1956;**72**(6):268-277. ISSN: 1478-4408

[6] Schwen G, Schmidt G. Some experiments on the effect of dye, fibre and atmosphere on light fastness. J.S.D.C. 1959;**75**(2):101-105. ISSN: 1478-4408

[7] Graff JH. The constitution of detergents in concentration with the cleaning of ancient textiles. Studies in Conservation. 1968;**13**(3):122-141. ISSN: 2047-0584

[8] Valko EI. Textile auxiliaries in dyeing. Review Progress in Coloration. 1972;**3**(1):50-62. ISSN: 1478-4408

[9] Jones F. The theory of dyeing. Review Progress in Coloration. 1967;**1**(1):15-22. ISSN: 1478-4408

[10] Wu FC, Tseng RL, Juang RS. Role of pH in metal adsorption from aqueous solutions containing chelating agents onto chitosan. Industrial and Engineering Chemistry Research. 1999;**38**:270-275. ISSN 0888-5885

[11] Aspland JR. A series on dyeing, Chapter 8: Disperse dyes and their application to polyester. Textile Chemist and Colorist. 1992;**24**(12):18-23. ISSN 0040-490X

[12] Ahmed NSE. The use of sodium edate in the dyeing of cotton with reactive dyes. Dyes and Pigments. 2005;**65**:221-225. ISSN: 0143-7208

[13] Dawson TL, Roberts BP. The essential role of pH control in the dyeing of carpet fibres. J.S.D.C. 1979a;**95**(2): 47-54. ISSN: 1478-4408

[14] Rice JW. Principles of textile conservation science. Textile Museum Journal. 1966;**2**:15-37

[15] McCarthy JP, Drozdowski RE. Household fabric softeners: The chemistry and characteristic of cationic softeners. Soap, Cosmetics, Chemical Specialties. 1989;**65**(91):33-34. ISSN: 1478-4408

[16] Duin MV, Peters JA, Kieboom APG, Bekkum HV. A general coordination-ionization scheme for polyhydroxy carboxylic acids in water. Recueil des Travaux Chimiques des Pays-Bas. 1989;**108**:57-60

[17] Tomasino C. Chemistry & Technology of Fabric Preparation & Finishing. North Carolina: North Carolina State University, Raleigh; 1992. ISBN 0840002297

[18] Chen XY, Lin Q , Luo YM, He YF, Zhen SJ, Yu YL, et al. The role of citric acid on the phytoremediation of heavy metal contaminated soil. Chemosphere. 2003;**50**(6):807-811. ISSN: 0045-6535

[19] Tandy S, Bossart K, Mueller R, Ritschel J, Hauser L, Schulin R, et al. Extraction of heavy metals from soils using biodegradable chelating agents. Environmental Science & Technology. 2004;**38**(3):937-944. ISSN: 0013-936X

[20] Broadbent AD. Basic Principles of Textile Coloration. Bradford, UK: SDC; 2001. ISBN 0-901956-76-7

[21] Kabir SMM, Koh J. Effect of chelating agent in disperse dye dyeing on polyester fabric. Fibers and Polymers. 2017;**18**(12):2315-2321. ISSN: 1229-9197

[22] Biedermann W. Effect of crystal modification on dyeing behaviour of disperse dyes. Coloration Technology. 1971;**87**(4):105-111. ISSN: 1478-4408

[23] Ingamells W. Colour for Textiles: A User's Handbook. Bradford, UK: SDC; 1993. ISBN 0-901956-56-2

[24] Johnson A. The Theory of Coloration of Textiles. Bradford, UK: SDC; 1989. ISBN 0-901-95648-1

[25] Murray A, Mortimer K. Carrier dyeing. Review Progress of Coloration. 1971;**2**(1):67-72. ISSN: 0557-9325

[26] Wolf, Bauer. Melliand textilber. 1968;**48**:464. ISSN: 0341-0781

[27] Heimanns S. Textile auxiliaries: Dispersing agents. Review Progress in Coloration. 1981;**11**(1):1-8. ISSN: 1478-4408

[28] Dawson TL, Roberts BP. Progress towards a greener textile industry. Coloration Technology. 2011;**128**:1-8

[29] Broadale G, Giauque WF. The heat of hydration of sodium sulfate. Low temperature heat capacity and entropy of sodium sulfate decahydrate. Journal of the American Chemical Society. 1958;**80**:2042-2044

[30] Ann A. Handbook of chemistry and physics. In: 71st edition. Michigan: CRC Press; 1990

[31] Neale, Hanson, Stringfellow. The absorption of dyestuffs by cellulose. Part VI. The effect of modification of the cellulose, and a theory of the electrolyte effect. Transactions of the Faraday Society. 1935;**31**:1718

[32] Iyer SRS, Jayaram R. Studies on the Donnan potential in systems containing cellulose and aqueous electrolyte solution. J.S.D.C. 1970;**86**(9):398-402. ISSN: 1478-4408

[33] Hamada K, Haizuka H, Yamazaki A. Effects of a bolaform electrolyte on the sorption of acid dyes by a nylon 6 film. Journal of Applied Polymer Science. 1998;**69**:2247-2254

[34] Dawson TL, Roberts BP. The essential role of pH control in the dyeing of carpet fibres. J.S.D.C. 1979b;**95**(2): 47-54. ISSN: 1478-4408

[35] Kamat SY, Borkar HS. pH and buffer systems in dyeing. Colourage. 1993;**40**(1):19. ISSN: 0010-1826

[36] Hannay RJ, Major WH. The use of hydrolysable esters in the control of dye baths. J.S.D.C. 1953;**69**(6):195-201. ISSN: 1478-4408

[37] Koh J, Sim G, Kim J. pH control in the dyeing of polyamide with acid dyes. Coloration Technology. 2001;**69**(6):195-201. ISSN: 1478-4408

[38] Schroder FR. Concentrations of anionic surfactants in receiving riverine water. Tenside, Surfactants, Detergents. 1995;**32**(6):492-497

[39] Samanta AK. Efficacy of selective surfactants/detergents as washing agents on soiled white and dyed cotton fabrics. Indian Journal of Fibre & Textile Research. 2004;**29**:223-232. DOI: 10.1051/matecconf/20171080 3001

[40] Zana R. Specialist Surfactants. London: Blackie Academic and Professional; 1997

[41] Rosen MJ. Surfactants and Interfacial Phenomena. 2nd ed. New York: John Wiley & Sons; 1989. p. 1989

[42] Jha BK, Patist A, Shah DO. Effect of antifoaming agents on the Micellar stability and foam ability of sodium dodecyl sulfate solutions. Langmuir. 1999;**15**:3042-3044

[43] Kekevi B, Berber H, Yildirim H. Synthesis and characterization of silicone-based surfactants as anti-foaming agents. Journal of Surfactants and Detergents. 2012;**15**:73-81

[44] Saikhao L, Setthayanond J, Karpkrid T, Suwanruji P. Comparison of sodium dithionate and glucose as a reducing agent for natural indigo dyeing on cotton fabrics. In: ICMAA, MATEC Web of Conferences. Vol. 108. 2017

[45] Fono A, Montclair. New process of color stripping dyed textile fabrics. US Patent No. 4227881; 1980

[46] Chavan RB. Stripping of Dyestuff from Textiles. US Patent No. 3591325; 1969

[47] Shukla SR, Pai RS. Sulphur dyeing using non-sulphide reducing agents. Indian Journal of Fibre & Textile Research. 2004;**29**:454-461. ISSN: 0975-1025

[48] Hassan MS. Crease recovery properties of cotton fabrics modified by urea resins under the effect of gamma irradiation. Radiation Physics and Chemistry. 2009;**78**:333-337

[49] Fikentscher R, Bille H. Chemical Aftertreatments of Textiles. New York: Willey InterScience; 1971. p. 357

Chapter 11

Synthesis and Optical Properties of Near-Infrared (NIR) Absorbing Azo Dyes

Sharad Rohidas Patil and Amol S. Chaudhary

Abstract

This chapter provides a general overview and information on near-infrared (NIR) absorbing azo dyes. In this work, we have developed an efficient and simple protocol for the synthesis of novel A-π-D-π-A NIR azo dyes. The near-infrared absorbing azo dyes were synthesized by using 2-hydroxy-1,4 naphthoquinone (Lawsone) and different substituted aromatic primary amines. Furthermore, author developed push-pull chromophores of A-π-D-π-A type containing an electron-withdrawing azo core, phenazine moieties, and a hydroxyl group as electron donor. The benzo[a]quinoxalino[2,3-i]phenazin-5-ol moiety was introduced to make the system planer as well as to increase the π-conjugation. The optical properties of these dyes were studied in *N,N*-dimethylformamide (DMF).

Keywords: NIR absorbing azo dyes, 2-hydroxy-1,4-naphthoquinone, Lawsone, optical properties, phenazine core

1. Introduction

In 1856, William Henry Perkin accidentally discovered the world's first commercially successful synthetic dye. Dyes are defined as colored organic or inorganic compounds, mixtures, complexes, or substances that once applied to the fibers/substrate, it imparts a permanent color to the fibers/substrate. This is able to resist fading upon exposure to sweat, light, water, and many chemicals, including oxidizing agents and microbial attack. Over 10,000 synthetic dyes were developed and used in manufacturing by the end of the nineteenth century. Azo dyes are a class of compounds containing a *N*=*N* double bond and, due to their ability to absorb visible to near-infrared (NIR) light and ease of synthesis, they have been extensively used in the textile, fiber, leather, paint, and printing industries for more than a century. The azo group is substituted with benzene or naphthalene groups, which can contain many different substituents, such as chloro (–Cl), methyl (–CH_3), nitro (–NO_2), amino (–NH_2), and hydroxyl (–OH), carboxyl (–COOH), which give different types of azo dyes. Azo dyes account for the majority (more than 3000 different varieties) of all textile dyestuffs produced because of the ease and cost-effectiveness of their synthesis, their stability, and the variety of colors available compared to natural dyes. They are extensively used in the textile, paper, food, leather, cosmetics, and pharmaceutical industries [1–3].

Azo colorant represents the greatest production volume in dyestuff chemistry due to simplicity of coupling reaction, adaption to the needs of most diverse applications such as textile dying, coloring of plastics and polymer, in liquid crystal displays (LCD), optical data storage, nonlinear optics, biological and medical field, and advanced application in organic synthesis [4–6]. Azo chromophores have versatile applicability ranging from textile dyeing [7], leather dyeing [8], coloring of plastics, and polymer [9] to advanced applications such as liquid crystal displays [10], biological and medical studies, and advanced application in the organic synthesis [11]. Also, they contribute greatest production volume of the dyestuff industry due to simplistic mode of their synthesis with high yield.

The infrared wavelength ranges of the electromagnetic spectrum cover the area between the visible region and the microwave region. The infrared radiation spectrum is generally accepted and can be subclassified into the near-, medium-, and far-infrared regions. An important challenge in creating molecules with desired optical properties lies in achieving a well-defined architecture while maintaining adequate electron density throughout the whole molecule [12]. Organic materials with intense absorptions in the near-IR region (i.e., 750–1300 nm) are particularly important in number of applications including thermal imaging, optical data storage, automatic identification, etc. NIR absorbing dyes have been attracting increasing interest due to the rapid progress achieved in their high technology applications.

In this region, infrared radiation provides sufficient energy to the NIR dyes, which impart a $\pi \rightarrow \pi^*$ transition of electrons within dye chromophore molecules. Despite the enormous synthetic versatility of azo dyes, which accounts for representing 50–70% of all commercially available dyes, little effort has been devoted to bring their absorption into the long wavelength region, and only a few examples of NIR absorbing azo dyes have till now been disclosed [13, 14]. Near-infrared absorbing dyes have attracted significant interest for their uses in the development of functional materials for high-technology end uses. Considerable interest is evinced in recent years in the design of near-infrared absorbing organic azo dyes. They find manifold applications in the high technology areas such as dye sensitized solar cells (DSSCs) [15], laser optical storage systems [16], clear coatings [17], organic

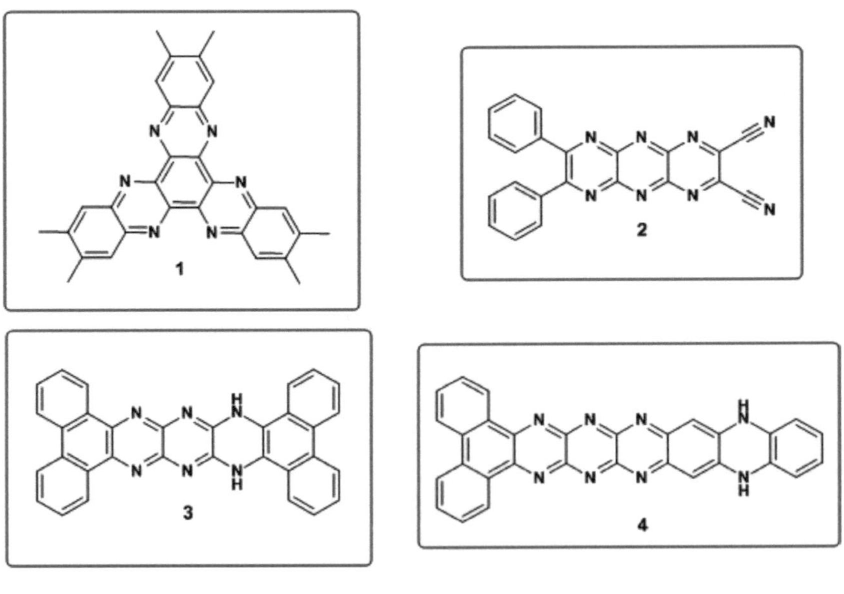

Figure 1.
Chemical structure of pyrazinacenes for organic thin film transistors.

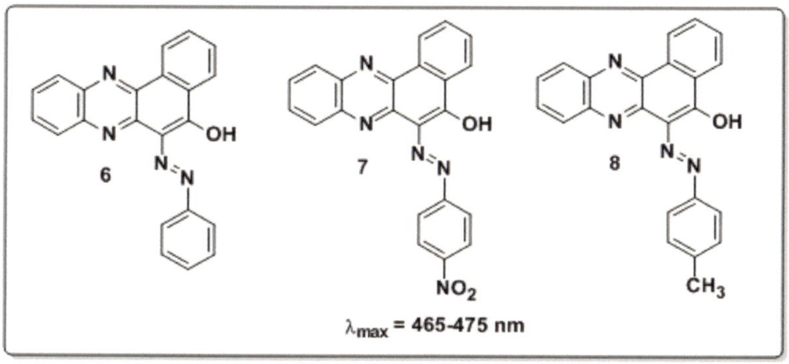

Figure 2.
*Reported azo derivatives of lawsone (**6–8**).*

photovoltaics [18], greenhouse claddings [19], molecular switches [20], NIR-active materials [21], colorimetric sensors [22], mesomorphic [23], solvatochromic [24–26], and electrochemical materials [27] and surface relief grating (SRG)-forming azo polymer films [28], laser-printing systems, laser filters, photography, photonics, and in telecommunication devices operating at wavelengths in the NIR region [29].

The planar structure of phenazine with a heterocyclic pyrazine nucleus and a fully conjugated aromatic π-system imparts to that special optical and redox properties. The optical properties give it functional for applications in molecular imaging as fluorescent tracers and stains for subcellular components and biological events. Highly electron-rich and redox-active molecules like phenazines might be functional for preparation of organic materials that includes donor-π-acceptor design. Pyrazine compounds are electron-deficient species. Hence, they are used in the design of n-type organic semiconductors; examples

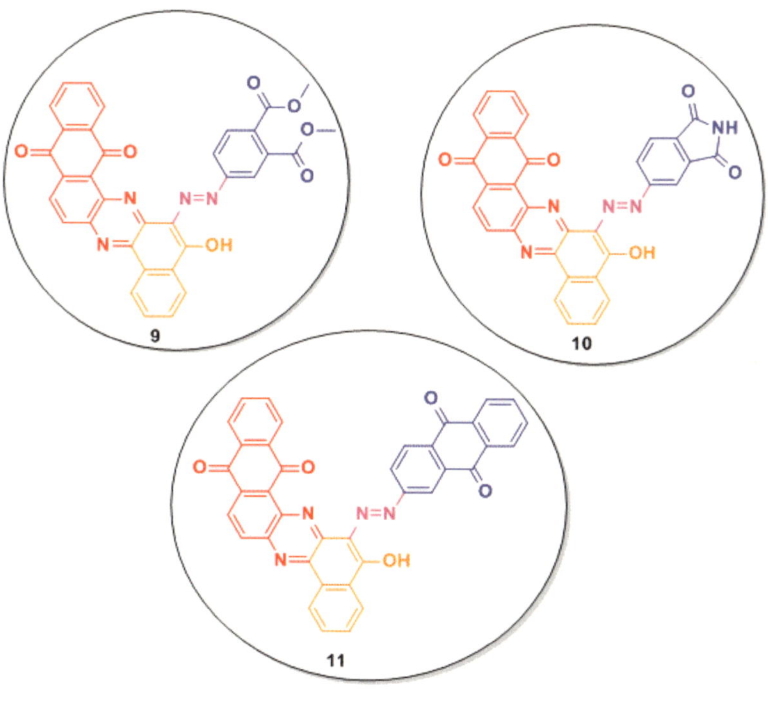

Figure 3.
Structures of the dyes **9–11**.

are pyrazinoquinoxaline derivatives, hexaazatriphenylenes, diquinoxalino, phenazine, and quinoxalinophenanthro phenazine [30]. Richards et al. have shown that pyrazinacenes are good candidates as materials for organic thin film transistors as shown in the **Figure 1** [31]. The existence of a donor-π-acceptor (D-π-A) alternating structure has been proved to be an efficient way to reduce band gap energies of conjugated molecules. The phenazine core consists of the electron-withdrawing pyrazine ring which might be an excellent charge acceptor.

Romanyuk et al. has reported azo derivatives of 2-hydroxy-1,4 naphthoquinones **5** as shown in **Figure 2** [32].

There are no further reports in the use of 2-hydroxy-1,4-naphthoquinone in the synthesis of azo colorants. It is known that the molecule with a quinoid fragment shows keto-enol tautomerism when a hydroxyl group is presenting adjacent to one of the quinoid carbonyl groups: the p-quinoid structure can transform into an *o*-quinoid or dihydro structure.

The reactions of 2-hydroxy-1,4-naphthoquinone in the synthesis of azo colorants for NIR-absorbing dyes have been explored in our previous work. Three different kinds of near-infrared (NIR) absorbing pull-push acceptor-π-donor-π-acceptor (A-π-D-π-A) V-shaped chromophoric dyes have been synthesized. These dyes are architect by doing substitution with 5-hydroxy and 6-arylazonium on the benzo[a] naphtho[2,3-h] phenazine dione core. Phenazine and anthraquinone moieties were envisaged in a single molecule with a view to enhance the photophysical properties in NIR region as shown in **Figure 3** [33].

This work is aimed at synthesizing novel molecules containing both phenazine and azo cores that were envisaged in a single molecule with a view to enhance the photophysical properties in the NIR region. In this study, we have extended our contribution to synthesis NIR absorbing dyes (**12–14**) as shown in **Figure 4**.

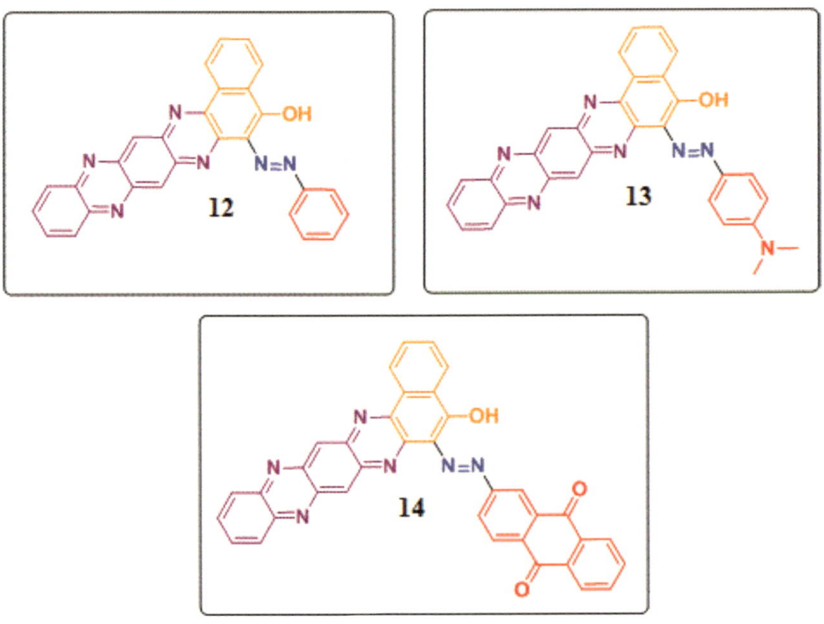

Figure 4.
*Structures of the near-IR absorbing dyes (**12–14**).*

2. Result and discussion

2.1 Synthesis strategy

In the present study, the use of 2-hydroxy-1,4-naphthoquinone (Lawsone) **5** coupling components and diazo components in the synthesis of azo disperse dyes followed by condensation with *o*–phenylenediamine OPDA to obtain the desired NIR absorbing azo dyes (**12–14**).

In this context, authors have extended our contribution to synthesis of three different kinds of near infrared (NIR) absorbing pull-push acceptor–π–donor–π acceptor (A–π–D–π–A) chromophoric dyes. The NIR absorbing dyes (**12–14**) were synthesized from derivatives of 2-hydroxy-3-(phenyldiazenyl)naphthalene-1,4-dione (**18–20**) and 2,3-diaminophenazine core (**21**) (**Figure** 5).

The scaffold of phenazine, *N,N*-dimethylaniline, and anthraquinone moieties was envisaged in a single molecule with a view to enhance the photophysical properties in NIR region. The UV-Vis absorption spectra of synthesized azo-azine dyes have been studied. The intermolecular interaction occurring in solvents of differing polarities as well in solution of different pH have been studied. The structures of the dyes are given in **Figure 4**. Synthetic route for the dyes **44–46** and their intermediates are depicted in **Figure** 5.

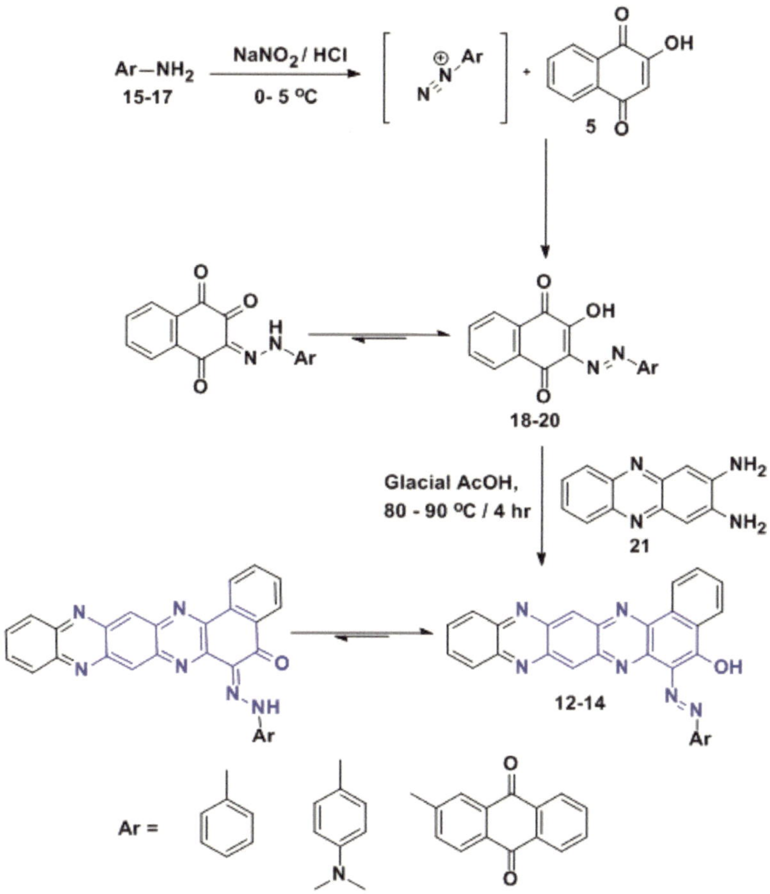

Figure 5.
Schematic representation of NIR azo dyes (12–14).

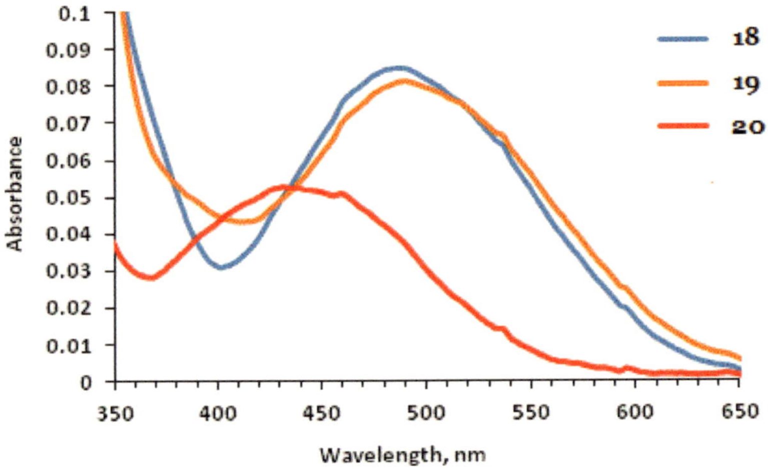

Figure 6.
Visible absorption of the dyes (18–20) in DMF solvent.

Dye	λ_{max} (nm)	(log ε)	ε_{max} (dm^3 mol^{-1} cm^{-1})
18	487	4.93	84,830
19	490	4.91	81,275
20	439	4.72	52,260
12	620	4.60	40,000
13	710	4.63	43,000
14	680	4.69	49,000

Table 1.
Visible absorption of the dyes (18–20 and 12–14) in DMF solvent.

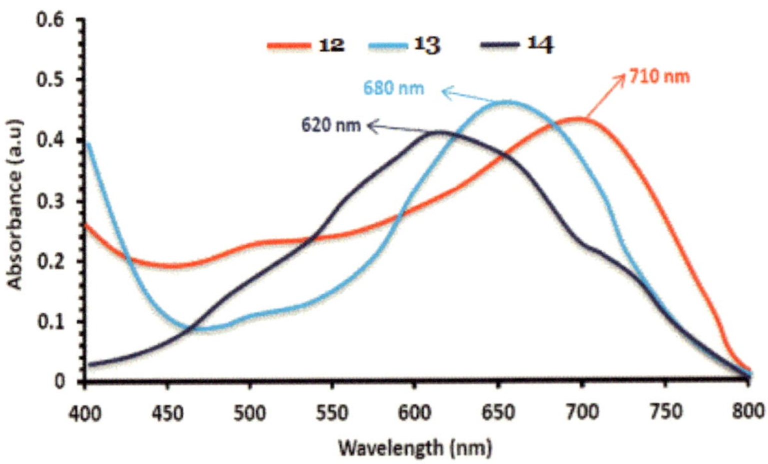

Figure 7.
Visible absorption of the dyes (12–14) in DMF solvent.

2.2 Optical properties

The UV-visible absorption spectra of the dyes **(18–20 & 12–14)** were measured in *N,N*-dimethylformamide (DMF). These dyes do not show solvatochromism significantly. The dyes (**18–20**) showed absorption at 439–490 nm (**Figure 6** and **Table 1**).

The red-shifted absorptions owing to the π-conjugation framework are present in these dyes. The dyes (12–14) showed absorption at 620–710 nm due to the intramolecular charge transfer transition (ICT) between donor-acceptor pair (**Figure 7** and **Table 1**).

3. Experimental sections

3.1 General experimental procedure of the azo dyes (18–20)

The aromatic primary amines (5 g) was partially dissolved in conc. HCl (5 mL) in 80 mL water and cooled to 0–5°C [34]. The solution of $NaNO_2$ (2 g) in 5 mL water was added slowly to the reaction mixture, and the reaction temperature was kept below 5°C. The resulting aryl diazonium salt was added to the alkaline solution of Lawson (5 g) and NaOH (1.15 g) in 60 mL of water, and reaction temperature was maintained at 0–5°C. The reaction mass was stirred for 2.0–2.5 h, after which mixture was acidified with dil. HCl. The precipitate was settled down and filtered, washed with water and then with aqueous EtOH, and was dried.

3.2 General experimental procedure of the azo-phenazine dyes (12–14)

Diketo-azo derivatives (**18–20**) and diaminophenazine (**21**) were mixed together then refluxed in acetic acid for 4 h. The AcOH was removed under reduced pressure in a rotary evaporator, and water was added to the residue. The precipitate was let out and filtered, washed with water and then with aqueous ethanol, and dried.

3.3 General experimental procedure of 2,3-diaminophenazine (21)

Finally powdered *o*-phenylenediamine (54.0 g, 0.5 mol) was dissolved in conc. hydrochloric acid (83.3 ml) and distilled water (2.5 L). A filtered solution of ferric chloride (400 g) in (750 ml) water was added slowly, and the mixture was stirred mechanically. After standing overnight at room temperature, the red-brown colored crystalline product was filtered off, washed with cold dilute 0.3 N HCl until free from ferric ions, then dissolved in hot water (2.5 L), and 2,3-diaminophenazine was precipitated by the addition of a concentrated solution of potassium hydroxide. The product was filtered off, washed with water, and dried at 100–110°C. The strongly alkaline filtrate was heated and acidified (pH 4.5) with glacial acetic acid. After cooling, 2-amino-3-hydroxy phenazine was collected, washed with water, and dried at 100–110°C. The crude products were used without further purification.

3.4 Characterization

*3.4.1 6-(Phenyldiazenyl)benzo[a]quinoxalino[2,3-i]phenazin-5-ol (**12**)*

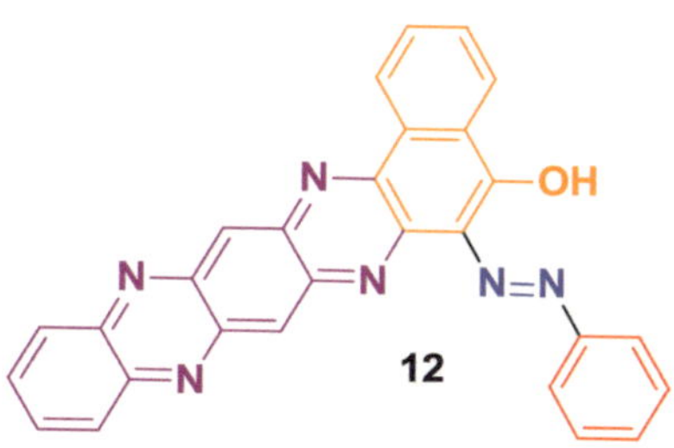

Yield 72%; melting point (measured) >300°C.
FT-IR (KBr, cm^{-1}) = 3057 (O–H), 1630 (C=N), 1589 (arom.), 1534 (arom. ring).

^{1}H-NMR (DMSO, 500 MHz) = δ 8.30–8.24 (m, 4H, aromatic), 8.26–8.16 (m, 3H, aromatic), 7.92–7.76 (m, 8H, aromatic), 5.2 (1H, s, OH).

*3.4.2 6-((4-(Dimethylamino)phenyl)diazenyl)benzo[a]quinoxalino[2,3-i]phenazin-5-ol (**45**)*

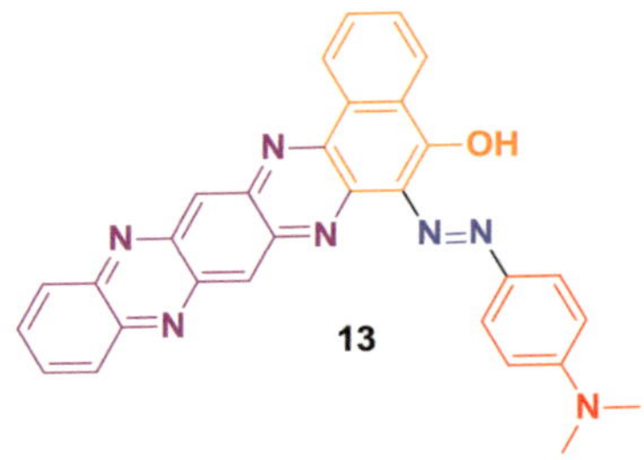

13

Yield 80%, melting point (measured) >300°C. FT-IR (KBr, cm^{-1}) = 3169 (O–H), 1674 (C=N), 1633 (arom.), 1514 (arom. ring). ^{1}H-NMR (DMSO, 500 MHz) = δ 8.25 (s, 2H, aromatic), 8.23–7.78 (m, 12H, aromatic), 3.86 (s, 6H, CH_3–N–CH_3), 6.18 (1H, s, OH).

*3.4.3 2-((5-Hydroxybenzo [a]quinoxalino[2,3-i]phenazin-6 yl)diazenyl)anthracene-9,10-dione (**46**)*

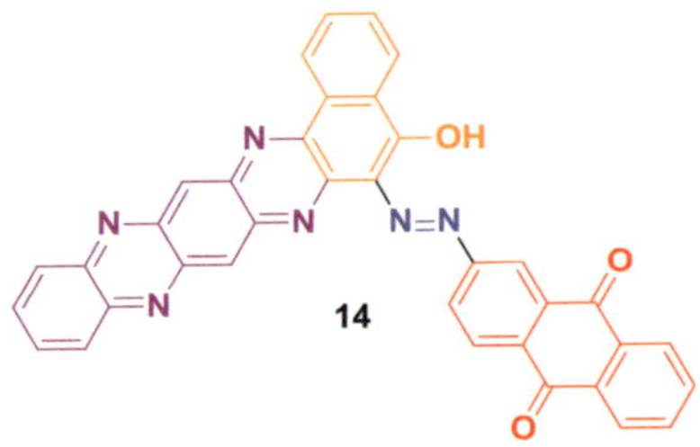

14

Yield 67%, melting point (measured) >300°C. FT-IR (KBr, cm^{-1}) = 3059 (O–H), 1629 (C=N), 1590 (arom.), 1529 (arom. ring). ^{1}H-NMR (DMSO, 500 MHz) = δ 8.304 (s, 2H, aromatic), 8.292 (m, 12H, aromatic), 7.77–7.76 (m, 3H, aromatic), 5.2 (1H, s, OH).

4. Conclusions

Azo colorants represent the greatest production volume in dyestuff chemistry due to simplicity of coupling reaction, adaption to the needs of most diverse applications such as textile dying, coloring of plastics and polymer, in liquid crystal displays (LCD), optical data storage, nonlinear optics, biological and medical field, and advanced application in organic synthesis. In this context, authors have designed and developed push-pull chromophores of A-π-D-π-A type containing an electron-withdrawing azo core, phenazine moieties, and a hydroxyl group as electron donor. The benzo[a]quinoxalino[2,3-i]phenazin-5-ol moiety was introduced to make the system planer as well as to increase the п-conjugation. These synthesized chromophores were confirmed by FT-IR, ^{1}H-NMR, ^{13}C-NMR, and MS spectral analysis. The optical properties of these dyes were studied in DMF. NIR absorbing dyes have been considered in numerous recent hi-tech applications like optical device, optical recordings, laser printings, laser thermal writing displays, infrared photography, and biological/medical applications. In optoelectronic application areas, especially,

NIR absorbing dyes are being developed and new structural designs of the dye chromophore molecules are being studied. Azo dyes are the most important class of organic dyes, but many of azo dyes suspect to be carcinogenic in nature. Therefore, the research of the dyes is aimed at development of new type of chromophores. The NIR azo dyes may probably prove to be a solution to this problem.

5. Future trends

There is an excellent range of organic compounds that absorb intensively higher than 700–1300 nm owing to electronic excitation. The basic acyclic NIR chromophores (polymethine dyes) as well as the basic cyclic NIR chromophores (annulenes, porphyrines, and tetraazaporphyrins) are well recognized. Structurally, more advanced NIR dyes with some meropolymethinic, quinonoid, and/or indigoid character will eventually be found. Color structure relationships are at present better well-known and more reliably predictable for closed shell than for open shell systems.

On account of their special optical properties, there exists scope for further research into designing small molecules bearing donor and acceptor functional groups directly attached to the phenazine nucleus that could possess NIR optical spectra, which would be sensitive to presence of solution analytes such as cations/anions as well as to pollutants such as polycyclic aromatic hydrocarbons and nitroaromatics. Additionally, preparation of more examples of phenazine-based low-band gap-containing molecules will also enhance its use in the field of organic photovoltaics.

Acknowledgements

Author is grateful to IntechOpen Publishers for being given the chance for contributory to the present book. This work was supported by Institute of Chemical Technology, Mumbai, India.

Notes

The authors declare no competing financial interest.

Author details

Sharad Rohidas Patil[1,2]* and Amol S. Chaudhary[3]

1 Department of Chemistry, SPDM Arts, SBB and SHD Commerce and SMA Science College, Dhule, Maharashtra, India

2 Department of Chemistry, Uka Tarsadia University, Gopal-Vidyanagar, Maliba campus, Bardoli, Gujarat, India

3 R&D Division, Rextone Dyes Ltd, Vadodara, Gujarat, India

*Address all correspondence to: sharad.omd.patil@gmail.com

References

[1] Li Y, Patrick BO, Dolphin D. Near-infrared absorbing azo dyes: Synthesis and X-ray crystallographic and spectral characterization of monoazopyrroles, bisazopyrroles, and a boron–azopyrrole complex. The Journal of Organic Chemistry. 2009;**74**:5237-5243. DOI: 10.1021/jo9003019

[2] Singh RP, Singh PK, Singh RL. Role of azoreductases in bacterial decolorization of azo dyes. Current Trends in Biomedical Engineering & Biosciences. 2017;**9**:1-3. DOI: 10.19080/CTBEB.2017.09.555764

[3] Telke A, Kalyani D, Jadhav J, Govindwar. Kinetics and mechanism of reactive red 141 degradation by a bacterial isolate *Rhizobium radiobacter* MTCC 8161. Acta Chimica Slovenica. 2008;**55**:320-329

[4] Arslanoglu Y, Sevim AM, Hamuryudan E, Gul A. Near-IR absorbing phthalocyanines. Dyes and Pigments. 2006;**68**:129-132. DOI: 10.1016/j.dyepig.2005.01.019

[5] Saratale RG, Saratale GD, Chang JS, Govindwar SP. Bacterial decolorization and degradation of azo dyes: A review. Journal of the Taiwan Institute of Chemical Engineers. 2011;**42**:138-157. DOI: 10.1016/j.jtice.2010.06.006

[6] Gregory P. In: Hunger K, editor. Industrial Dyes: Chemistry, Properties and Applications. Weinheim: Wiley-VCH; 2002. pp. 543-585

[7] Bamfield P, Hutchings MG. Chromic Phenomena: Technological Applications of Colour Chemistry. 2nd ed. Cambridge: RSC; 2010

[8] Koh J, Greaves AJ. Synthesis and application of an alkali-clearable azo disperse dye containing a fluorosulfonyl group and analysis of its alkali-hydrolysis kinetics. Dyes and Pigments. 2001;**50**:117-126. DOI: 10.1016/S0143-7208(01)00028-6

[9] Vijayaraghavan R, Vedaraman N, Surianarayanan M, MacFarlane DR. Extraction and recovery of azo dyes into an ionic liquid. Talanta. 2006;**69**:1059-1062. DOI: 10.1016/j.talanta.2005.12.042

[10] Fleischmann C, Lievenbrück M, Ritter H. Polymers and dyes: Developments and applications. Polymer. 2015;7:717-746. DOI: 10.3390/polym7040717

[11] Gouda MA, Eldien HF, Girges MM, Berghot MA. Synthesis and antitumor evaluation of thiophene based azo dyes incorporating pyrazolone moiety. Journal of Saudi Chemical Society. 2016;**20**:151-157. DOI: 10.1016/j.jscs.2012.06.004

[12] Nejati K, Rezvani Z, Seyedahmadian M, Seyedahmadian M. The synthesis, characterization, thermal and optical properties of copper, nickel, and vanadyl complexes derived from azo dyes. Dyes and Pigments. 2009;**83**:304-311. DOI: 10.1016/j.dyepig.2009.05.007

[13] Dong M, Babalhavaeji A, Hansen MJ, Kalman L, Woolley GA. Red, far-red, and near infrared photoswitches based on azonium ions. Chemical Communications. 2015;**51**:12981-12984. DOI: 10.1039/C5CC02804C

[14] Salvador MA, Reis LV, Almeida P, Santos PF. Delocalized cationic azo dyes containing a thiazole moiety. Tetrahedron. 2008;**64**:299-303. DOI: 10.1016/j.tet.2007.11.004

[15] Mikroyannidi J, Tsagkournos D, Sharma S, Kumar A, Vijay Y, Sharma G. Efficient bulk heterojunction solar cells based on low band gap bisazo dyes containing anthracene and/or pyrrole units. Solar Energy Materials and Solar Cells. 2010;**94**:2318-2327. DOI: 10.1016/j.solmat.2010.08.001

[16] Fuxin H, Yiqun W, Donghong G, Fuxi G. Synthesis, spectroscopic and thermal properties of nickel (II) eazo complexes with blue-violet light wavelength. Dyes and Pigments. 2005;**66**:77-82. DOI: 10.1016/j.dyepig.2004.08.013

[17] Knischka R, Lehmann U, Stadler U, Mamak M, Benkhoff J. Novel approaches in NIR curing technology. Progress in Organic Coating. 2009;**64**:171-174. DOI: 10.1016/j.porgcoat.2008.09.015

[18] Sekar N, Raut R, Umape P. Near infrared absorbing iron-complexed colorants for photovoltaic applications. Materials Science and Engineering B. 2010;**168**:259-262. DOI: 10.1016/j.mseb.2010.01.018

[19] Lamnatou C, Chemisana D. Solar radiation manipulations and their role in greenhouse claddings: Fresnel lenses, NIR- and UV-blocking materials. Renewable and Sustainable Energy Reviews. 2013;**18**:271-287. DOI: 10.1016/j.rser.2012.09.041

[20] Li N, J L, Li H, Kang E. Nonlinear optical properties and memory effects of the azo polymers carrying different substituents. Dyes and Pigments. 2011;**88**:18-24. DOI: 10.1016/j.dyepig.2010.04.010

[21] Salvador M, Almeida P, Reis L, Santos P. Near-infrared absorbing delocalized cationic azo dyes. Dyes and Pigments. 2009;**82**:118-123. DOI: 10.1016/j.dyepig.2008.12.003

[22] Kaur P, Sareen D. The synthesis and development of a dual-analyte colorimetric sensor: Simultaneous estimation of Hg2+ and Fe3+. Dyes and Pigments. 2011;**88**:296-300. DOI: 10.1016/j.dyepig.2010.07.012

[23] So B, Kim H, Lee S, Song H, Park J. Novel bent-shaped liquid crystalline compounds: IV. Dimesogenic compounds containing 2-hydroxy-1,3-dioxypropylene and azobenzene mesogens. Dyes and Pigments. 2006;**70**:38-42. DOI: 10.1016/j.dyepig.2005.04.006

[24] Tsai P, Wang I. A facile synthesis of some new pyrazolo[1,5-a]pyrimidine heterocyclic disazo dyes and an evaluation of their solvatochromic behavior. Dyes and Pigments. 2007;**74**:578-584. DOI: 10.1016/j.dyepig.2006.03.022

[25] Karcı F, Karcı F. The synthesis and solvatochromic properties of some novel heterocyclic disazo dyes derived from barbituric acid. Dyes and Pigments. 2008;**77**:451-456. DOI: 10.1016/j.dyepig.2007.07.009

[26] Yazdanbakhsh R, Mohammadi A, Abbasnia M. Some heterocyclic azo dyes derived from thiazolyl derivatives; synthesis; substituent effects and solvatochromic studies. Spectrochimica Acta. Part A, Molecular and Biomolecular Spectroscopy. 2010;**77**:1084-1087. DOI: 10.1016/j.saa.2010.08.079

[27] Ucar M, Polat K, Solak A, Toyc M, Aksu M. The electrochemical behaviour of 20-halogenated derivatives of 4-methoxyazobenzene at a mercury electrode. Dyes and Pigments. 2010;**87**:55-61. DOI: 10.1016/j.dyepig.2010.02.004

[28] Wang D, He Y, Deng W, Wang X. The photoinduced surface-relief-grating formation behavior of side-chain azo polymers with narrow Mr distribution. Dyes and Pigments. 2009;**82**:286-292. DOI: 10.1016/j.dyepig.2009.01.012

[29] Zollinger H. Color chemistry. In: Syntheses, Properties, and Applications of Organic Dyes and Pigments. 3rd ed. Zurich: Verlag Helvetica Chimia Acta, Wiley VCH; 2003

[30] Banerjee S. Phenazines as chemosensors of solution analytes and as sensitizers in organic photovoltaics.

ARKIVOC. 2016;**i**:82-110. DOI: 10.3998/ark.5550190.p009.347

[31] Richards G, Hill JP, Subbaiyan N, D'Souza F, Paul A, Mark K, et al. Pyrazinacenes: Aza analogues of acenes. The Journal of Organic Chemistry. 2009;**74**:8914-8923. DOI: 10.1021/jo901832n

[32] Romanyuk AL, Polishchuk OP, Litvin BL, Ganushchak N, Russ I. Synthesis and transformations of 2-hydroxy-3-arylazo-1,4-naphthoquinones. Journal of General Chemistry. 2002;**72**:251-254. DOI: 10.1023/A:1015429803432

[33] Choudhary A, Patil S, Sekar N. Solvatochromism, halochromism, and azo-hydrazone tautomerism in novel V-shaped azo-azine colorants—Consolidated experimental and computational approach. Coloration Technology. 2016;**132**:1-12. DOI: 10.1111/cote.12226

[34] Mei L, Tai LS, Tao FH, Jie S, Rong QL. A novel synthesis of 2,3-diaminophenazine. Research on Chemical Intermediates. 2012;**38**:499-505. DOI: 10.1007/s11164-011-0366-z

Chapter 12

Removal of Methylene Blue Dyes from Aqueous System Using Composite Polymeric-Apatite Resins

Nasser S. Awwad, Adel A. El-Zahhar and Jamila A.M. Alasmary

Abstract

Removal of cationic dyes from industrial effluents is still a big and challenging subject in the field of environmental purification. Millions of tons of cationic dyes are consumed by the textile, rubber, paper, and plastic industries. These dyes have thousands of different chemical structures. Most of them have special properties, such as high hydrophilicity and stability to light or heat. Adsorption is commonly used as a technique for removing dyes. Removal of cationic dyes by adsorption is a promising approach because of its low performance cost and easy technical access. The amount adsorbed of the dye onto the polymeric resin is studied with time for estimating the adsorption mechanism. The adsorption of dye with time shows that mixing period of 10 min is optimum for attaining equilibrium with respect to R1 and R2, while attaining equilibrium with R3 takes 60 min. This findings represent a rapid kinetic for adsorption of MB, particularly R1, on the prepared resins. Different kinetic models were applied on the obtained results and the kinetic parameters were determined. The kinetic models correlate the amount adsorbed of dye with time. The values of calculated adsorption capacity q_e and the linear regression coefficient clarify that the studied kinetic model could not fit with the experimental results for adsorption of MB onto R1, R2, and R3. The results of the studied kinetic model clarify that the experimental results for adsorption of MB onto R1, R2, and R3 could be described by kinetic model supporting chemical adsorption. The sorption of MB could be favorably described by the pseudo-second-order kinetic model onto the composite resins. This finding refers to the participation of chemical adsorption within the adsorption mechanism for MB onto R1, R2, and R3.

Keywords: blue dyes, composite polymeric-apatite resins, cationic dyes, environmental purification

1. Introduction

The environmental issues associated with residual dye content or residual color in treated textile effluents are always a concern for each textile operator that directly discharges, sewage treatment works as well as commercial textile operations to meet the requirements of color and residual dye imposed on the discharge of treated effluent [1]. In watercourses higher than 1.0 mg/L, dye concentrations induced by direct discharges of textile effluents, treated or not, can give rise to public enforcement. High concentrations of fabric dyes in water bodies halt the reoxygenation potential of the receiving water and cutoff sunlight, thereby disrupting biological activity in aquatic life as well as the aquatic plant or algae photosynthesis process [2]. Presence of dye is accepted as an esthetic issue in watercourses rather than as an eco-toxic hazard. The public appears to know the blue, green and brown color of the rivers, but the color 'non-natural' as red and purple usually causes the most concern. Due to their long-term existence in the environment (i.e., half-life of several years), accumulation in sediments and particularly in fish or other aquatic life forms, decomposition of contaminants in carcinogenic or mutagenic compounds, and also their low aerobic biodegradability, the polluting effects of dyes on the aquatic environment may also result in toxic effects. The majority of dyes are nonbiodegradable due to their artificial origin or chemical composition, possess carcinogenic activity, and induce asthma, dermatitis, inflammation of the skin, and various tissue changes. In addition, various azo dyes, primarily aromatic compounds, show toxicity both acute and chronic. High potential health risk is caused by adsorption of azo dyes and their breakdown products (toxic amines) through the gastrointestinal tract, body, lungs, and hemoglobin adduct formation as well as blood formation disturbance [3].

There are about 700,000 tons of colors, approaching 10,000 forms, often used as color operators in most of projects. The use of colors in characteristic media is disturbing considering their heavy workload, poisonous content, and bioaccumulative ability in living beings. In particular, the azo dyes that are most advertised and that cause cancer, need to be taken seriously. The current status of cationic and anionic dyes is audited here. For this reason, common adsorbents are commonly used, such as activated carbon, chitosan, composite, and natural waste. Various dangerous engineered dyes (cationic and anionic) have been fabricated with high creation rate. To dispose of their negative effects, the broad utilization of composite was watched for oxidative debasement/expulsion of colors from wastewater.
In cationic dye, methylene blue, precious stone violet, Splendid Blue-R, and Rhodamine-B, while in anionic dye, Methyl orange, Congo red, Alizarin red S and Eosin Y, are broadly treated with composite.

Adsorption methods are used as procedures of high quality for the removal of dissolved organic contaminants from industrial wastewater, such as dyes. The fiber, pulp and paper industries are stated to use large quantities of a number of colors; such chemicals can be used in many wastewater factories that generating vast amounts of sprayed, toxic and even cancerous wastewater, causing serious hazard to aquatic living organisms. It is well known that dye effluents from the production of dyestuffs and fabric industries may have toxic effects on microbial organisms and may be toxic and/or cancerous to mammalian animals. Most of the dyes used in textile industries are resistant to light and not biodegradable. These are also resistant to aerobic digestion [4]. Because of simple design, adsorption has advantages over other methods and can entail low investment in terms of both initial and necessary land costs. The adsorption method is commonly used to treat industrial wastewater from organic and inorganic contaminants, and the researchers are paying close attention to it. The search for low-cost adsorbents with pollutant-binding capabilities has intensified in recent years. Locally available materials such as natural materials, agricultural waste, and industrial waste can be used as adsorbents at low cost. For groundwater and wastewater treatment, the activated carbon produced from these materials can be used as an adsorbent [5]. Natural hydroxyapatite derived from bio-waste, bovine and camel bones in general. To remove the natural hydroxyapatite, three separate methods are applied: thermal decomposition, subcritical water, and alkaline hydrothermal processes. Results from many physiochemical analyses have shown that all the methods used are capable of removing the organic compounds present in bovine bones and producing pure bioceramic hydroxyapatite with an average yield of 65% [6]. A composite material is made by consolidating at least two materials—frequently ones that have altogether different properties. The two materials cooperate to give the composite novel properties. In any case, inside the composite you can without much of a stretch differentiate the distinctive materials one from the other as they do not break up or mix into each other [7]. One of the biggest advantages of the composite is it can be formed into entangled shapes more effortlessly than most different materials. This gives fashion designers the versatility to build any form or structure. In addition, composites are light in weight, contrasted with most woods and metals, possess high strength, dimensional stability, and are nonmagnetic. Composites contain no metals; hence, they are not attractive. They can be utilized around touchy electronic gear. The absence of attractive impedance permits huge magnets utilized as a part of MRI (Magnetic Resonance Imaging) hardware to perform better. Composites are utilized as a part of both the hardware lodging and table. Also, the development of the room utilizes composite rebar to strengthen the solid dividers and floors in the healing center [8].

In this chapter, our team will investigate the performance of (polymer/hydroxyapatite) composite for removal of methylene blue as cationic dye from aqueous system.

2. Materials and methods

2.1 Synthesis of composite resins

Polyacrylonitrile co-acrylic acid apatite resins, P(AN-co-AA)-HAP, were synthesized by radical copolymerization in DMF solution in the presence of (BPO) as follows: In three-neck round flask equipped with nitrogen, thermometer, and mechanical shaker, AN and AA (mass ratio of 4:1) were dissolved in the DMF (mass ratio of total monomer to DMF 4:11) solution 1. Different types of as-prepared natural HAP produced from camel bone with appropriate amount were dispersed in DMF and sonicated for 30 min then added to solution 1 in the three-neck flask and stirred under nitrogen purging. The initiator BPO was transferred to the reaction mixture with weight ratio of total monomers of 1:450. The solution was stirred until all substances were completely dispersed and monomers dissolved. After passing nitrogen to the solution for 1 h, the polymerization was performed in an airtight device at 60°C for 3 h till complete polymerization. After complete polymerization, the obtained composite was washed several times with water and acetone to remove the residual monomers and initiators thoroughly. The final co-polymer-HAP composites P(AN-AA)-HAP were dried at 60°C till complete dryness and ground to the specified particle size.

Different types of the as-prepared natural HAP camel bone samples were used as follows:

Resin	HAP preparation temp. °C	HAP preparation gas	HAP content % in the composite
R1	700°C for N_2 and 900°C for CO_2 with surface area 122.56 m^2/g	Passing N_2 gas for 2 h (1 h for rising temperature to reach to desired temperature and 1 h for holding) then passing CO_2 gas for 1 h	14%
R2	500°C for N_2 and 900°C for CO_2 was surface area 94.88 m^2/g		
R3	900°C for N_2 and 900°C for CO_2 was surface area 124.35 m^2/g		

2.2 Sorption studies

The sorption experiments were carried out on solutions containing methylene blue dye with the prepared sorbents in batch as well as column techniques.

2.2.1 Batch experiments

Batch experiments were conducted with 100 ml of methylene blue solution of concentration 1×10^{-5} M in 250-ml stoppered bottles containing 0.1 g of the composite material at pH 5.5. The mixture was shaken by a mechanical water shaker thermostated at 25 ± 1°C (except when studying the effect of temperature). The parameters affecting the sorption process were studied by varying any one of the parameters and keeping the other parameters constant. These factors include composite particle size, equilibrium time, dose, temperature, and the initial metal ion concentration. The solid material was separated from solution by centrifuge and the dyes concentration in the solution was determined. Methylene blue was determined spectrophotometrically using UV-visible spectrophotometer.

The amount of dye retained in the solid phase (q_e) (mg/g) was calculated using this relation:

$$q_e = \frac{v}{m}(C_0 - C_e)\ \text{mg/g} \quad (1)$$

where v is the volume of aqueous solutions (ml), m is the weight of solid material used (g), and C_0 and C_e are the initial and equilibrium concentration (M) of MB dye.

The sorption of dye solution was calculated using the relation:

$$\text{Sorption\%} = \frac{C_0 - C_e}{C_0} \times 100 \quad (2)$$

The distribution coefficient (K_d) of the dye between the aqueous phase and the solid phase is calculated from the relation:

$$K_d = \frac{v}{m}\left(\frac{C_0 - C_e}{C_e}\right) \quad (3)$$

$$K_d = \frac{q_e}{C_e} \quad (4)$$

2.2.2 Column studies

Column studies were conducted using glass column of dimensions 20 cm length and 1.2 cm of internal diameter and 0.4 g of sorbent of particle size 30 mesh was introduced as slurry in the column. A 1000-ml solution containing 1×10^{-5} M of MB added sample was passed through HAP. All the experiments were carried out at room temperature. The sample solution was collected after different volumes. Amounts of MB in each volume were determined using UV/VIS spectrophotometry. The breakthrough sorption capacity of HAP was obtained in column using the equation:

$$Q_e = \left[\frac{C_i - C_e}{m}\right] bv \quad (5)$$

where C_i and C_e denote the initial and equilibrium (at breakthrough) dye concentration (M) respectively. *bv* was the breakthrough volume of the MB solution in liters, and *m* was the mass of the adsorbent used (g).

Desorption of solutes from the loaded column was carried out by elution using (H_2SO_4 + H_2O_2), HCL, NaOH, CH_3COOH, $C_2O_4H_2$, and $C_6H_8O_7$. From the start of the experiment, effluent samples at different volumes were collected at the bottom of the column for analysis. The percentage desorption of solutes was obtained in column using the equation:

$$\text{Desorption \%} = \frac{C_e}{C_i} \times 100 \quad (6)$$

3. Results and discussions

This chapter is divided into four parts. The first one deals with characterization of the synthesized sorbent materials, R1. The second and third parts are concerned with the removal of MB dye respectively from aqueous solution by the investigated sorbents. Capacities of the sorbent materials for removal of the studied dye were investigated. The fourth part deals with column studies (**Figure 1**).

3.1 Characterization of the sorbent materials

3.1.1 X-ray diffraction (XRD), Fourier transform infrared (FTIR), and scanning electron microscopy (SEM) measurements

The IR-spectra revealed wide bands of H—O—H lattice water in the regions 1600–1700 cm^{-1} and 3200–3600 cm^{-1} and an [OH]— stretching mode band at 3350 cm^{-1} for HAP at physical activation at 500°C. The bands at 1097 cm^{-1} and 1029 cm^{-1} were assigned for $(PO_4)^{3-}$. Also the bands at 960, 603, and 562 cm^{-1} are due to $(PO_4)^{3-}$. The $(PO_4)^{3-}$ vibrational bands depend on the activation temperature. The sharpness of the peaks at 603 cm^{-1} and 562 cm^{-1} indicates a well crystallized HAP. Characteristic vibration bands of the C—O in the carbonate group were observed at 1418–1456 cm^{-1}. New peaks were observed for composite resin (R1) at 2006 cm^{-1} and peak of phosphate is slightly shifted to 1042, 592, and 560 cm^{-1}, while the loaded resin R1 with MB dye showed a broad peak of the phosphate at 560 cm^{-1} and the stretching mode of vibration for OH group was observed at 3340 cm^{-1} as shown in the spectra. As can be seen from the morphologies of particles in the SEM images, there is a distribution of small particles and large agglomerates. These agglomerates are consisted of very fine particles that are welded together and the powders displayed a significant level of agglomeration. Forty-point BET surface area analyses were done to study the effect of polymer on the surface area of the natural hydroxyapatite. The natural HAP surface area was found to be 122.56 m^2/g, while composite shaped reaches 218.48 m^2/g. The XRD analyses revealed straight base lines and sharp peaks confirming that hydroxyapatite was formed in all samples, and the produced calcium phosphate was well

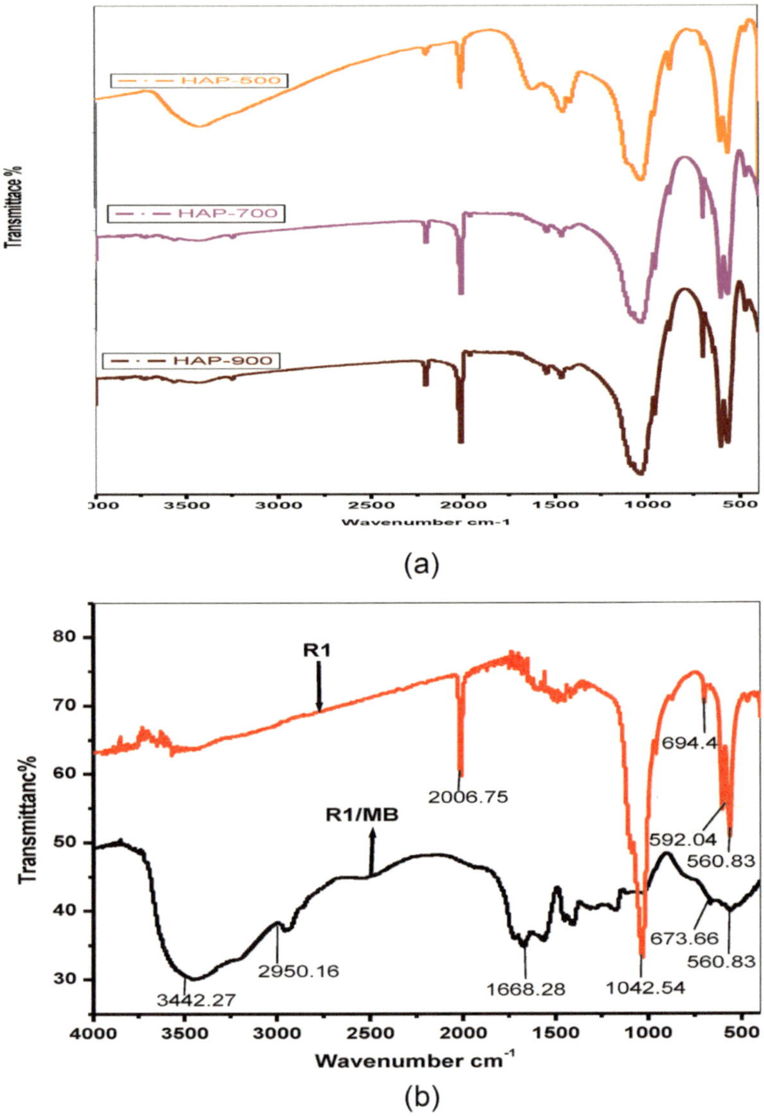

Figure 1.
(a) Infrared spectrum of the HAP samples and (b) infrared spectrum of the synthesized R1 and R1/MB.

crystallized. On the other hand, **Figure 2(a–c)** XRD analysis of HAP samples indicates that the samples are in the crystalline phase. In addition, XRD pattern shows a broad reflection peak at the range of 31.7–32.8 of 2 values, which correspond to the characteristic peak of hydroxyapatite. The XRD patterns of polymer/supported HAP composite resins with 14% HAP (R1) show that the characteristic peaks of HAP disappeared after the interaction between the polymer and HAP. This finding explains the interaction of the polymer within the HAP active groups (**Figure 3**).

3.1.2 Thermogravimetric analysis (TGA) measurements

The thermogravimetric analysis (TGA) measurements were performed in flowing nitrogen gas up on the prepared resins R1 as show in **Figure 4**. The thermal

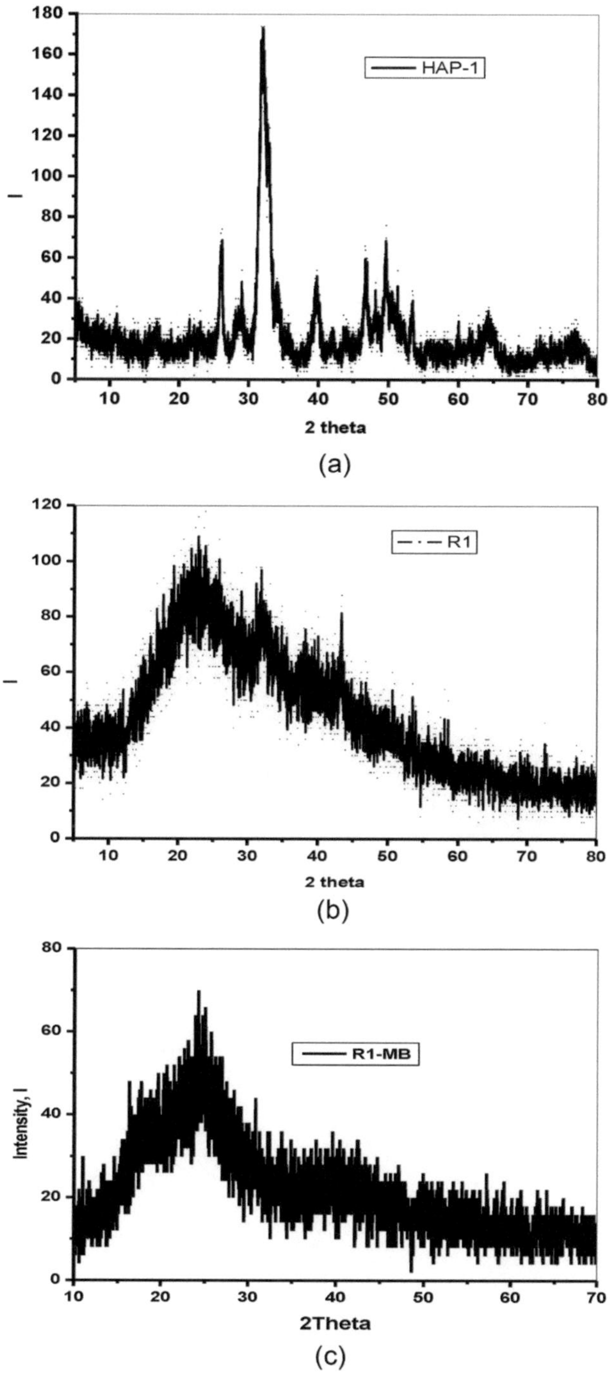

Figure 2.
(a) X-ray diffraction of the synthesized HAP, (b) X-ray diffraction of the synthesized R1, and (c) X-ray diffraction of R1 after sorption the of MB.

decomposition of the HAP-polymer composites illustrates that up to 350°C, 5 % weight loss occurred followed by three steps up to 800°C. It is clear that the presence of HAP with the polymer increases the thermal stability and delays the polymer thermal decomposition.

Figure 3.
SEM micrographs for the natural HAP and prepared resins before and after MB dye loading.

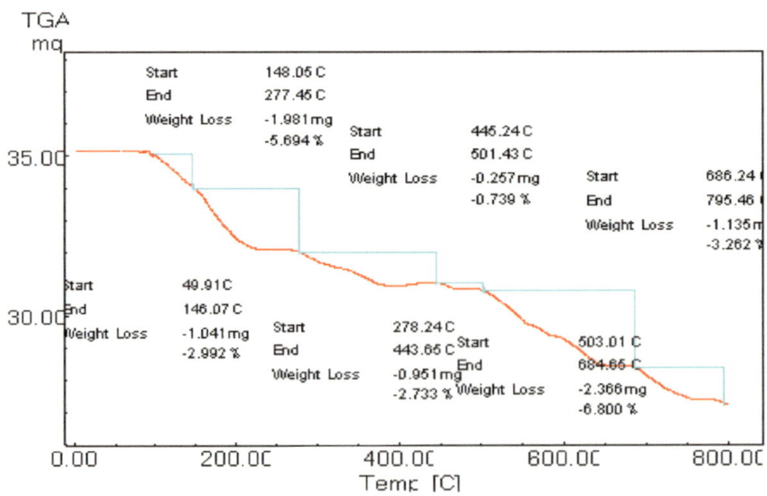

Figure 4.
TGA thermogram for R1.

The thermal decomposition of apatite content depends on its hydroxide and phosphate content; both are related to the change in apatite mass upon heating. After 800°C a replacement of $—OH^-$ occurred, resulting in water evolution.

The weight loss after 800°C was attributed to the —OH^- content [9]. The calcium phosphates containing HPO_4 undergo weight loss between 400 and 700°C due to formation of pyrophosphate and evolution of water. It is reported that after 700°C the calcium pyrophosphate reacts with HAP to produce TCP and water.

3.2 Sorption investigations of methylene blue dye

3.2.1 Batch investigations

Batch experiments were carried out to find out the optimum conditions for the removal of MB dye from aqueous solution by the synthesized sorbents. Different parameters affecting the sorption of MB dye were separately studied, such as particle size of sorbent, shaking time, sorbent weight, temperature, and MB dye concentration.

3.2.1.1 Effect of shaking time

The effect of shaking time on the removal of 1×10^{-5} M MB dye from aqueous solution using 0.1 g of each the sorbents used was investigated in the range 1–120 min at 25 ± 1°C. The results illustrated in **Figure 5** show an increase in the removal percentage (uptake) of MB dye from aqueous solution with shaking time up to 60 min, which then remains constant with further increase of shaking time. A plateau is seen to be reached for all curves representing that the adsorbent is saturated at this time. Therefore, equilibrium time of 60 min was chosen in the subsequent studies.

3.2.1.2 Effect of particle size

The effect of particle size on the removal efficiency of 1×10^{-5}M MB dye from aqueous solution using 0.1 g of selected sorbents was studied at different sorbent particle sizes in the range (0.063–4) mesh at 25 ± 1°C, the removal of MB dye was found to increase from 22.39 to 82.5% as the sorbent particle size increased from (4–0.063) mesh, see **Figure 6**. The higher sorption with smaller sorbent particles may be attributed to the fact that smaller particles provide a larger surface area. Hence, selected sorbents with particle size of 0.063 mesh has been chosen for further experiments.

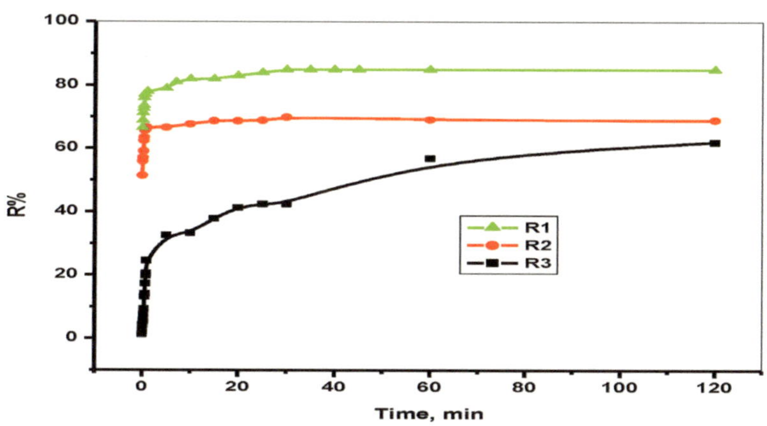

Figure 5.
Effect of shaking time on the adsorption of MB from aqueous solutions using apatite resins.

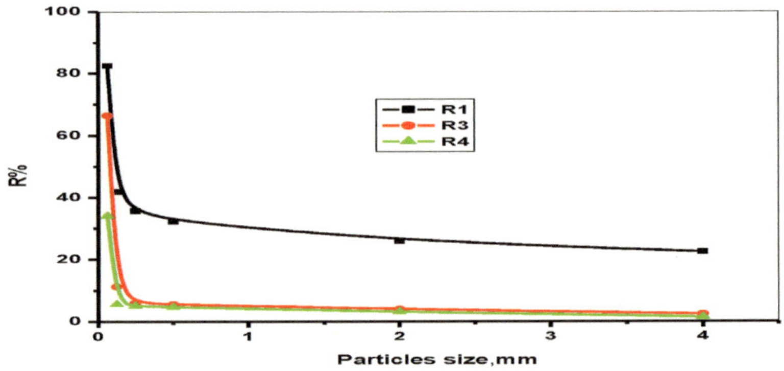

Figure 6.
Effect of particle size on the adsorption of MB from aqueous solutions onto apatite resins.

3.2.1.3 Effect of sorbent weight

The results for the removal of 1×10^{-5} M MB dye from aqueous solution using the investigated sorbents with respect to sorbents' weight are shown in **Figure 7** in the range 0.01–0.3 g/50 ml of 1×10^{-5} M MB dye solution at 25 ± 1°C. The results obtained show that when increasing sorbents' weight from 0.01 to 0.3 g, the removal of MB dye from aqueous solution increases and remains constant with further increase of sorbents' weight up to 0.2 g. This can be due to the availability of more surface functional groups and surface area at higher sorbent weight. The removal percentage of MB dye becomes constant from 0.2 to 0.3 g. This may be referred to the aggregation of the sorbent particles at high concentration of MB dye. Such aggregation would lead to decrease in the total surface area available for metal ion sorption.

3.2.2 Kinetics of adsorption

The amount adsorbed of dye onto the polymeric resin was studied with time for estimating the adsorption mechanism. The adsorption of dye with time shows that mixing period of 10 min is optimum for attaining the equilibrium with respect to R1and R2, while attaining the equilibrium with R3 takes 60 min. These findings

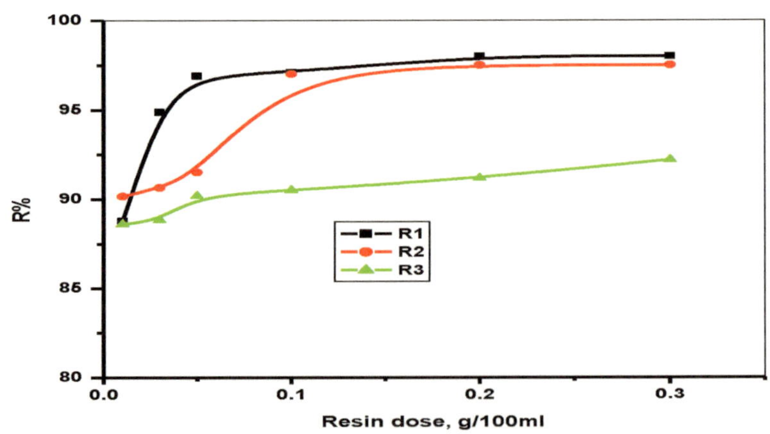

Figure 7.
Effect of resin dose on the adsorption of MB from aqueous solutions onto apatite resins.

reflect a fast kinetic for adsorption of MB onto the prepared resins, especially R1 and R2.

Different kinetic models were applied on the obtained results and the kinetic parameters were determined. The kinetic models correlate the amount adsorbed of dye with time. Lagergren presented the following equation for pseudo-first-order reactions [10]:

$$\frac{dq_t}{dt} = k_1(q_e - q_t) \tag{7}$$

where q_e and q_t are the dye concentration in solid phase at equilibrium and at time t, respectively, and k_1 is the model constant (min^{-1}). The linear form of the above equation was obtained by integration at the borders (q_t = 0 to q_t = q_t and t = 0 to t = t) as:

$$\log(q_e - q_t) = \log q_e - \frac{k_1 t}{2.303} \tag{8}$$

The rate constant k_1 was determined from the plot of $\log(q_e - q_t)$ with t as shown in **Figure 8** while the value of q_e was determined from the intercept. The model variables with the coefficient are given in **Table 1**.

The plots in the figures above show linear fit with correlation coefficient (R) of 0.994 for R1. The values of calculated adsorption capacity q_e and the linear regression coefficient clarify that the studied kinetic model kinetic model could not fit with the experimental results for adsorption of MB onto R1.

Second-order kinetic model, which describes the chemical adsorption is given by [11].

$$\frac{dqt}{dt} = k_2(q_e - q_t)^2 \tag{9}$$

where k_2 is the model constant (g/mg min). The above equation could be integrated at the border (q_t = 0 to q_t = q_t at t = 0 to t = t) to give:

$$\frac{t}{q_t} = \frac{1}{k_2 q_e^2} + \frac{t}{q_e} \tag{10}$$

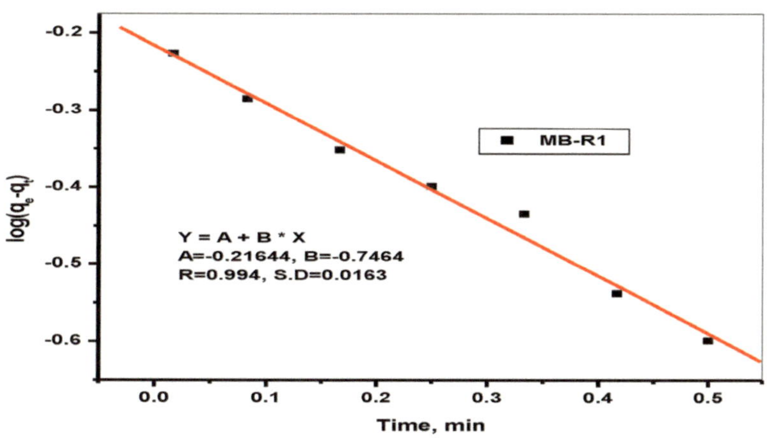

Figure 8.
Pseudo-first-order model kinetic plot for the sorption of MB onto the synthesized sorbents R1.

Adsorption system	Kinetic model	Parameters	R^2	SD
MB-R1	Pseudo first order	$k_1 = 1.718$ $q_e = 0.607$	0.994	0.0163
	Pseudo second order	$K_2 = 2.18$ $q_e = 2.73$	0.987	0.133
	Elovich	$\beta = 5.98$ $\alpha = 8.72$	0.978	0.0376
	Intra-particle	$k_{id} = 3.168$ $C = 2.93$	0.991	0.834

Table 1.
Kinetic models' variables for adsorption of MB onto R1.

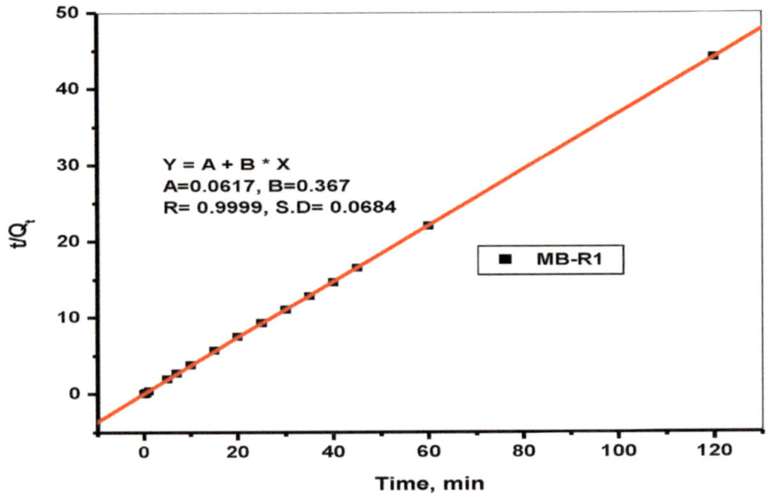

Figure 9.
Pseudo-second-order model kinetic plot for the sorption of MB onto the synthesized sorbents R1.

The model variables were calculated from the plot of t/q_t with t as shown in **Figure 9**. The plot showed a linear relation, and the model parameters with the correlation coefficient R^2 are given in **Table 1**.

The results of the studied kinetic model clarify that the experimental results for adsorption of MB onto R1, R2, and R3 could be described by kinetic model supporting chemical adsorption. The MB sorption could be represented favorably on the composite resins by the pseudo-second order kinetic model. This finding refers to the participation of chemical adsorption within the adsorption mechanism for MB onto R1.

Elovich kinetic model was applied upon the results to explain mainly the chemisorptions onto heterogeneous solid surfaces. The linearized form of Elovich model equation is given in [12].

$$q_t = \left(\frac{1}{\beta}\right) \ln\left(\alpha\beta\right) + \left(\frac{1}{\beta}\right) \log t \tag{11}$$

where α and β are model parameters representing the starting sorption rate (g mg^{-1} min^{-2}) and the leaching constant (mg g^{-1} min^{-1}), correspondingly. The model parameters were calculated from the linear fit of q_t vs. log(t) plot, shown in **Figure 10**, and are presented in **Table 1**.

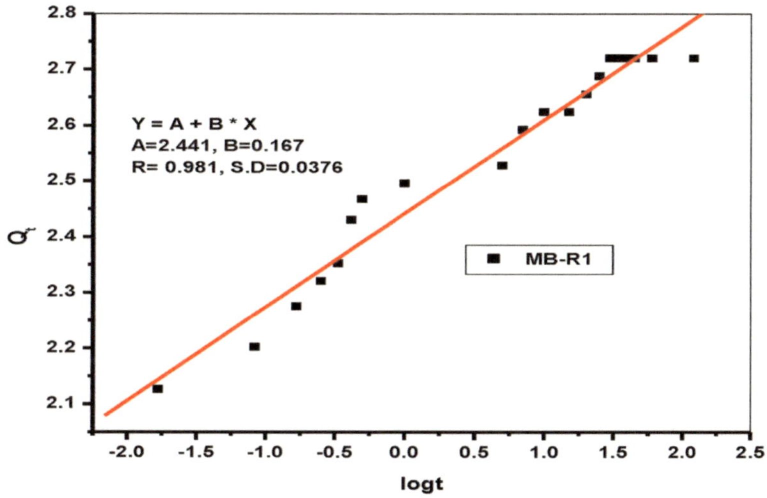

Figure 10.
Elovich model kinetic plot for the sorption of MB onto the synthesized sorbents R1.

The value of the Elovich constant (α and β) for the adsorption of MB on R1 predicate the effect of adsorbent dose and the possibility of performing sorption-desorption regeneration cycles of adsorbent. The value of correlation coefficient (R) reflects a poor fit of Elovich model with the experimental results. It could be inferred that both chemical and surface adsorption mechanisms are participating in the studied systems.

Intraparticle diffusion model was studied to explain the influence of transfer of dye from solution to solid surface of adsorbent on the reaction. The adsorption reaction could be affected by film diffusion, pore diffusion, surface diffusion, and/or adsorption on pore surface. The studied batch experiment was performed with shaking; therefore, the transfer of adsorbate particles could be described by diffusion coefficient that gives a considerable fitting with the experimental results. Weber model could be studied to explain the intraparticle diffusion's influence on the reaction by the following equation [13]:

$$q_t = k_{id}t^{0.5} + C \tag{12}$$

where k_{id} is the Webber model constant (mg g^{-1} min$^{-0.5}$) and C is a constant (mg g^{-1}) connected to the depth of the boundary layer, reflecting the boundary layer effect.

If the adsorption takes place within multilayer adsorbent, the adsorbate particles have to spread within the interior pores of solid materials. The model parameters were obtained from the plot of q_t vs. $t^{1/2}$, shown in **Figure 11,** and are presented in **Table 1**.

The results in **Figure 11** show two linear regions referring to the participation of at least two steps in the reaction. The linearity in the first region refers to a diffusion of dye into macro-pores of adsorbent, while the second linear region shows that the adsorbate particles diffuse within a micro-pore of adsorbent. The third region refers to mesoporous of adsorbent. The data obtained show that the synthesized sorbent has different pores.

The results in the figures reflect a variation of particle migration rate between different stages of sorption. The deviation of straight lines from the origin (when extrapolated), reflects that the diffusion within pore is not only the rate

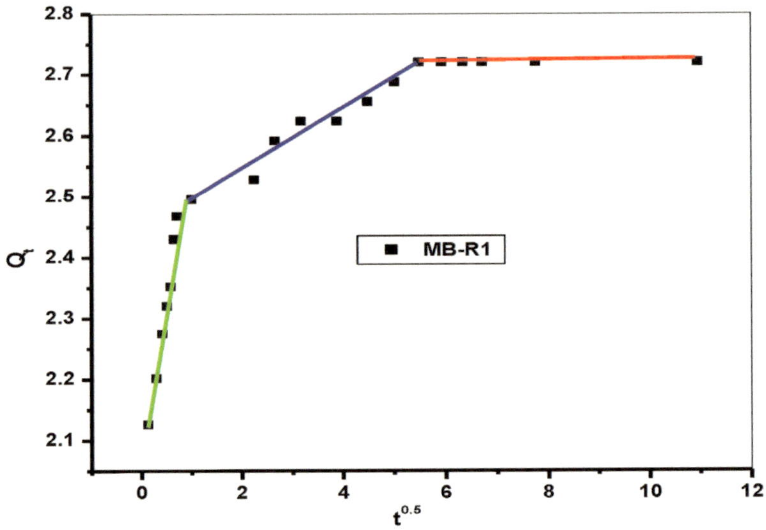

Figure 11.
Intraparticle model kinetic plot for the sorption of MB onto the synthesized sorbent R1.

determining step. The model variables for adsorption of MB using both R1, were given in **Table 1**. These results indicate that Webber diffusion model could not be considered as the controlling mechanism in the adsorption reaction.

3.2.3 Adsorption isotherm

The adsorption of MB onto R1 was studied at different initial dye concentrations (within the range 1–50 mg/L). The removal percentage and the adsorbed amount of MB are presented against the starting dye concentration in **Figure 12**. The results show that the amount of MB retained on the solid adsorbent increases with increasing dye concentration.

Different isotherm models were studied for describing the adsorption mechanism controlling the reaction. Langmuir isotherm model [13] was studied for adsorption of MB onto R1, which is expressed as:

$$\frac{C_e}{q_e} = \frac{1}{K_L} + \frac{C_e}{q_{max}} \tag{13}$$

where C_e is the dye concentration in solution after experiment (mg/L), q_e is dye concentration on the solid resin (mg/g), and q_{max} and K_L are model parameters connected to maximum adsorbed amount (mg/g) and adsorption energy, correspondingly. A plot of C_e vs. C_e/q_e is presented in **Figure 12** and the model variables were determined from the plot and are given with coefficient R in **Table 2**.

R values for the adsorption systems were found to be 0.975, indicating less compatibility with Langmuir isotherm. This finding prove that monolayer chemical adsorption on the homogeneous surface is not participate in the adsorption process.

Freundlich isotherm model [13] was applied on the experimental results, which is described by the equation:

$$\log q_e = \log k_f + \frac{1}{n} \log C_e \tag{14}$$

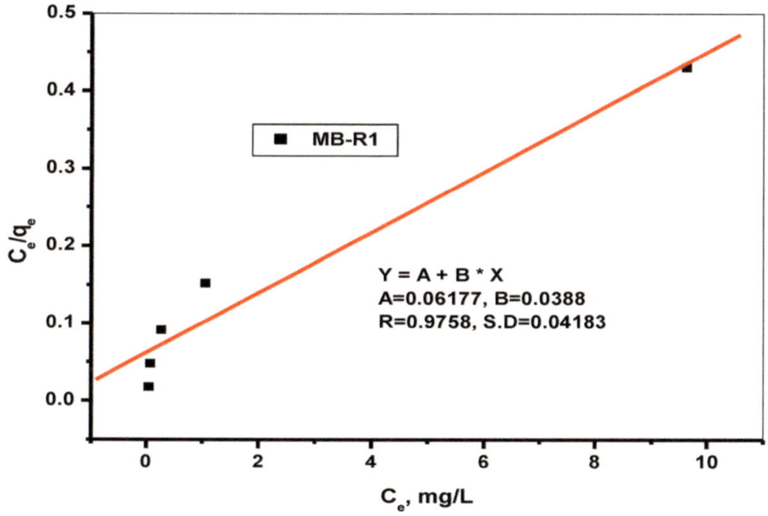

Figure 12.
Langmuir-1 isotherm plot for the sorption of MB onto the synthesized sorbent R1.

Model	Resin	Parameter	R	SD
Langmuir	R1	q_{max} = 25.77 K_L = 16.18	0.975	0.0418
Freundlich	R1	$1/n$ = 0.617 K_f = 6.102	0.988	0.104
Dubinin-Radushkevich	R1	$\beta = -0.093$ q_{max} = 10.39 E_S = 2.318	0.979	0.359

Table 2.
Adsorption isotherm models' parameters for MB-R systems.

where k_f (mg/g) and n are model constants, indicating the adsorption capacity and favorability nature of the adsorption process. Freundlich model constants were determined from the linear fit of log q_e vs. log C_e plot in **Figure 13** and are given with the correlation coefficient in **Table 2**.

The values of R of Freundlich plots for MB-R system showed bad fit of the experimental results with Freundlich isotherm model.

Dubinin-Radushkevich (D-R) adsorption isotherm model was studied; it describes adsorption onto porous solid surfaces, and is described by the following equation [13]:

$$\ln q_e = \ln q_{max} - \beta\varepsilon^2 \tag{15}$$

where β is the D-R model constant (mol^2/kJ^2), q_{max} is the constant referring to the maximum adsorbed amount (mg/g), and ε is Polanyi potential (ε = RTln$(1 + 1/C_e)$). The D-R model constants were obtained from the linear fit of the plot of ln q_e vs. ε^2 **(Figure 14)** and are given with correlation coefficient in **Table 2**. The adsorption free energy (E_S) is calculated as:

$$E_S = (-2\beta)^{-1/2} \tag{16}$$

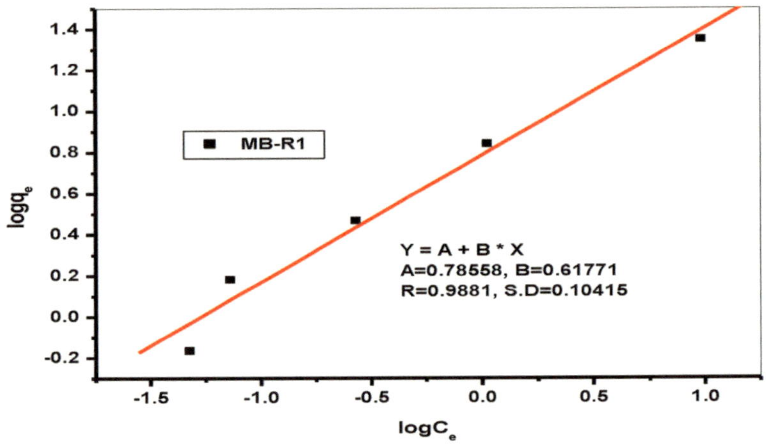

Figure 13.
Freundlich model isotherm plot for the sorption of MB onto the synthesized sorbent R1.

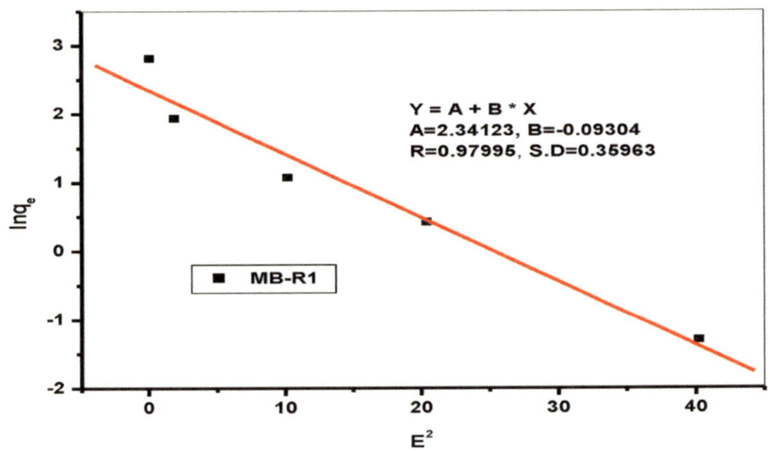

Figure 14.
Dubinin-Radushkevich model isotherm plot for the sorption of MB onto the synthesized sorbent R1.

The calculated mean adsorption free energy (E_S) from D-R model for adsorption of MB using R1 was found to be 8.03 kJ/mol. These values reflect that physical adsorption is a participating mechanism.

3.2.4 Effect of temperature

The effect of temperature on the removal efficiency of 1×10^{-5} M MB dye from aqueous solution using 0.1 g of each investigated sorbents was studied at different temperatures ranging from 25 to 50°C. It is observed from the results that the removal of MB dye increases slightly with increasing temperature, see **Figure 15.** This behavior indicates that the sorption process of MB dye from aqueous solution using all selected sorbents is an endothermic process. The increase in the sorption efficiency of MB dye by the sorbents used at high temperature may be attributed to the increase of MB dye mobility and decrease of the retarding forces acting on the diffusing dye.

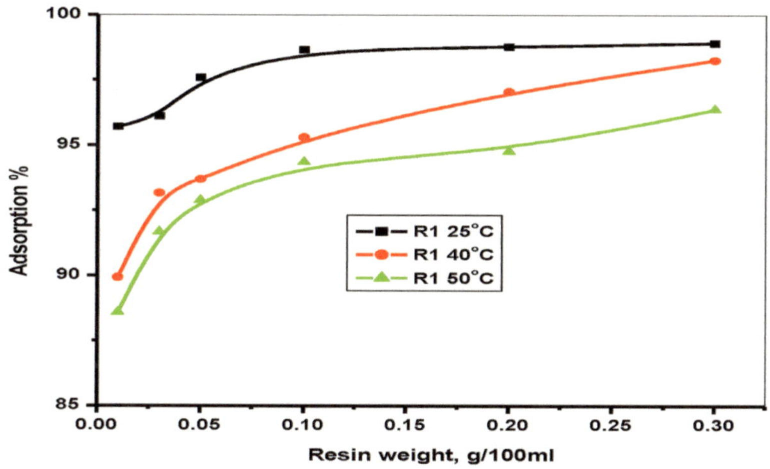

Figure 15.
Effect of temperature on the adsorption of MB from aqueous solution onto R1.

3.2.5 Adsorption thermodynamics

The thermodynamic parameters corresponding to dye sorption on the prepared resins were assessed using Van't Hoff equation [14] (**Table 3**).

$$\log K_d = \frac{\Delta S^o}{2.303R} - \frac{\Delta H^o}{2.303RT} \quad (17)$$

where k_d is the distribution coefficient of the solute ions, ΔS° is the entropy change (J mol^{-1} K^{-1}), R is the ideal gas constant (8.314 J mol^{-1} K^{-1}), and T is the absolute temperature. A plot of log k_d vs. $1/T$ was constructed as shown in **Figure 16** from which the slope of the straight line equal $-(\Delta H^o)/2.303$; consequently, the value of apparent enthalpy change (ΔH^o) for the overall system was calculated. The values of other thermodynamic parameters are calculated at different temperatures, using the following equations:

$$\Delta G^o = -2.303RT \ \log \ k_d \quad (18)$$

$$\Delta G^o = \Delta H^o - T\Delta S^\circ \quad (19)$$

3.2.6 Column studies

Sorption dynamics of MB in a fixed bed flow through a sorption column is eventually conducted for multiple reuse of the sorbent. Column sorption studies of MB on the R1 beads at room temperature were investigated using aqueous solution

Adsorption system	Temperature (°K)	ΔH^o (kJ mol^{-1})	ΔG^o (kJ mol^{-1})	ΔS^o (J K^{-1} mol^{-1})	SD
R1-MB	298	44.991	6.961	172.81	0.257
	313		−7.850		
	323		−11.626		

Table 3.
Thermodynamic parameters for sorption of MB onto R1.

of 1×10^{-5} M influent concentrations (C_i). Experimental breakthrough curve was studied using flow rate 0.2 ml/min. It is obtained by plotting the ratios of effluent concentration to initial concentration versus the volume of the effluent, see **Figure 17**. It was observed that the column gets saturated after passing 1 l of MB solution. The estimated breakthrough sorption capacity (Q_e) was 79.65 mg/g for MB.

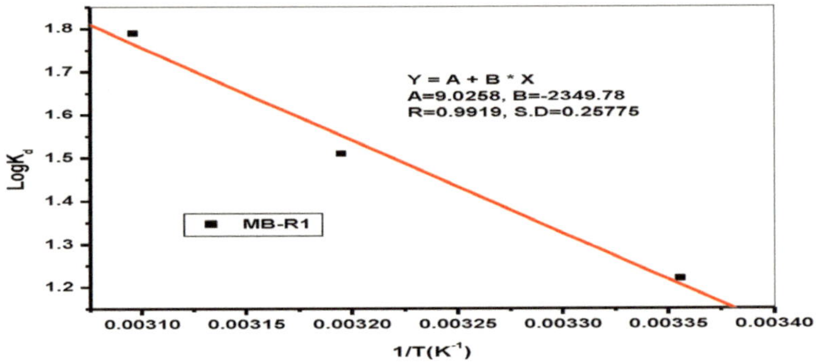

Figure 16.
Effect of temperature on the sorption of MB by the synthesized sorbent R1.

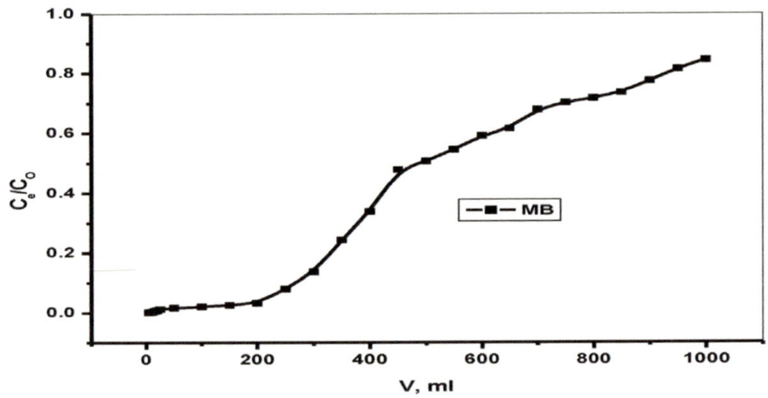

Figure 17.
Breakthrough curve of MB dye with R1.

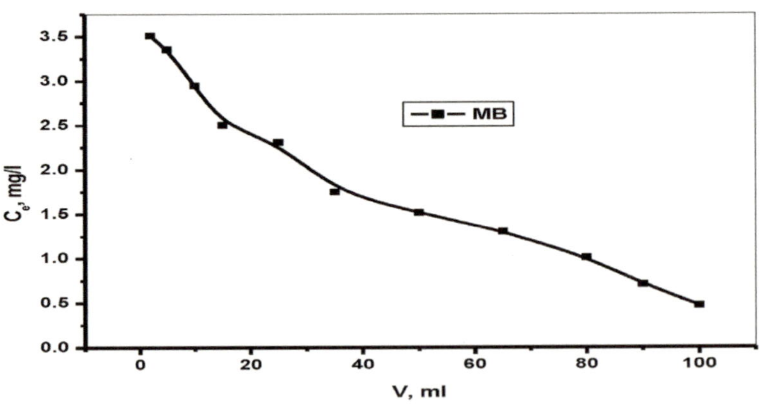

Figure 18.
Dynamic desorption curve of MB from R1 by H_2O_2 and H_2SO_4.

Once the column reached exhaustion, efficient elution of adsorbed solute from resin in column is essential to ensure the recovery of dye as well as the reuse of resin for repeated adsorption/desorption cycles. Desorption of MB from R1 was studied using different concentrations of sulfuric acid and hydrogen peroxide.

Desorption curve shown in **Figure 18** was obtained by plotting the effluent concentration (C_e) versus elution volume from the column at a flow rate of 0.2 ml/min. The result obtained shows that 93.79% recovery was achieved for MB by 1:1 W/W H_2SO_4 and H_2O_2.

4. Conclusions

In the current study, the removal of MB dye from aqueous solution was investigated by three prepared composite resins and the following conclusions can be drawn:

- The prepared composite resins can potentially be applied for removal of MB dye from aqueous solutions.
- The adsorption mechanism was suggested based on applying different isotherm models and kinetic models.
- From the data obtained for the uptake of MB dye by practical experiment, good match was found with calculated values obtained from isotherm models.
- Working on the composite is more advanced than working on HAP alone because:
  - The spontaneous reaction was carried out at low temperatures while HAP was carried out with spontaneous reaction at high temperatures.
  - The surface area and mechanical strength for composite are higher than the surface area and mechanical strength of HAP alone.
- Different inorganic and organic solutions were studied for regeneration of composite.
- Aqueous solution containing mixture of sulfuric acid and hydrogen peroxide of 1:1 w/w showed the maximum release for the adsorbed dyes with 87.46% for methylene blue dye.

Acknowledgements

The authors extend their appreciation to the Research and Development Grants Program for National Research Institutions and Centers (GRANTS) at King Abdulaziz City for Science and Technology (KACST) for supporting this work through research groups program under grant number 1-18-01-010-0002.

Author details

Nasser S. Awwad[1*], Adel A. El-Zahhar[1] and Jamila A.M. Alasmary[2]

1 Faculty of Science, Department of Chemistry, King Khalid University, Abha, Saudi Arabia

2 Faculty of Science for Girls, Department of Chemistry, King Khalid University, Abha, Saudi Arabia

*Address all correspondence to: aawwad@kku.edu.sa

References

[1] Zaharia C, Suteu C, Muresan A. Proceedings of International Conference on Environmental Engineering and Management. 2011;**4**:121

[2] Zaharia C, Suteu D, Muresan A, Muresan R. Environmental Engineering and Management Journal. 2009;**6**:1359

[3] Börnick H, Schmidt TC. J. Anal. Environ. 2006;**9**:181

[4] Doulati AF, Badii K, Yousefi Limaee N, Mahmoodi NM, Arami M, Shafaei SZ, et al. Dyes and Pigments. 2007;**73**:178

[5] Hassan SSM, Awwad NS, Aboterika AHA. Journal of Hazardous Materials. 2008;**154**:992-997

[6] Barakat NAM, Khila MS, Omrand AM, Sheikhd FA, Kima HY. Journal of Materials Processing Technology. 2009;**209**(2):3408

[7] El-Zahhar AA, Awwad NS. Journal of Environmental Chemical Engineering. 2016;**4**:633-638

[8] Awwad NS, Alshahrani AM, Saleh KA, Hamdy MS. Molecules. 2017, 1947;**22**(12). DOI: 10.3390/molecules22121947

[9] TÕnsuaadu K, Agris Gross K, Plūduma L, Veiderma M. Journal of Thermal Analysis and Calorimetry. 2012;**110**:647

[10] Lagergren S. Kungliga Svenska Ventenskapsakademiens Handlingar. 1898;**24**:1

[11] Ho YS, McKay E. Canadian Journal of Chemical Engineering. 1998; **76**:822

[12] Ho YS, McKay G. Science and Technology. 2002;**20**:797

[13] Dada AO, Olalekan AP, Olatunya AM, DADA O. Journal of Applied Chemistry. 2012;**3**:38

[14] El-Zahhar AA, Abdel-Aziz HM, Siyam T. Journal of Radioanalytical and Nuclear Chemistry. 2006;**267**(3):657

Section 8

Miscelleneous Other Review on Dyed Textiles and Clothing

Chapter 13

Review on Analysis of Seam Quality

Suprit Borse, Kamlesh Sonawane, Madhuri V. Kakde and Tushar A. Shinde

Abstract

The apparel quality largely depends on fabric quality and also on seam quality. Seam performance and appearance largely correlate with fabric characteristics, sewing-thread properties, selection of stitches and seams, and also with sewing conditions such as size of needle, sewing-thread tension, number of stitches per inch, and maintenance and operation of sewing machine. In good quality garments, functional requirements of the seam are very essential for apparel serviceability. The seam performance deteriorates the quality of garment by reducing seam strength and seam-slippage strength, seam puckering, and seam damage. Apparel durability and its esthetic performance are defined by seam quality. Seam performance is influenced by a selection of seam type with respect to stitch classes, selection of sewing threads irrespective to fabric characteristics, and different sewing process parameters.

Keywords: apparel quality, apparel serviceability, seam appearance, seam performance, seam quality

1. Introduction

The basic constituent of quality apparel is stitches and seams. The different components of apparel joint together by means of stitching operation and seams are responsible for giving the apparel shape for wear [1, 2]. The overall performance of the apparel in use mainly depends on the quality of seam [3]. Seam is the joining of two or more pieces of fabrics by means of stitching. The quality of seam greatly influences on the quality of apparel. The quality of seam is assessed by means of its efficiency, elongation, bending, stiffness, abrasion resistance, seam-slippage strength, puckering, tightness, boldness, and seam damage [4]. A good seam has functional and aesthetic requirements. The functional performance of the seam is evaluated by efficiency, elongation, density, slippage, bending stiffness, and abrasion resistance of seam [5]. Whereas seam puckering, tightness, boldness, and seam damage are assessed for better elegant performance of the seam [6]. The basic raw material of garmenting is fabric and sewing thread. The apparel seam quality depends on raw material characteristics [7]. Seam performance by means of durability is based on the strength and efficiency of seam as well as seam appearance along the seam line attributed by the seam pucker [8]. These seam parameters are very important for garment manufacturers and customers, as it affects the customers' opinion regarding quality of garment [9]. Some of the manufacturers select stitch classes, stitch density, and sewing threads without consideration to their influence on the overall performance of the apparel being made, resulting in failure of the seam during use [10]. It is therefore very important to select appropriate seam for better performance of apparel. Seam types are one of the factors influencing the overall performance of the seam in terms of durability, comfort, and potential for alteration. The appearance of a seam is generally lead by the proper relationship between the size and type of thread, the stitch density and its types, and the texture and weight of the fabric [11]. Many researchers found that fabric quality parameters, such as fabric density, fabric thickness, tensile strength, extensibility, bending rigidity, and shear rigidity have extreme effect on quality of seam [12]. Different researcher gives a subjective ranking for evaluation of seam performance [13]. Experts revealed that seam is evaluated by its analytical dimensions, such as strength, puckering property, efficiency, slippage strength, and seam boldness. High-speed sewing machine exerts high tensions and needle penetrates in high forces through the fabric. In this stage, damaging of sewing threads and warp and weft in the fabric result in poor seam performance by means of reducing seam strength and seam slippage [14, 15].

2. Influence of sewing thread on seam quality

Sewing thread is a crucial factor that not only has effect on durability of garment but also on overall design [16]. Tensile properties of sewing thread influence on the productivity of apparel manufacturing [17]. Sewing thread should have sufficient tensile strength, should be uniformly twisted throughout the length, smooth and elastic, and plied yarn treated by special finishes to increase its abrasion resistance during stitching operation [18]. It can be spun from staple yarn, filament, or core spun yarns. The thread must be withstand the several wash and hold the seams together for the life of the garment. Cotton and polyester sewing thread are most commonly utilized by the clothing industry [19, 20]. Sewing thread plays a very crucial role for improving the quality of garment. Poor quality of sewing thread increases the production cost rapidly since they cause frequent stoppages of sewing machines. Types of sewing thread used for stitching depend upon the fabrics being used. Polyester sewing thread is more durable than the cotton sewing thread [14, 21]. Higher strength of thread is prone for giving higher seam strength shows better functional performance of seam [22]. The other parameters of sewing thread also decide the seam performance, such as type of sewing thread, number of ply, type of finish, twist, and size of thread. For all types of apparel spun polyester thread, three-ply thread with optimum twist and standard finish sewing thread quality is widely used [23]. Seam efficiency of polyester textured thread lies in between cotton and spun polyester threads, but more closer to apparel stitched with spun polyester threads [21]. A few research works emphasis on influence of thread finishes on performance of seam. It revealed that as per the specific end use of apparel, different types of finishes are used on thread [22]. The size of sewing thread and number of ply decide the seam stability in apparel. The seam strength and seam puckering also depend on the thread size and ply [24]. During stitching, greater friction occurs with higher sewing-thread size, which ultimately results into less seam strength and increased seam puckering. While more number of ply assists for increasing the thread strength resulting higher seam strength and seam efficiency with less seam slippage. But the diameter of thread increases with number of ply leading to structural jamming along the seam line, which results into more seam puckering [16, 25]. The draping characteristics of apparel are also affected due to the increasing stiffness by means of number of ply in sewing thread [20]. Type of sewing thread and fabric weave influences on seam elongation in both directions. Plain cotton fabrics show higher seam elongation in warp direction, whereas twill weave of cotton fabrics shows higher seam elongation in weft direction. This is attributed due to the fact that during stitching, stitching threads may be subjected to stretching and bending, results into shrinkage in relaxation process after sewing operation [26]. One of the performance property of seam is seams abrasion resistance, which is determined by sewing thread. The abrasion resistance of core-spun threads is better than other fiber threads [19, 27].

3. Influence of fabric parameters on seam quality

Apparel manufacturers considered the specification of fabrics in terms of basic and supplementary characteristics for apparel quality. The basic qualities are related to the mechanical dimensions of fabric, and supplementary qualities are related to the fabric performance on an applied dynamic force [28]. The clothing manufacturers are mainly focused on the supplementary qualities of the material and emphasis on seam quality during fabrication. On contrary to this, consumer are mainly focused on comfort, look, and wear ability of fabric and assess seam quality on the basis of seam appearance [4, 10]. The good quality fabrics gives comforts as well as evenly working of apparel processes and leads to fault-free garments [15, 29]. Apparel seam performance depends on the fabric cover factor, fabric weight, thickness, tensile strength, fabric shrinkage along the fabric length and width, resiliency, bending rigidity, flexural rigidity, and shear rigidity [8]. Fabric selected for making the garment should possess good sewability [1]. A good sewable fabric has the propensity and the ease with which two-dimensional fabric components can be qualitatively and quantitatively be seamed together to three-dimensional garment [22]. The ease of formation of shell structures on the fabric surface indicates that the fabric possesses good sewability producing pucker-free garment. A resulting sewability is due to the integration of various variables interconnected with stitching thread, material, and stitching machine settings at their optimum level [27]. Many researchers have been studied on fabric sewability of various types of fabrics. Research studies revealed that both stitch length and stitch density together with knitted fabric relaxation methods has found significant influence on seam elongation. But, it shows negative impact on seam strength and seam efficiency. Increasing stitch length and yarn twist factor, decreases seam strength and efficiency. But, stitch density is directly proportional to the seam strength and seam efficiency. If stitch length increases, decreases the strength of knitted fabric which attributed to decreasing the seam strength. But, increasing stitch length will increases fabric elongation, which shows higher seam elongation. More twist factor leads to an increase in yarn elongation, which attributed to increase of the knitted fabrics elongation. Therefore, as the yarn twist factor increases, causes the increase in seam elongation [1, 8]. The study on sewing performance of stretch denim revealed that with the increase in Lycra percentage in denim seam efficiency increases. This is due to the higher flexibility and better compatibility of the sewing thread with denim. More Lycra percentage in denim increases the flexibility thereby improving the seam performance [16]. Many researchers studied on the effect of fiber type and its percentage on the seam performance. The fiber percentage significantly influences on the seam stiffness. The study revealed that the change in blending of polyester and cotton, the seam stiffness increases due to decreasing polyester fiber content in fabric which has low bending stiffness. On the other hand if polyester percentage is more in fabric show less resistance during sewing operation. However, the linear density of sewing thread shows insignificant effect on seam stiffness [12, 24]. Different blended fabrics show significant effect on drape coefficient. Polyester dominating fabric shows low drape coefficient due to higher low-stress extensibility. On the other, hand cotton dominating fabric shows high drape coefficient, since cotton has the more bending stiffness than polyester [28].

4. Seam strength

Poor seam performance makes the garment inappropriate even though the material strength is high. Therefore, it is very crucial to divine the seam strength in terms of the performance of apparels during use [7]. It is defined as a load required for breaking seam and it results from the breakage of either textile material or sewing thread or sometimes both simultaneously [29]. Different types of garments have different seam strength requirements. Many factors affect the level of seam strength [30]. Seam strength depends on the fabric quality, thread strength, and stitch density along with the stitch and seam selection and sewing conditions. But, it has been found that the fabric strength is not expected to cause any adverse effect on seam strength. Seam strength correlates with thread types, different classes of seam and stitch [16]. Many previous studies reported that seam strength depends on the thread strength and stitch density. Stitch length is indirectly proportional to the seam strength. It is due to the fact that, as the stitch length increases, the number of loops that shares the load decreases, resulting in decreasing seam strength [9, 19, 21]. The fabric speed during stitching operation also decreases with lower stitch lengths, which leads to a uniform distribution of stitches [27]. The seam strength is also higher with polyester core spun sewing thread is used as the bobbin thread rather than polyester-cotton core spun sewing thread [9]. Seam strength of higher density fabric is found better than that of lower density. This is attributed due to the higher tensile strength of fabric and more stripping of the needle thread through the fabric [19]. Sewing-thread parameters such as type of thread, its size, and its constructional features and finishing influence on seam strength. The linear density of sewing thread is expected to affect the seam strength, the thicker thread gives a better seam strength, and nevertheless it requires the use of a thicker needle which may damage the fabric [8]. Seam strength is also the product of the sewing thread loop strength at a standard gauge length and the number of loops in a given width of seam which depends on stitch length. It is therefore reveled that loss in the strength of the needle and bobbin threads significantly affects the seam strength of the fabric. The loss of needle thread strength is higher during sewing operation than bobbin threads strength [21]. The fabric types show significant influence on the needle thread strength loss [22].

5. Seam strength efficiency

Seam efficiency plays a very crucial role for obtaining desirable seam performance with the help of the ratio of sewn fabric strength to unsewn fabric strength [31]. The tensile behavior of fabrics and sewing thread, the sewing-thread properties, the combination of fabric and thread, and machine and process parameters together decide the seam efficiency. Seam efficiency of 85 and 90% is difficult to achieve. The apparel manufacturer expected seam efficiency in the range of 60–85% [28, 31]. Many researchers were studied on the effect of different fabric blend percentage and thread tension on seam efficiency. Higher fabric strength gives less seam strength efficiency for a given sewing thread. Seam efficiency is also depends on the blend composition in fabric. Polyester dominating fabric shows low seam strength efficiency. This is due to the fact that polyester gives higher flexibility in the low load region, hence, reducing fabric dimensional stability. The decreasing seam efficiency is due to the less uniform fiber matrix causing fiber slippage and yarn failure. Higher linear density sewing thread has better gripping with the textured polyester in polyester dominating fabric resulting high seam strength efficiency. But, seam strength efficiency increases with cotton dominating fabric due to the uniform fiber matrix [1, 29]. The seam strength efficiency also depends on the needle-thread tension. During stitching operation, all low-thread tension considerably improves the seam efficiency. This is attributed to the fact that some flexibility at the seam occurs at low needle thread tension under sudden stress. In contrast at high tension, the fabric has been pulled along the seam line leading to puckered and unstable seam [17]. The linear density of sewing thread also shows considerably effect on seam efficiency. Seam efficiency increases with increasing the linear density of sewing thread due to the more number of fibers is incorporated in the coarser sewing thread [14, 19]. Researchers concluded that core-spun polyester thread show higher seam efficiency. Stitch density is directly proportional to the strength and efficiency of seam [8].

6. Seam puckering

It is the increments in the seam thickness over the fabric under a constant compressive load, which appears along the seam line. This particular effect appears when the sewing variables and fabric properties are not correlated [24]. It has been revealed from many studies that seam puckering depends on the sewn fabric [6]. Seam puckering is assessed in terms of seam thickness strain percentage and it is measured by the thickness difference in fabric and seam [18]. Seam pucker is an unacceptable seam appearance found due to seam gathered either just after stitching or after finishing and the contractive forces introduced in seam during stitching operation. Seam puckering increases with increasing fabric weight irrespective of type of sewing thread [24]. Many research studies reported that seam puckering is highly contingent on linear density of sewing thread [16]. The coarser thread leads to fabric structural jamming and less flexibility at the seam causing more seam puckering. Also, polyester fabrics show high seam pucker than cotton fabrics. This is due to the fact that polyester dominating fabric gives higher extensibility in the low load region, hence, reducing fabric dimensional stability resulting into high seam puckering [18]. In spite of this, fabric weave and weft setting show significantly effect on seam pucker [31].

7. Seam slippage

Seam slippage is identified by the load need to discrete the seam by a certain distance. Seam slippage is less if higher load is required to separate the seam and indicates the textile is more resistant to seam slippage. The fabric stitched with core spun sewing thread showed higher slippage strength. This can be attributed to high extensibility of core spun sewing thread. With polyester filament, the seam slippage is lower because of less flexibility and compact structure of polyester thread. Seam slippage is also increased with silicone finish, since silicone is participating in mobilizing the threads and at less force the desired seam opening is possible [16, 28].

8. Seam damage

In garments, a constant occurring phenomenon is a seam damage. It is also termed as a sewing damage. Most of the complaints arise from customers are due to the seam damage [17]. Seam damage is evaluated by the needle cutting index. It not only depends on the fabric weave and cover factor but also on the thread diameter, surface properties of thread and stitch density [32]. The results from many studies found that with increasing Lycra percentage in denim and fabric weight rises needle cutting index. It is attributed due to the voluminous structure of Lycra yarn inside the fabric structure mainly at interlacing points and leads to higher frictional force between yarn and needle [3, 32, 33].

9. Conclusion

In this chapter, seam quality parameters have been discussed. During the production of the apparel, manufacturers only deals with fabric characteristics and more prominence about seam quality. On the other hand, consumers are assessing seam quality on the basis of seam appearance, comfort, and wear ability. Seam performance largely depends on the interrelationship between fabric structural properties, sewing threads, sewing process conditions, and the number of washes the garment will receive. Different stitching threads will show different results on seam standard and seam performance. Types of fabric and their structural properties have significant effect on seam strength, seam efficiency, seam elongation, and seam puckering. The seam quality expectations for various apparels will be different. It can be therefore concluded that, it will be advantageous to investigate the sewing performance of any type of fabrics fit for various applications.

Author details

Suprit Borse, Kamlesh Sonawane, Madhuri V. Kakde* and Tushar A. Shinde*
Centre for Textile Functions, Mukesh Patel School of Technology Management and Engineering, Narsee Monjee Institute of Management Studies (NMIMS), Shirpur, Dhule, India

*Address all correspondence to: madhutext.kakade@gmail.com
and tushar.shinde176@gmail.com

References

[1] Behera BK, Sharma S. Low-stress behavior and sewability of suiting and shirting fabrics. Indian Journal of Fiber and Textile Research. 1998;**23**(4):233-241

[2] Behera BK. Evaluation and selection of sewing thread. Textile Trends. 1997;**39**(12):33-42

[3] Behera BK, Chand S, Singh TG, Rathee P. Sewability of denim. International Journal of Clothing Science and Technology. 1997;**9**(2):128-140

[4] Brain DH. The prediction of strengths of lockstitch seams in woven fabrics. Journal of the Textile Institute. 1970;**61**(1):493-505

[5] Dobilaite V, Juciene M. The influence of mechanical properties of sewing threads on seam pucker. International Journal of Clothing Science and Technology. 2006;**18**(5):335-345

[6] Ebrahim' FFS. Influence of mechanical properties of cotton fabrics on seam quality. Journal of American Science. 2012;**8**(5):831-836

[7] Ferreira FBN, Harlock SC, Grosberg P. A study of thread tensions on a lockstitch sewing machine (part I). International Journal of Clothing Science and Technology. 1994;**6**(1):14-19

[8] Germanova-Krasteva D, Petrov H. Investigation on the seam's quality by sewing of light fabrics. International Journal of Clothing Science and Technology. 2008;**20**(1):57-64

[9] Giorgio Minazio P. The fabric pressing performance and its role in predicting the appearance of men's wool suit jackets. International Journal of Clothing Science and Technology. 1998;**10**(3/4):182-190

[10] Gribaa S, Ben Amar S, Dogui A. Influence of sewing parameters upon the tensile behavior of textile assembly. International Journal of Clothing Science and Technology. 2006;**18**(4):235-246

[11] Gurarda A, Meric B. Sewing needle penetration forces and elastane fiber damage during the sewing of cotton/elastane woven fabrics. Textile Research Journal. 2005;**75**(8):628-633

[12] Haghighat E, Mohammad Etrati S, Shaikhzadeh Najar S. Evaluation of woven denim fabric sewability based on needle penetration force. Journal of Engineered Fabrics and Fibers. 2014;**9**(2):47-60

[13] Hu J, Chung S. Bending behavior of woven fabrics with vertical seams. Textile Research Journal. 2000;**70**(2):148-153

[14] Hui CL, Ng SF. Predicting seam performance of commercial woven fabrics using multiple logarithm regression and artificial neural networks. Textile Research Journal. 2009;**79**(18):1649-1657

[15] Hui PC, Chan KC, Yeung KW, Ng FS. Application of artificial neural networks to the prediction of sewing performance of fabrics. International Journal of Clothing Science and Technology. 2007;**19**(5):291-318

[16] Kang TJ, Lee JY. Objective evaluation of fabric wrinkles and seam puckers using fractal geometry. Textile Research Journal. 2000;**7**(6):469-475

[17] Mandal S, Abraham N. An overview of sewing threads mechanical properties on seam quality. Pakistan Textile Journal. 2010;**1**:40-43

[18] Meric B, Durmaz A. Effect of thread structure and lubrication ratio on seam

properties. Indian Journal of Fibre and Textile Research. 2005;**30**(3):273-277

[19] Midha VK, Mukhopadhyay A, Chatopadhyay R, Kothari VK. Studies on the changes in tensile properties of sewing thread at different sewing stages. Textile Research Journal. 2009;**79**(13):1155-1167

[20] Mukhopadhyay A, Sikka M, Karmakar AK. Impact of laundering on the seam tensile properties of suiting fabric. International Journal of Clothing Science and Technology. 2004;**16**(4):394-403

[21] Nayak R, Padhye R, Dhamija S, Kumar V. Sewability of air-jet textured sewing threads in denim. Journal of Textile and Apparel Technology and Management. 2013;**8**(1):1-11

[22] Nergis B. Thread lubrication. Canadian Textile Journal. 1997;**114**(3):42-44

[23] Padhye R, Nayak R. Sewing performance of stretch denim. Journal of Textile and Apparel Technology and Management. 2010;**6**(3):1-9

[24] Pasayev N, Korkmaz M, Baspinar D. Investigation of the techniques decreasing the seam slippage in chenille fabrics (part I). Textile Research Journal. 2012;**82**(9):855-863

[25] Pavlinic DZ, Gersak J, Demsar J, Bratko I. Predicting seam appearance quality. Textile Research Journal. 2006;**76**(3):235-242

[26] Rengasamy RS, Kothari VK, Alagirusamy R, Modi S. Studies on air-jet textured sewing threads. Indian Journal of Fiber and Textile Research. 2003;**28**(3):281-287

[27] Stjepanovic Z, Strah H. Selection of suitable sewing needle using machine learning techniques. International Journal of Clothing Science and Technology. 1998;**10**(3/4):209-218

[28] Stylios G, Lloyd DW. Prediction of seam pucker in garments by measuring fabric mechanical properties and geometric relationships. International Journal of Clothing Science and Technology. 1990;**2**(1):6-15

[29] Sundaresan G, Hari PK, Salhotra KR. Strength reduction in sewing threads during high speed sewing in industrial lockstitch machine. Part-II: Effect of thread and fabric properties. International Journal of Clothing Science and Technology. 1998;**10**(1):64-79

[30] Gurarda A. Seam performance of garments. DOI: 10.5772/intechopen.86436. Available from: Intechopen.com

[31] Topalbekiroğlu M, Kübra Kaynak H. The effect of weave type on dimensional stability of woven fabrics. International Journal of Clothing Science and Technology. 2008;**20**(5):281-288

[32] Zervent Ünal B. The prediction of seam strength of denim fabrics with mathematical equations. Journal of the Textile Institute. 2012;**103**(7):744-751

[33] Zunic-Lojen D, Gersak J. Study of the tensile force of thread in relation to its pre-tension. International Journal of Clothing Science and Technology. 2001;**13**(3/4):240-250